THE COLLECTED PAPERS OF

Albert Einstein

VOLUME 13

THE BERLIN YEARS:
WRITINGS & CORRESPONDENCE,
JANUARY 1922–MARCH 1923

DIANA KORMOS BUCHWALD

GENERAL EDITOR

THE COLLECTED PAPERS OF ALBERT EINSTEIN

English Translations Published to Date

VOLUME 1
The Early Years, 1879–1902
Anna Beck, translator; Peter Havas, consultant (1987)

VOLUME 2
The Swiss Years: Writings, 1900–1909
Anna Beck, translator; Peter Havas, consultant (1989)

VOLUME 3
The Swiss Years: Writings, 1909–1911
Anna Beck, translator; Don Howard, consultant (1993)

VOLUME 4
The Swiss Years: Writings, 1912–1914
Anna Beck, translator; Don Howard, consultant (1996)

VOLUME 5
The Swiss Years: Correspondence, 1902–1914
Anna Beck, translator; Don Howard, consultant (1995)

VOLUME 6
The Berlin Years: Writings, 1914–1917
Alfred Engel, translator; Engelbert Schucking, consultant (1997)

VOLUME 7
The Berlin Years: Writings, 1918–1921
Alfred Engel, translator; Engelbert Schucking, consultant (2002)

VOLUME 8
The Berlin Years: Correspondence, 1914–1918
Ann M. Hentschel, translator; Klaus Hentschel, consultant (1998)

VOLUME 9
The Berlin Years: Correspondence, January 1919–April 1920
Ann M. Hentschel, translator; Klaus Hentschel, consultant (2004)

VOLUME 10
The Berlin Years: Correspondence, May–December 1920
and
Supplementary Correspondence, 1909–1920
Ann M. Hentschel, translator; Klaus Hentschel, consultant (2006)

VOLUME 12
The Berlin Years: Correspondence, January–December 1921
Ann M. Hentschel, translator; Klaus Hentschel, consultant (2009)

VOLUME 13
The Berlin Years: Writings & Correspondence, January 1922–March 1923
Ann M. Hentschel and Osik Moses, translators; Klaus Hentschel, consultant (2012)

THE COLLECTED PAPERS OF

Albert Einstein

VOLUME 13

THE BERLIN YEARS:
WRITINGS & CORRESPONDENCE,
JANUARY 1922–MARCH 1923

ENGLISH TRANSLATION
OF SELECTED TEXTS

Diana Kormos Buchwald, József Illy, Ze'ev Rosenkranz, and
Tilman Sauer, Editors

Ann M. Hentschel and Osik Moses, TRANSLATORS
Klaus Hentschel, CONSULTANT

Princeton University Press
Princeton and Oxford

CONTENTS

PUBLISHER'S FOREWORD

We are pleased to be publishing this translation of selected documents of Volume 13 of *The Collected Papers of Albert Einstein,* the companion volume to the annotated, original-language documentary edition. As we have stated in all earlier volumes, these translations are not intended for use without the documentary edition, which provides the extensive editorial commentary necessary for a full historical and scientific understanding of the source documents. The translations strive first for accuracy, then literariness, though we hope that both have been achieved. The documents were selected for translation by the editors of *The Collected Papers of Albert Einstein.*

We thank Ann M. Hentschel, Osik Moses, and Klaus Hentschel for their continuing good work and dedication to making Einstein's writings and correspondence available to the English-speaking world; Diana Kormos Buchwald for her attention to the translation project; Rosy Meiron and Jennifer E. James for stylistic suggestions; and A.J. Kox for typesetting the manuscript and producing the camera-ready copy, assisted by Rudy Hirschmann, Terri O'Prey, and Linny Schenck. Many thanks to Alice Calaprice for her thorough copyediting of the manuscript.

Finally, we are most grateful to the National Endowment for the Humanities, Grant RQ-50267-7, and to the National Science Foundation, Grant No. SES-1058125, for their support of our project.

Princeton University Press
June 2012

LIST OF TEXTS

In this list, writings are indicated by a **bold** running number.

SELECTED TEXTS

Vol. 3, 10a. "On Boltzmann's Principle and Some of Its Direct Consequences"

[Zurich, 2 November 1910][1]

~~Thermodynamics~~ is ⟨known to be⟩ based on two principles, the energy principle (also called the 1st law) ~~and the~~ principle of the irreversibility of natural events (also called the 2nd law). ⟨This latter principle ~~by states,~~⟩ The substance of this latter principle can be expressed ⟨in the Planckian sense⟩ thus, ~~according to Planck~~⟩

All natural science[2] is founded on the presumption of an ⟨unbroken⟩ entirely causal relation for all occurrences. Let us assume that Galileo had found from his pendulum experiments that the period of one oscillation of this pendulum changed in a very irregular way. Let us assume furthermore that this change could not be connected with any change in other observable relations. Then it would have been impossible for Galileo to gather his observations under a law. If all phenomena accessible to us had as irregular a character as we have just pictured in this fictitious case, people would certainly never have resorted to scientific endeavors.

Which characteristic must phenomena have in order for science to be possible? ⟨To this, one might first want to reply somewhat as follows: If we put a system into a particular state, then provided this system is separated from other systems—such as by a large spatial distance—then over time the course of the states of this system is completely determined; i.e., if we put ⟨two⟩ arbitrarily many equally composed isolated systems into exactly the same state and leave these systems alone, then the evolution of the phenomena in time is exactly the same for all these systems.⟩

Now, how about the ⟨unbroken⟩ entirely causal connection between events according to our knowledge today? This question has to be specified more precisely before it can be solved. Let us do this right away by means of an example. Take a cube of copper of a given size. Within this cube ⟨we imagine⟩ we establish by external influences a very specific temperature distribution and then, after having enveloped it in a thermally insulating shell, we leave it be. We know that in the course of time a temperature equilibrium will then set in by the process of thermal conduction. The temperature gradient at all points of the cube will thereby prove to be "uniquely determined" by the initial state; by the expression "uniquely determined" we mean that we are always going to perceive the same temperature gradients, no matter how often we may repeat the experiment, that is, no matter how often we may set the starting temperature distribution and then leave the cube be. Does this unique determinability of the process, this entirely causal connection of events really exist? In order to address a pertinent objection but of no interest to us here, we rather pose this question in the following form: Do we always establish an entirely causal connection between the events to closer approximation, the more exactly we realize the initial state and the more exactly we follow the process in time by measurement?

The point of view of physicists toward this problem changed considerably in the last century. If we leave aside for the time being Brownian motion, radioactive fluctuations, and a few other phenomena, which came into the focus of scientific interest in the past few years, we arrive decidedly at the verdict that according to [p. 3] experience[3] an entirely causal connection in the last sense indicated does exist. Nevertheless, physicists, more specifically, heat theoreticians, managed to deny the entirely causal connection between events, more precisely speaking, between events insofar as they can be the objects of observation. Let us throw a fleeting glance at this development! From the simple idea that gases are composed of material points (molecules) that essentially only act upon one another mechanically by contact (collision), Clausius was able to derive a relation between the specific heats and the constants of the equations of state of monatomic gases as well as a ratio between thermal conduction, internal friction, and the diffusion of gases, which magnitudes or phenomena, resp., were entirely unrelated without Clausius's theory.[4] This major success prompted physicists to attribute heat phenomena to irregular motions by molecules. This kinetic theory of heat required, however, that the laws of conduction, etc., be taken as only approximately valid laws; according to this theory a precisely valid law of conduction is absolutely impossible, being instead just a law of averages. That deviations from these averaging laws must normally be very small is unimportant in principle.

The kinetic theory so broadly supported by experience is not just incompatible with the presumption that observable occurrences be related ⟨precisely⟩ completely causally, however. The analyses carried out by ⟨Cla⟩ Maxwell, Boltzmann, and Gibbs[5] also show that randomly large deviations from those averaging laws would have to occur within the range of observation, even though this happens so rarely in most ⟨groups of phenomena accessible to observation⟩ systems according to the theory that we are not really able to detect such deviations.

[p. 4]   The following well-known consideration shows most concisely that the laws of conduction, just as all other laws concerning irreversible processes, cannot be precise. According to the kinetic theory of heat the reversal in time of any process of molecular motion is likewise a possible process of motion; hence no such thermal process exists that could not also run in [the] opposite direction. Therefore, from the point of view of the molecular theory of heat, it has to be regarded as possible that by mere thermal conduction heat would flow from a colder body into a warmer one. Why do we not observe this? Doesn't this consideration show that the kinetic theory of heat has to be abandoned?

This question was answered by Boltzmann, specifically, in [the] following way: Observe some closed physical system having a particular given amount of energy. We signify by $Z_1, Z_2 \ldots Z_l$ all observable states that this system can assume at the given energy value. In the example of the copper cube, each Z_ν would thus mean a particular temperature distribution, where in total l distinguishable temperature distributions are possible. But now it is assumed that these states Z be of entirely different probabilities, such that from among all the states differing very little from a given state Z_a, one (Z_b) be far more probable than all the others, at least provided that Z_a differ substantially from the so-c[alled] state of thermodynamic equilibrium. Then, if brought into the state Z_a and then left to itself, the system is far more likely to change into the state Z_b than into any other states neighboring state Z_a. The probability that this would occur can come as close to unity (i.e., certainty) as you like, although it is excluded in principle that this transition be entirely certain. T[his] m[eans]: If we bring the system very frequently into the state Z_a, then in the great majority of cases, but by no means always, state Z_b will follow state Z_a; a transition into every other neighboring state to state Z_a will also occasionally occur, even if only extremely rarely. What has been said about the transition from state Z_a into neighboring state Z_b is again valid for the change a system experiences from state Z_b in the following little segment of time. Thus one arrives at a conception of (apparently) irreversible processes.

[p. 5]

This sketch of the Boltzmann conception is incomplete. Still needing to be answered are the questions: "How should the probability of individual states $Z_1, Z_2 \ldots$ be understood?" and "Why is a transition from one state Z_a to the most probable neighboring state Z_b more probable than a transition to other neighboring states?"

For the first of these questions we note the following: According to the ⟨molecular th⟩ kinetic theory of heat, there cannot be a thermal equilibrium in the strict sense. The state we call a thermal equilibrium is one that a system left to itself for enormously long most frequently has. However, it is a consequence of the kinetic theory that over long periods of time the system will take on all possible states; in particular, the further a state is away from thermodynamic equilibrium, the more rarely does the system assume it. The copper cube left infinitely long to itself incessantly changes its temperature distribution, whereby it extremely seldom assumes temperature distributions differing considerably from the temperature distribution of thermal equilibrium. If we imagine a system under observation for an immensely long time T, there will be for most states Z_ν an abnormally small portion τ of this total time during which the system is just assuming the state Z_v. We shall call the ratio $\frac{\tau}{T}$ the probability W of the state concerned.

[p. 6]   If this definition of the probability of a state is taken as a basis, it is generally understandable that a system changes on average from one state Z_a such that from this state follows with the greatest probability the neighboring state Z_b. I only have to mention this without going into the proof. ⟨This is the answer to the second of the questions posed above.⟩

It is essential that the definition of the probability of a state be definable independent of the kinetic picture; probability W is a magnitude in principle accessible to observation, even though in most cases direct observation of it is excluded, owing to the brevity of the time at our disposal.

If a system in a state substantially differing from thermodynamic equilibrium is left alone, it successively assumes states of ever greater W. A state's probability W shares this property with the entropy S of a system, and Boltzmann found out that the relation between W and S

$$S = k \lg W$$

holds, where k is a universal constant, i.e., independent of the system chosen. ⟨This is the important equation that ... the mathematical expression of the Boltzmann conception⟩

This Boltzmann equation can be applied in two different ways. There can be a more or less complete picture of molecular theory on the basis of which one can calculate the probability W. The Boltzmann equation then yields the entropy S. This was how Boltzmann's equation has mostly been applied hitherto.

Example.[6] In a volume V, let there be N molecules, i.e., one gram molecule of a particular type. The volume is large enough compared to the eigenvolume of the N molecules and of the other existing matter besides the N molecules—provided such matter is there—distributed evenly over V_0 so that the various points of V_0 are equivalent for each of the N molecules. This is an incomplete expression of how we visualize an ideal gas or a dilute solution. How large is the probability W that at a randomly chosen instant all N molecules are within the partial volume V of volume V_0?

[p. 7]

A simple consideration yields

$$W = \left(\frac{V}{V_0}\right)^N.$$

From this, using the Boltzmann ⟨constant⟩ equation, we find

$$S = kN\lg\left(\frac{V}{V_0}\right) = kN\lg V + \text{const.,}$$

where the constant "const." can depend on the temperature but not on the volume. From this we immediately obtain the force that the N molecules are able to exert on a wall that forces them to remain within volume V. For, if the energy of the system is independent of V, and if G signifies the work received upon infinitesimally enlarging volume V along a reversible path, then

$$pdV = G = +TdS = + kNT\frac{dV}{V}, \text{ holds,}$$

hence $$pV = kNT.$$

We thus have the equation of ideal gases and osmotic pressure. At the same time it is revealed that the universal constant kN of this equation is equal to constant R of the gas equation.

In my opinion, the main importance of the Boltzmann equation does not lie in that with its help one is able to calculate the entropy for a known molecular scenario. Rather, the most important application is, conversely, that by means of Boltzmann's equation one can get the statistical probabilit[ies] of the individual states from the empirically established entropy function S. Thus it is possible to assess how much the systems' behaviors differ from the behavior required by thermodynamics.

[p. 8]

Example.[7] A particle that is slightly heavier than the liquid in which it is suspended and which it displaces.

Such a particle should, according to thermodynamics, sink to the bottom of the vessel and stay there. According to Boltzmann's equation, however, a probability W is ascribed to each height z above the bottom; the particle incessantly changes its height in an irregular way. We want to determine S and from it W. If μ is the particle's mass and μ_0 is that of the liquid it displaces, then the work $A = (\mu - \mu_0)gz$ has to be expended to raise the particle to height z above the bottom. In order for the energy of the system to stay constant, the amount of heat $G = A$ has to be removed from the system, whereupon the entropy diminishes by $\frac{G}{T} = \frac{A}{T}$. Therefore,

$$S = \text{const} - \frac{1}{T}(\mu - \mu_0)gz.$$

From the Boltzmann equation it follows, if one substitutes the value $\frac{R}{N}$ for k:

$$W = \text{const}\, e^{-\frac{N}{RT}(\mu - \mu_0)gz}.$$

If there are many particles present in a single liquid, then the right-hand side of the equation indicates the distribution density of the particles as a function of depth. Perrin tested this relation and found it confirmed.[8]

From this relation the law of Brownian motion can very easily be deduced. For, from it, it immediately follows that the mean height $\bar{z}$ of a particle above the vessel bottom is equal to

$$\frac{\int z e^{-\frac{N}{RT}(\mu - \mu_0)gz}\,dz}{\int e^{-\frac{N}{RT}(\mu - \mu_0)gz}\,dz} = \frac{RT}{N} \cdot \frac{1}{g(\mu - \mu_0)} \quad .$$

Now, however, because of its greater density the particle drops downward, according to Stokes's law, by $D = \dfrac{g(\mu - \mu_0)}{6\pi\eta P}\tau$

in the time τ if η signifies the liquid's viscosity coefficient and P, the radius of the

[p. 9] (spherically shaped) particle. But in the same time τ, as a consequence of the irregularity of the molecular thermal process, [it] is also shifted a distance Δ upwards or downwards, where positive and negative values for Δ appear equally frequently; so $\overline{\Delta}$ is $= 0$.

A particle that, before time τ has elapsed, is located at height z, is, after τ has elapsed, at the height $z - D + \Delta = z'$. As the distribution law of all the particles should not depend on time, the mean value of z^2 must be equal to z'^2, therefore,

$$\overline{(z - D + \Delta)^2} = \overline{z^2},$$

or for sufficiently small τ, D^2 is negligible and $\overline{z\Delta} = \overline{D\Delta} = 0$

$$\overline{\Delta^2} = 2\bar{z}D = \frac{RT}{N} \cdot \frac{1}{3\pi\eta P}\tau \quad .$$

This is the familiar law of Brownian motion, which has likewise been confirmed by experience.[9]–

The just-described example of a particle suspended in a liquid offers a fitting depiction of Boltzmann's conception of irreversible processes. For, if we imagine a particle suspended in such a tall vessel, and that it is so much heavier than the displaced liquid that the expression for probability W is very small, even at a height z just barely above the bottom of the vessel when compared to the value W_0 for $z = 0$, then very rarely will the particle rise much from the bottom, once it has reached the bottom (thermodynamic equilibrium). If we raise the particle to a considerable height z, then, obviously, it will with the greatest probability sink back down to the bottom (irreversible process) in order then to dance up and down, as before, in its proximity. If this sinking back did not occur in the overwhelming majority of cases, a probability function of the assumed quality could not be valid.–

Before I go into other applications of Boltzmann's equation, I would like to draw [p. 10] a general conclusion about it regarding the mean size of the fluctuations that the parameters of a system perform around the values for ideal thermodynamic equilibrium.[10] $\lambda_1 \ldots \gamma_n$ are parameters defining the state of a system. The null values for the λ's are chosen so that at thermal equilibrium $\lambda_1 = \lambda_2 \ldots$ is $= 0$. The work that, according to thermodynamics, would have to be performed in order to bring the system from the state of thermodynamic equilibrium into the state very close to thermodynamic equilibrium characterized by the values $\lambda_1 \ldots \gamma_n$, is

$$A = \sum A_v = \sum_1^n \frac{a_v}{2} \lambda_v^2.$$

In order that the system's energy be the same as before, after the state has been established, the amount of heat $G = A$ must be removed from it, which corresponds to a reduction in the system's entropy by $\frac{G}{T} = \frac{A}{T}$. Thus, if the system has assumed the considered state on its own, its entropy is

$$S = \text{const} - \frac{1}{T}\sum_1^n \frac{a_v}{2}\lambda_v^2.$$

If this is plugged into the Boltzmann equation, one obtains

$$W = \text{const } e^{-\frac{N}{RT}\sum_1 \frac{a_v}{2}\lambda^2{}_v}$$

In this case, therefore, Gauss's law of error distribution applies to the deviations of the individual parameters from the values for thermodynamic equilibrium. For the mean work $\overline{A_v}$ that according to thermodynamics would have to be expended to bring the parameter λ_v from equilibrium to the temporal mean $\sqrt{\overline{\lambda_v^2}}$ in a reversible process, one obtains the value

$$\overline{A_v} = \frac{RT}{2N}.$$

[p. 12]

This result can be expressed thus: Provided A can be described in the above-indicated manner in the vicinity of thermodynamic equilibrium, deviations from the state of ideal thermodynamic equilibrium set in of their own accord; these deviations are, on average, for each parameter so large that the work required by thermodynamics for arbitrary generation of the deviation is equal to one third of the mean kinetic energy of the propagating motion of a gas molecule at the same temperature. Perceptible deviations from the state of ideal thermodynamic equilibrium occur everywhere where the performance of work, however small, can achieve a perceptible effect. Measurement of each such deviation provides us with a determination of the energy of the monatomic gas molecule, hence also a determination of the absolute size of the atom.

Smoluchowski has indicated a very interesting application of this general result. According to classical thermodynamics, the individual components of a phase in the case of thermodynamic equilibrium are distributed evenly over the volume of the phase. From what was said before, irregularities must, on the contrary, occur in the spatial distribution of the matter, which are greater, the weaker are the forces opposing a change in the even distribution of the matter, or, resp., the separate, independent components. So the phase is in reality inhomogeneous, which makes itself noticeable by an optical opacity (opalescence). This opalescing is particularly strong in the vicinity of critical states (for uniform substances and for solutions), because in these cases only slight forces oppose a change in density or concentration. A short while ago I demonstrated that on the basis of the outlined interpretation by Smoluchowski an exact calculation of the light diffracted by opalescence is possible.[11]

[p. 13]

Finally, I would not like to leave unmentioned that by means of the Boltzmann equation the statistical properties of thermal radiation are derivable in a simple way from the laws of thermal radiation, more precisely speaking, without having to avail oneself of electromagnetic and kinetic heat theory. The problem is the following: In a cavity surrounded by opaque bodies at temperature T, there is radiation whose quality is determined solely by the temperature. At time τ a defined radiant energy E passes through a surface σ imagined to lie somewhere within the cavity; its directional range is indicated by a definite elementary cone $d\Omega$ and its frequency range is $d\nu$. If this radiant energy is imagined to occur often and moreover is measured very precisely, then one would not always find the same value E but a magnitude $E = E_0 + \varepsilon$, slightly deviating from

a mean E_0. One would seek the quadratic mean $\overline{\varepsilon^2}$ of this magnitude ε. This problem is of essential interest because its solution contains a statement about the structure of thermal radiation.

I would just like to suggest the way this problem can be solved. If some arbitrary body K is thermally related with another one of relatively infinitely large thermal capacity, then according to thermodynamics K will take on the temperature of this second body and continue to maintain it. According to Boltzmann's principle, however, the temperature of K will constantly change, albeit rarely substantially far removed from the temperature of thermal equilibrium; the Boltzmann equation yields the mean of those temperature fluctuations. The temperature fluctuations thus obtained are entirely independent of the way in which the thermal exchange between K and the relatively infinitely large body takes place; the temperature fluctuation is of the calculated magnitude even when this thermal exchange takes place exclusively by way of radiation. One then only has to investigate the question: Which would have to be the statistical properties for the radiation to really generate the calculated temperature fluctuations? If this analysis is carried out as suggested, the result one obtains is that the temporal fluctuations of the thermal radiation at a low radiation intensity and great frequency are far greater than would be expected according to our current theory.–[12]   [p. 14]

If we now in closing again ask the question: "Are observable physical phenomena connected entirely causally with one another?" we definitely have to negate this question. To even the most conscientious observer, the positions a particle undergoing Brownian motion in two time values separated by one second must always seem completely independent of each other; and the greatest mathematician will never succeed in calculating in advance in a specific case, even approximately, the path traveled in one second by such a particle. According to the theory, in order to be able to do so, one would have to know precisely the positions and velocities of all the individual molecules, which in principle appears to be out of the question. Nevertheless, the ever-reliable ⟨exact⟩ averaging laws as well as the statistical laws valid for these fields of the subtlest effects lead us by means of these fluctuations to the conviction that we must retain in the theory the assumption of a completely causal connection between events, even if we may not hope ever to attain by refined observations direct confirmation by Nature of this conception.

Vol. 5, 315a. From Heinrich Zangger[1]

[Zurich, after 28 November 1911][2]

[Not selected for translation.]

Vol. 5, 505a. To Paul Langevin

[Zurich], 19 January 1914

[Not selected for translation.]

Vol. 8, 86a. From Heinrich Zangger

[Zurich, after 28 May 1915][1]

[Not selected for translation.]

Vol. 8, 95a. From Heinrich Zangger

[Zurich, 9 July 1915][1]

[Not selected for translation.]

Vol. 8, 95b. From Heinrich Zangger

[Zurich, 12 July 1915][1]

[Not selected for translation.]

Vol. 8, 113a. From Elsa Einstein

[Berlin, 31 August 1915]

D[ea]r Albert,

I just sent a long letter to you at the Glockenhof in Zurich; now your postcard arrives with the news that you're staying there until Wednesday. I think that's entirely right of you, after not having seen your mother for so long.[1] You're surely having a very nice time over there. Don't think so rarely of me and don't be miserly with your news. Warm greetings also for your dear mother.

I'm visiting the Mendels at Wannsee today.[2] Yours,

Elsa.

Vol. 8, 113b. From Elsa Einstein

[Berlin, 1 September 1915][1]

Dear Albert,

Yesterday I wrote you at length to Glockenhof Hotel.[2] Meanwhile, I've become worried; maybe a hotel by that name doesn't exist there at all. Please telegraph me immediately whether that address is correct. I'm not able to write you before I receive the telegram; I'm too worried that the letter will fall into the wrong hands.

While here, you didn't know for certain whether the hotel had that name. Einstein Haberland St. suffices for the telegram. I expect to hear from you as soon as poss. Affectionate greetings!

Elsa.

Vol. 8, 113c. From Elsa Einstein

[Berlin, 4 September 1915][1]

Dear Albert,

Tomorrow it will have been a week since we've been apart. Before me lie three *very meager* postcards. That's all I've heard from you, in eight days. This third postcard just arrived and therefore I'm able to write you. I was uneasy about your address. Maybe a "Glockenhof" doesn't exist there anymore; and I was all jittery at the thought: Who will get their hands on that letter?[2] With the same surname, this conjecture was just too close at hand. Thank God! Now, at least my first letter reached you. It's so incomprehensible to me that you didn't have your children come and see you immediately in the hotel. How could you stay there even for just an hour and tolerate not having seen them? She really couldn't have refused you that.[3]

I can't tell you how much I yearn for the time when you're here again. It's such an unpleasant feeling, not being able to reach you by letter for days on end. I would have had *so much* to report. But the censors do let unduly long letters lie around for a while; in any case, they do suffer a delay. I'm recuperating visibly, you'll be

pleased about that. At the end of next week, perhaps Thursday, I want to go to Heilbronn, from there to Hechingen.[4] I'd like to meet you along the way in order to
accompany you then on your return trip via Frankfurt. But these are preliminary
plans; whether they will be realized is a second question. Yesterday I went into your
apartment;[5] I conscientiously looked through your mail. Some things could be
among them that needed to be immediately dealt with. Weinstein wrote you that
he'd like to speak with you. You may set a time.[6] Miss Sidy Fischer, chem. stud.,
argues in a finely composed letter that she would like to become your secretary; she
requested an answer. You had supposedly already written to her three years ago
from Zurich. I answered both very shortly and correctly: Prof. E. is traveling in
Switzerland for the time being. That way these people don't have to wait in vain.
You have me as secretary; I am entirely at your disposal in this regard!–

I imagine you'll be staying in Zurich; you do want to use the time available to
you to see the children daily? You'll get much more out of that than if you went
hiking. Follow my advice. Greet Mr. Zangger for me. Don't forget to buy a gift for
the child.–[7]

A warm hug from me! In response to your "best regards"!

Elsa.

Vol. 8, 113d. From Elsa Einstein

[Berlin, 5 September 1915]

Dear Albert,

Sunday today, without any news from you. It's been a week, and just three slim
cards the whole time.[1] So, could you keep your promise? You did want to please
me so much with your letters.–

Meanwhile you saw the children.[2] How did they welcome you? I'm always
thinking about that, for days. And otherwise? There are so many things I'd like to
ask you; do pull yourself together and write a proper letter.–

At home everything's taking its usual course. Last night I was visiting the F's in
Grunewald; they're really true friends of mine. The children are dear. Healthwise
I'm "on top of it" again. No more news from the "[New Fatherland] League" anymore, what a pity![3] I'm counting the days until you're back home again. I'm hoping to be able to depart on Thursday.[4] Affec. gr. and a kiss,

Elsa

In case I do travel, I'll go first to Heilbronn.

Vol. 8, 113e. From Pauline Einstein

Heilbronn, 7 September 1915

My dear Albert,

I was hugely pleased about your postcard; I received it yesterday & I hope that you soon give me more news & then have more to tell me.

So now you're surely staying with your friend Zangger;[1] I'm so anxious to know whether you saw your children & how. Were they just as happy to see you? Or had their mother already totally poisoned them?

You're surely spending the time quite nicely; how did you get on with your various friends? Did you write to Maja right away? & when will you be going to see her?[2] I'm also eager to know how the reunion with Uncle Jacob went.[3] He was letting off quite some steam about you, but that'll have dissipated long ago, I hope.

I still often think about our fine time together; it was just a little short![4] I still do talk a lot about it with Guste;[5] & Mr. O[ppenheimer] greatly enjoyed having his dear guest;[6] he sends his cordial greetings. Guste is leaving tomorrow for a few weeks; I don't like that at all; maybe her absence will be shortened by my also going away for a time; Mr. O. is going on about Baden-Baden, you know; hopefully something will come of it.

From Aunt Fanny[7] I received a big letter, but there's nothing noteworthy in it; a postcard from Elsa arrived after you had departed; I enclose it.

Do write me in detail & my love & kisses from your

Mama.

Vol. 8, 113f. From Elsa Einstein

[Berlin, 7 September 1915]

My d[ea]r Albert,

Today, a sign of life, at last; but again only in the form of a paltry postcard. And you could be telling me so much from there. Throughout your 10-day trip, four scant little postcards: That costs 15 minutes time. You really could have been a little more generous. Or do you think I'd become too "uppity"? In the meantime you saw the boys[1] and I'm dying to hear how the children met you. I really don't understand why you didn't summon them immediately after you had arrived. Despite everything! I intend to travel to Heilbronn the day after tomorrow.[2] I'm hurrying up now, even though this time is very inconvenient for me to be able to be back here again at the same time as you. I'm looking forward to that like a child. Do please send me a daily note; I beg you. Countless questions I have to ask you. Yet a post-card like this is something "official," and a letter, under the circumstances, likewise. Heartfelt gr. & kiss also from the children.[3]

Elsa.

Vol. 8, 113g. From Elsa Einstein

[Berlin, 7 September 1915]

Dear Albert,

Still no message today and I'm in such suspense about your report about the children. Until now I've had to content myself with those sparsely written post-cards. I imagine that some of my letters got lost because you *never made any mention of them*; and yet I've been writing quite long letters so frequently already. Tomorrow I'm traveling to Heilbronn; and the Hechingen trip weighs down on me like a ton. If only I had that fatal mission behind me, I'd breathe freely again. It's embarrassing for me there for other reasons as well; the family regards me with great distrust.[1] I'll not write more today, I'm too jittery; and you do only want to get cheerful letters.

Most affectionately!

Elsa.

Vol. 8, 116a. From Elsa Einstein

[Konstanz, 12 September 1915]

Dear Albert,

I'm on my way to Constance. My plan is to stay at the Waldhaus Jakob for about 5 days.[1] What a pity that we can't see each other! 2 hours distance from Zurich!

Stuck in Singen right now;[2] I'll have to pass by here again on the way back. Now it's been two weeks already; you didn't want to stay longer! I'm looking forward to this magnificent autumnal stay by Lake Constance. Yours,

Elsa.

Vol. 8, 177a. From Paul Ehrenfest

Leyden, 1 January 1916

[Not selected for translation.]

Vol. 8, 493a. To Heinrich Zangger

[Berlin, after 26 March 1918][1]

My dear friend Zangger,

In greatest distress I gather from my *Züricher Zeitung* that you lost your beloved little daughter Trudi after a very brief illness.[2] My heartfelt condolences to you and your wife. It's terrible how destiny treats the best of us. How much I regret, now more than ever, having annoyed you recently![3] I'll try with all my might to make amends. I hope at least little Gina is well again,[4] so that you can regain your emotional balance somewhat.

Politics has settled in my stomach and is grumbling there. The eye searches in vain for something to look at with joy. I seek refuge in the objective, in articles and proofs. Weyl wrote a brilliant book about the general theory of relativity;[5] his departure from Zurich would be a great loss,[6] it seems to me. Albeit I did hear that his lectures were virtually incomprehensible to the students. In any event, he is outstandingly talented.

My wife and I now have a quite satisfactory relationship, despite my wanting to divorce. I'm very satisfied that she, and as it seems, Tete[7] also are feeling reasonably well. There's a lively exchange of letters between me and her; and now I believe that it works best if I discuss all matters openly with her.

Dear Zangger! A person as valuable to others as you are should not abandon himself to grief. All the Swiss are your brothers and your children; and you perhaps do not even know with how much joy and sympathy many of those in your nearer or farther surroundings look upon you and your work.

Fond greetings from your

Einstein.

Vol. 8, 510a. From Heinrich Zangger

[Zurich, after 16 April 1918][1]

[Not selected for translation.]

Vol. 9, 35a. To Luise Karr-Krüsi[1]

[Berlin,] 6 May 1919

[Not selected for translation.]

Vol. 9, 140a. To Albert Karr-Krüsi

[Berlin,] 17 October 1919

[Not selected for translation.]

Vol. 7, 33a. Statement on the Hebrew University

[Berlin, 18 February 1920][1]

The thought that the dream of a Jewish university is now close to materialization elates me. Considering the general interest among Jews in academic things and the great hurdles impeding the Eastern European Jews from studying, establishing the Jewish university would be a necessity even if the development of Palestine had no need for an intellectual center. However, we also need the university to train Palestine's academic youth so that the country can be motivated to ⟨a⟩ develop its own cultural life. Interest in the new university among Jews living abroad will assure that a lively exchange be maintained between the new university and the civilized nations of Europe and America so we do not need to fear crippling isolation. May the university become a new shrine for our nation!

Vol. 7, 39a. Page proofs for "Propagation of Sound in Partly Dissociated Gases"[1]

[Berlin, before 29 April 1920][2]

[Not selected for translation.]

Vol. 10, 80a. From Mileva Einstein-Marić

[Zurich, before 23 July 1920][1]

[Not selected for translation.]

Vol. 7, 45a. Opinion on Jakob Grommer's Textbook Project[1]

[Berlin, before 11 October 1920][2]

Expert Opinion by Professor A. Einstein on the Mathematical Physics Textbook by Dr. J. Grommer

Dr. J. Grommer has been working diligently in constant consultation with me on the textbook assigned to him. He has already developed part of the material using a modern scientific approach in as simplified and clear a form possible.

Basic arithmetic (elementary operations, irrational numbers, etc.), ca. 2 printed sheets

Algebra (linear equations) 1 ½ printed sheets

Mr. Grommer encloses herewith the outline of the entire work, of which I thoroughly approve.

I am convinced that Mr. Grommer, on the strength of his rare mathematical knowledge and talent, is very suitable for authoring an academically appropriate textbook for the university in Palestine of use to students in engineering as well as to aspiring mathematicians. He is working on it with dedication and zeal.

I would like to permit myself to advance the following for consideration. Mr. Grommer has been familiar with Hebrew since his youth. Nonetheless, it should be realized that the technical terminology in use in the exact sciences, in a certain sense, still needs to be devised in Hebrew. Therefore, in my view, the right thing would be for Mr. Grommer to be in direct contact with the mathematics teachers in Palestine so that the mentioned terminology evolve as uniformly as possible.

Transplanting Mr. Grommer to Palestine would be advantageous also because then the immediate teaching needs could guide his choice of material completely.

sig. A. Einstein.

Vol. 7, 50a. On the Present Situation in Theoretical Physics

[Vienna, 14 January 1921][1]

If I survey the present situation in theoretical physics, I find one point of the greatest importance that is not taken into account sufficiently. A theoretical system can only claim completeness when the relations between the concepts and the facts of experience are unambiguously established. It does not suffice, for example, to base the theory of relativity only on a fundamental mathematical invariant.

It also has to be clear how this invariant relates to the observable data, as has happened with the fundamental concepts of Maxwellian theory through Heinrich Hertz.

If this aspect is left out of consideration, one can only arrive at unrealistic systems.

Vol. 7, 52a. Opinion on Eggeling and Richter's Project

Berlin W. 30, 3 February 1921 Haberlandstr. 5
[Not selected for translation.]

Vol. 7, 56a. "International Relations in Science"

[2 April–10 August 1921?][1]

... In this agitated time of strife among nations and the social strata, one of the most precious goods of humanity seems to be under threat: the internationality of science.[2] Scholarly organizations in the individual countries have allowed themselves to be carried away by nationalistic passions to the point that they rival political bodies in the political contest; so much did they forget that they are made to foster and preserve endeavors that must stand high above all political struggles among people ... I believe the most important thing is to awaken in the younger generation a strong love for scientific truth and ambitions, so that the purer atmosphere thus created will gradually drown out the insensitive emotional motives that have brought so much misfortune upon our current generation ...

Vol. 7, 56b. Professor Einstein on the Proposed Hebrew University of Jerusalem

[before 3 April 1921][1]

[See documentary edition for English text.]

Vol. 7, 60a. Calculations on a Cooler

[July 1921– March 1922][1]

Given to Mr. Müller & Nernst[2]

In isothermal compression there must be supplied:

1) work

$$A = -pv + (p + \Delta p)(v + \Delta v) - p\Delta v = v\Delta p;$$

[3]

2) heat

$$W = \Delta u + p\Delta v.$$

Hence
$$A + W = \Delta(u + pv).$$

The negative thereof is the total energy (per unit mass of the working fluid[4]) discharged from the compressor, therefore, according to the equation of energy, also the heat extracted or cold discharged from the other component, if the effect of heat conduction between exchanger and compressor is neglected, which is not always permissible. (If heat flows from exchanger to compressor, then the amount of heat discharged outwards by the latter is conserved, that is, also the cold discharged by the other component.) If one ignores this, then

$$K = -(A + W) = \left(-\frac{\partial u}{\partial p} - v - p\frac{\partial v}{\partial p}\right)\Delta p.$$

According to the second law,

$$-\frac{\partial u}{\partial p} - p\frac{\partial v}{\partial p} = T\frac{\partial v}{\partial T}$$

hence also

$$K = \left(T\frac{\partial v}{\partial T} - v\right)\Delta p$$

and

$$\frac{K}{A} = \frac{T\,dv}{v\,dT} - 1 \quad \text{(taken at the temperature of the compressor).}$$

If there is also cooling between compressor and exchanger, then the total cold developed by the exchanger and evaporator is raised by an amount equal to the removed heat W'. The degree of efficiency thereby rises by $\frac{W'}{v\Delta p}$.

A more accurate theory must take into account the thermal conduction of the exchanger, as a completely insulated nonconducting exchanger could not have a stationary temperature distribution.

Comment: A lower limit for the attainable refrigeration is given in that at the lowest temperature (evaporator) $\frac{T\,\partial v}{v\,\partial p} - 1$ cannot become negative.

Vol. 7, 65a. Expert Opinion on Proposal by Heinrich Löwy[1]

Berlin, 12 October 1921

Expert Opinion[2]

I know about the suggestions by Dr. Heinrich Löwy, Vienna, regarding ground depth analysis through the influence of the ground's interior on an external oscillating electric circuit.[3] In my view this proposal deserves serious examination from the technical angle. From the standpoint of the theory it may be stated that Dr. Löwy's method rests on solid theoretical bases and that the anticipated effects do lie thoroughly within the range of modern measurement precision.

1. To Charlotte Weigert[1]

[Berlin, early 1922][2]

Dear Miss Weigert!

Elsa has been pestering me dreadfully about giving you a recommendation to Niels Bohr.[3] But I had to resist, I'm sorry. Bohr is the greatest genius in physics today, enormously busy and mentally exhausted for a long while. Now he is immensely productive again. Out of consideration for this person all other interests must give way, legitimate or serious though they may be. As long as he is fruitfully at work, I shall never be held to blame for his being disturbed. You must understand this and not think of it as an unwillingness to help.[4]

I am very glad that you are experiencing such fine success with your teaching. It is doubly difficult for a woman to be heard and win recognition. That you are offering lectures for workers I rate particularly highly. I also tried it but noticed that it is quite difficult to fathom the way of thinking of people of so entirely different a mental prehistory.[5]

With best wishes for 1922, I am yours,

Albert Einstein.

2. "Preface" to Bertrand Russell, *Political Ideals*

[*Einstein 1922d*]

PUBLISHED 1922

IN: Bertrand Russell, *Politische Ideale*. Berlin: Deutsche Verlagsgesellschaft für Politik und Geschichte m. b. H., 1922, p.[5]

The availability to the German public of this great English mathematician's[1] lucid discourse is very welcome. It is not some wavering professor vacillating between "on the one hand and on the other" speaking to us here, but one of those resolute, straightforward individuals who exist independent of the period into which they happened almost by chance to have been born. Unbending consistency and warm human sensitivity prescribe his path. He follows down this path unperturbed by the consequences his stance may bring. He did not play the martyr when he let himself be robbed of his professorship and found himself in prison for anti-militaristic propaganda.[2]

He wants military force to be completely eliminated and recommends consistent training of the population in organized passive resistance as the means of counteracting aggressive military force from abroad. This solution will not appear utopian to those who experienced the Kapp putsch[3] in Germany.

Russell furthermore addresses the problem of social policy. Driven by an ardent interest in the progress of human organization, he traveled through Bolshevist Russia to learn its lessons.[4] His ideal is the development of the free creative powers of the individual within a social order that banishes fear about one's livelihood without lapsing into a hypertrophic bureaucratism, the worst enemy of socialistic endeavors.

We may or may not agree with individual details of Russell's opinions. What a delight it is to become acquainted with the thoughts of a sharp-minded and truly noble man of our time on issues that touch all serious people today.

Let each person make up his own mind about this great Englishman.

A. Einstein

3. "The International Character of Science"

[Berlin, before or on 1 January 1922][1]

⟨Prof. Dr. A. Einstein⟩[2]

Contribution to the textbook by Graf Kessler and Ströbel. January 1922[3][4]

When during the war nationalist and political delusions had reached their zenith, Emil Fischer emphatically stated at an Academy meeting: "You can do nothing, gentlemen, science is—and shall remain—international!"[5] This the great scientists always knew and passionately felt, even if during times of political conflict they remained in isolation from their fellows of lesser stature. During the war this mass of eligible voters promoted in every camp the sacred wares entrusted to them. The internationalist association of academies was blasted asunder.[6] Conferences were—and still are—being organized to the exclusion of fellow professionals from former enemy countries.[7] Solemnly argued political considerations stand in the way of the supremacy of purely factual considerations so essential in fostering the great causes.

What can the well-intentioned do, who are not cast down by the emotional attacks of the moment, in order to win back what has been lost? Truly international large-scale conferences still cannot be organized owing to still existing agitation by the majority of intellectual workers; psychological resistance to the restoration of internationalist scientific labor associations is still too powerful to be overcome by the minority among them inspired by grander motives and feelings. They can serve the greater cause of the recovery of internationalist societies by maintaining close contacts with like-minded fellows from all the other countries and by persistently advocating international interests within their own spheres of influence. Overall success will take its time; but it is sure to come. I would not like to let this opportunity pass without pointing out with admiration that, particularly among a large number of our English fellow professionals, the effort to uphold the intellectual community has remained alive throughout all these difficult years.[8]

Everywhere, official proclamations are worse than the mentality of the individual. The well-intentioned should bear this in mind and not let themselves be irritated and led astray: *senatori boni viri, senatus autem bestia* [the senators are good men but the senate is a malicious animal].[9]

If I am full of confident hope with regard to the advancement of organized internationalism in general, this is based less on trust in reason and the nobility of convictions than on the despotic pressure of economic development. Because it is based to a high degree on the intellectual efforts even of backward-minded scientists, they too will involuntarily help make organized internationalism a reality.

4. From Max Born and James Franck[1]

Göttingen, 1 January 1922

Dear Einstein,

We, Franck and Born, are extremely crushed by the content of your letter, even though, in our stupidity, we cannot replicate the setup for the canal-ray experiment for ourselves.[2] We have 1,000 questions on our minds and all sorts of considerations for which we need you as a tranquilizer. As this letter cannot become 50 pages long and we are also afraid that we mustn't expect a 100-page reply, we had the brilliant idea of having you officially invited to visit us in Göttingen at the expense of the Wolfskehl endowment to deliver a talk in an informal manner. We have the secondary motive of having you here for Hilbert's 60th birthday, a thought that put the old gentleman in raptures.[3] The birthday is on January 23rd; the talk could be on Tuesday the 24th, while you would have to devote at least Sunday the 22nd to us. Perhaps your wife feels like coming along. It would be wonderful if you could make this possible; and we are so much looking forward to this prospect that you must not decline under any condition. Warm regards and good wishes for the New Year.

Born and Franck.

5. From Hermann Weyl[1]

Arosa, Villa Anita, 3 January 1922

Dear Colleague,

Many thanks indeed for your efforts regarding the printing permission; it is good that it turned out to be superfluous, as the printing had already been completed in the interim!—[2]

I do not know anything more specific about the canal-ray experiment; is it an experimental answer to the question of whether whatever it is that would be excited by a Bohr-like atomic jump is an ether wave?[3] (namely, the answer no)? Thus I do not know what to reply to your question: Now what??[4]—nor would I know, of course, even if I were fully informed about your experiment. This is yet another tidbit you found for yourself, reconciling apparently completely contradictory things within an overarching principle. I already feel sorry for the poor "field" in advance; but its theory has meanwhile acquired such a degree of harmony that it is—ripe for its demise; it almost seems to me that it is. Best wishes for the New Year and good luck along the new path! Yours,

H. Weyl.

6. To Max Born

[Berlin, 6 January 1922]

Dear Born,

I'll gladly come and visit you, partly to congratulate Hilbert personally, partly to tell you about the experiment, simple as it is.[1] The trick is this: According to the wave theory, the canal-ray particle emits contin[uously] variable color in different directions. Such a wave propagates in dispersive media at a velocity that is a function of the location. Therefore a bending of the wave planes would have to follow, as with terrestrial refraction. But the experimental outcome is reliably negative.

Cordial greetings also to Franck and your family, yours,

Einstein.

7. From Hedwig Born

[Göttingen,] 7 January 1922

Dear Mr. Einstein,

First of all, heartfelt thanks to you and your wife for the New Year's greeting to us that exuded such friendly warmth![1] May all your wishes be multiply fulfilled.– I hurried over to Hilbert with your postcard[2] from today, who at first hardly wanted to believe that you were really coming and then was enormously pleased.[3] He asks me to write you to definitely be here on his birthday, Monday the 23rd, and to appear among the large circle of guests in the evening. The talk could then take place on *Tuesday*, at a time to be determined by you. I hope you won't just flash by so meteorically but will be our guest for a couple of days. You should see how "nourishing" living here is and should get all kinds of light things to eat. If your wife feels like coming along, she is cordially invited and heartily welcome. Max is visiting Blaschke today and tomorrow, who is unfortunately feeling *very* unwell.[4] Perhaps you will also delight him with another visit? It just should not look too improvised. Warm regards to you and yours, from your

Hedi Born

8. From Paul Ehrenfest

On the homeward trip Christiania–Copenhagen. 8 January 1922

Dear Einstein,

I took (together with Tanichka and van Aardenne)[1] a magnificent trip: Christiania (where the Goldschmidts had much to tell me about you)[2] and then into the mountain snow—*Tofte* (Gudbrandsdalen)[3]—on my way back now, unfortunately a little over-hastily.–

It is easily possible that in the coming months I will be needing your very strong personal support (not in the material sense but purely in moral respects).

= : =

In connection with a circular invitation that various physicists have now received about collaborating on a publication in honor of Lenard, I would like to ask you please to let me know *immediately* at Bohr's address whether you also have been invited to do so.[4]

Warm regards to you and all your family, yours,

P. Ehrenfest

9. To Hermann Anschütz-Kaempfe[1]

[Berlin,] 9 January 1922

Dear Mr. Anschütz,

So you really meant seriously to what you alluded, as I see from a message by my children.[2] I regard this more as a gift, not as anything merited by me and feel obliged to tell you explicitly that I view that statement by you as the expression of a fleeting emotion but not as an agreement, especially considering that the times are boding even more ill economically. I welcome the conversion to water filling, as no significant solid residue accumulates from the electrolysis. If only we succeed in making the electrodes sufficiently resistant.[3] Wouldn't an isolating layer with large capacity come into consideration for them?

I am curious about Mr. Schuler's results. Even if nothing is likely to come of it, a positive result would be hugely interesting; so an attempt very certainly does seem justified. About the how, I and Dr. Schuler are in agreement.[4] The light experiment is finished now, with a securely negative result. *The undulatory theory is therefore certainly refuted in the very area of optics.*[5] Do tell Mr. Sommerfeld about it, who had been expecting the opposite result full of conviction.[6]

It is a noble deed of yours to create the vacation asylum.[7] We shall see that we visit you there during the summer, if only for a short while. Albert has to cram for his finals during the summer vacation.[8]

With best wishes to both of you for the year 1922 from me and my wife and with cordial greetings, I am yours,

A. Einstein.

10. To Friedrich Vieweg & Sohn

[Berlin,] 9 January 1922

Dear Sir,

By the same post, registered, I send you the manuscript of my Princeton lectures with the request that it be printed as soon as possible.[1] Because no individual headings could be attached so as not to interfere with the style of a lecture, I have a few additions to make in the form of marginalia to facilitate orientation. For this reason I ask you please to leave room for these marginalia not only on the correction proofs but also in the booklet itself. Furthermore, please send me a number of copies of the correction proofs so that I can give them to the translators.[2] Publication of this booklet in Germany can only take place after the issuance in America, of course. Consequently, I request that you await my instructions in this regard.

I anticipate receipt of the contract. I insist on the compensation of 20% of the sales price[3] because not only the paper and printing costs but also everything that the publishing house needs for its existence and the author needs for living has risen proportionately and is still rising.

In utmost respect.

11. From Richard B. Haldane[1]

Cloan, Auchterarder, Perthshire, 9 January 1922

Dear and highly esteemed Professor,

It was a very great pleasure for me to receive the very kind New Year's greeting from your family on Saturday.[2]

I have the most vivid memories of the days you and your gracious wife gave my sister and me in London during the summertime.[3] And London itself received and welcomed you as a personality not just from one nation but from all.

Today, so it seems to me, public affairs are developing better than in those days. And science, which knows no borders, has achieved quite a lot in this respect. You yourself, I believe, have accomplished more than you know, more than by relativity theory alone.

Recently I met Mr. Rathenau, at Queen Anne's Gates, and heard things are well with you. I believe he is going to Paris to develop these new relations as far as is possible.[4]

I, myself, am busy with public affairs, particularly in the direction of national education, as well as with philosophy. I have read Cassirer's and Hans Reichenbach's books.[5] But I believe still more will come out of Germany re. the philosophy of relativity theory. Weyl's interpretation of the latest subjects in *Space Time and Matter* now satisfy me completely.[6] It is clear to me that this is just the beginning of such considerations.

I was, as you have observed, in Göttingen at the beginning of August.[7] However, I only had three days off and had to return immediately to Parliament in London.

I hope I may visit Germany (Berlin included) again later when I have more time. Nowadays there is so terribly much to do here.– I visited Klein in Göttingen. He is old but full of life and is occupied with the complete edition of Gauss's papers.[8]

Your doctrines are making good progress here. Eddington & Whitehead are writing and others are stepping forward.[9] The public is interested in relativity but mostly believes in a new kind of ghost.

With best wishes for the new era which, I think, the New Year will bring, ever yours,

Haldane

12. "Proof of the Non-Existence of an Everywhere Regular, Centrally Symmetric Field According to the Field Theory of Kaluza"

[*Einstein and Grommer 1923a, 1923b*]

RECEIVED 10 January 1922
PUBLISHED 1923

IN: *Scripta Universitatis atque Bibliothecae Hierosolymitanarum. Mathematica et Physica* 1 (1923), VII: 1–5; *Kitvei ha-Universita ve-Beth-ha-Sfarim bi-Yerushalayim. Mathematica u'Fisica.* A (5684), VII: 1–4 (Hebrew).[1]

[p. 1] Surely the most important current issue of the general theory of relativity today is the essential unity of the gravitational field and the electromagnetic field. Although the essential unity of both kinds of fields cannot, by any means, be required a priori, it would undoubtedly be a great advance in the theory if this dualism could be overcome. Until a short while ago the sole attempt in this direction has been Weyl's theory.[2] Considerable misgivings about it exist, however. It does not do justice to the independence of the measuring rods and clocks, or atoms, from their prehistories.[3] Furthermore, it does not remove this dualism to the extent that its Hamiltonian function is composed additively of two parts, an electromagnetic one and a gravitational one, which are not independent of each other. Furthermore, this theory leads to differential equations of fourth order while we have no indication that equations of second order would work out.

A short while ago a draft of a theory was presented to the Academy of Science in Berlin by Mr. Th[eodor] Kaluza that avoids all these troubles and is formally of astonishing simplicity.[4] Let us first sketch Mr. Kaluza's thoughts and then move on to the question we wish to examine.[5]

A five-dimensional manifold whose field variable does not depend on the fifth variable is (with a suitable choice of coordinates) equivalent to a four-dimensional
[p. 2] continuum. It therefore does not signify any special physical hypothesis if we interpret the four-dimensional space-time manifold of physical experience as such a five-dimensional manifold, which we will call "cylindrical" with reference to x_5.[6] This is what Kaluza does. He furthermore assumes that physical reality in this continuum is characterized by a quadratic line element

$$ds^2 = g_{\mu\nu}dx_\mu dx_\nu \tag{1}$$

whose coefficients $(g_{\mu\nu} = g_{\nu\mu})$

$$
\begin{matrix}
g_{11} & g_{12} & g_{13} & g_{14} & g_{15} \\
g_{21} & g_{22} & g_{23} & g_{24} & g_{25} \\
— & — & — & — & — \\
\\
— & — & — & — & — \\
g_{51} & g_{52} & g_{53} & g_{54} & g_{55}
\end{matrix}
$$

should not, according to what has been said, depend on x_5. The components g_{11} ... g_{44} should describe the gravitational field; g_{15}, g_{25}, g_{35}, g_{45} would be the electric potentials, g_{55} a field value that still awaits interpretation and may perhaps be related to the Poincaré pressure that played a kind of awkward stand-in role in the theory of the electron.

Kaluza's essential hypothesis, now, consists of the assumption that the laws of nature should be generally covariant in this five-dimensional world. Thus the ways and means by which the electromagnetic potentials occur in the laws of nature are necessarily connected with the ways and means by which the gravitational potentials occur, which signifies a trenchant limitation of the possibilities. Thus the possibility arises for us to construct the physical worldview on a *uniform* Hamiltonian function that does not contain heterogeneous terms superficially welded together by a plus sign. Mr. Kaluza introduces another tensor for the material current besides the magnitudes $g_{\mu\nu}$, however. But it is clear that the introduction of such a tensor only serves to give a preliminary, merely phenomenological description of matter, whereas the ultimate goal we envision today is a pure field theory in which the field variables represent the field of "empty space" as well as the electric elementary particles that make up "matter."

Nonetheless, the fundamental weak points of Kaluza's idea must not be left unmentioned. In the general theory of relativity, which operates with the four-dimensional continuum,

$$ds^2 = g_{\mu\nu}dx_\mu dx_\nu \qquad\qquad \text{[p. 3]}$$

means a directly measurable magnitude for a local inertial system using measuring rods and clocks, whereas the ds^2 of the five-dimensional manifold in Kaluza's extension initially stands for a pure abstraction that seems not to deserve direct metrical significance. Therefore, from the physical point of view, the requirement of general covariance of all equations in the five-dimensional continuum appears completely unfounded.[7] Moreover, it is a questionable asymmetry that the requirement of the cylinder property distinguish one dimension above the others and yet with reference to the structure of the equations all five dimensions should be equivalent.

If one asks whether the $g_{\mu\nu}$'s alone suffice for a description of the total field, one can set, with a selection of coordinates in which the determinant $|g_{\mu\nu}| = g$ is assigned the value 1, the Hamiltonian function H

$$H = g^{\mu\nu}\Gamma^{\alpha}_{\mu\nu}\Gamma^{\beta}_{\nu\alpha}.\qquad(2)$$

Thus to first approximation, i.e., provided the deviations of the $g_{\mu\nu}$'s from the constants are slight, the field equations take the form[8]

$$\frac{\partial\Gamma^{\sigma}_{\mu\nu}}{\partial x_{\sigma}} = 0.\qquad(3)$$

Mr. Kaluza already saw that in this way the laws of gravity and Maxwell's field equations in vacuo are correctly obtained to first approximation.

In this situation it is of interest to know whether the stringent equations (in three dimensions) corresponding to the Hamiltonian function (2) have centrally symmetric static solutions that are everywhere singularity-free [i.e., nowhere may the $g_{\mu\nu}$'s become infinite or their determinants vanish] and suitable for describing the elementary electric charges.[9]

The following approach corresponds to a centrally symmetric solution:

1. For the three spatial indices, $g_{\alpha\beta} = \lambda\delta_{\alpha\beta} + \mu x_{\alpha}x_{\beta}$ ($\delta_{\alpha\beta} = 1$ or 0, for $\alpha = \beta$ or $\alpha \neq \beta$, respectively) must be valid.

[p. 4]   2. $g_{14}, g_{24}, g_{34}, g_{15}, g_{25}, g_{35}$ should vanish throughout.

3. $g_{44}, g_{45}, g_{55}, \lambda, \mu$ should be functions of $r(= \sqrt{x_1^2 + x_2^2 + x_3^2})$ alone.[10]

If one now also sets the abbreviation

$$\gamma = g_{44}g_{55} - g_{45}^2,$$

then for the Hamiltonian function from (2) one obtains:

$$r^2 H = \frac{\lambda^2}{2}r^2(g'_{44}g'_{55} - g'^2_{45}) + \frac{1}{2}\gamma\lambda'^2 r^2 + \lambda\lambda'\gamma' r^2 - \frac{2\lambda'r}{\lambda^2} - 2\lambda\lambda'\gamma r\qquad(4)$$

The variation of the action integral

$$\int Hr^2\,dr$$

according to g_{44}, yields the equation[11]

$$[(g_{44}\lambda^2)'r^2]' - g_{44}\lambda'^2 r^2 + 4g_{44}\lambda'\lambda r = 0.\qquad(5)$$

The equations for g_{45} and g_{55} read analogously.

Through suitable combination of two of each of these equations, three similarly structured equations follow, of the type

$$\frac{[r^2\lambda^2(g_{44}g'_{45} - g'_{44}g_{45})]'}{r^2\lambda^2(g_{44}g'_{45} - g'_{44}g_{45})} = -\frac{(\lambda^2)'}{\lambda^2}. \tag{6}$$

From this, by integration,[12]

$$r^2\lambda^4(g_{44}g'_{45} - g'_{44}g_{45}) = \text{const.} \tag{7}$$

The constant on the right-hand side has to vanish because the left-hand side for $r = 0$ vanishes. That is why, after another integration,

$$\frac{g_{45}}{g_{44}} = \text{const.} \tag{8}$$

follows from (7). Likewise follows

$$\frac{g_{45}}{g_{55}} = \text{const.} \tag{8a} \qquad [\text{p. 5}]$$

Within infinite space-time the manifold must be Euclidean and the electrostatic potential must vanish there. Hence the equations (8) and (8a) require that $\frac{g_{45}}{g_{44}}$ and

$\frac{g_{45}}{g_{55}}$ vanish throughout. Therefore, no spatially variable electric potential exists and hence no electric field, either.

Thus it is proven that Kaluza's theory possesses no centrally symmetric solution dependent on the $g_{\mu\nu}$'s alone that[13] could be interpreted as a (singularity-free) electron.

13. To Paul Ehrenfest

[Berlin,] 11 January 1922

Dear Ehrenfest,

I know nothing about a Lenard *Festschrift*.[1] You need not attach any importance to such trifles, especially to purely interpersonal relations that aren't based on cordial feelings. It's all a mockery, whether good or bad. The light experiment came out reliably negative;[2] tell this to Bohr.[3] One can conclude from it that the emissive field has no component with both the properties

    a) undulatory

    b) isotropic.

I now ask myself whether an interaction of elementary processes cannot *after all* substitute for this phantom field. I'll try to decide this experimentally as well.

My article on superconductivity cites Haber.[4] He developed a similar interpretation a few years ago in an Academy paper, albeit without "snakes."[5]

Warm regards, now, to you and Bohr, Tanya, & von Aardenne[6] from your

                                              Einstein.

14. From Arnold Sommerfeld

Munich, 11 January 1922

Dear Einstein,

The boring circulars by the *Math. Ann*[*alen*] this time give me welcome occasion to write to you.[1]

It really is lamentable that you let yourself be prevented from coming here by the tactless article in the *Schaubühne*.[2] I finally managed to obtain that noble little paper and am repulsed by its mindset. This kind of internationalism appears to me really disgusting in our present situation, especially since it is dictated by sensationalism and irreverence.[3] As regards what the writer of the article says about you, he certainly is spreading tall tales. What he says about the committee meetings was new to me, of course, as the students otherwise tend to treat their consultations with discretion.[4] The rector,[5] to whom I showed the paper, naturally agreed with me that this affair should be passed over in silent disdain. You should compare our poor students of 1922 some day to the bloodcurdling impression you paint of them.[6] It is true, they did get a bit excited in 1919 and at the beginning of 1920;[7] but now they are meek as lambs, as long as a new Entente dictate doesn't kick them off again. You said that very well in an Italian interview that I read in the *Auslandspost* and should raise your voice very loudly abroad.[8] Regarding the infamous *Figaro* interview, I would like to send that out to the *Auslandspost* as well, with the remark: "lies from start to finish." But you are hardly going to grant me permission to do so. It is disgraceful that even your wife is implicated in this interview.[9] Please don't hold it against me that in this case I asked Anschütz to confirm your good national comportment;[10] a stain on my impression of you would otherwise have remained! Three other neutral colleagues besides you to whom I had sent my Lusitania article[11] received it with deathly silence. The world simply does not want to be awakened from its dogmatic slumbers regarding the German guilt; rather, Lloyd George still wants to make all sorts of capital out of it.[12]

It really is time that we talked together again so that the Entente's insane thieving policy does not end up separating the two of us. But let us turn to something positive:

So, you made another grand discovery, buried the wave theory, as Anschütz tells me.[13] I am glad if you can get a spy hole into some tip of it. The way things have been with the dualism in viewpoints cannot go on. If you say your experiment is decisive, I will gladly believe it, although I do not understand it yet, despite Geiger's explanations in Jena.[14]

I meanwhile clarified for myself some wonderful numerical laws for line combinations in connection with Paschen's measurements and presented them in the 3rd ed. of my book.[15] A student of mine (Heisenberg, 3rd semester!)[16] even interpreted these laws and those of the anomalous Zeeman effects using models (Z[eitschrift] für Ph[ysik], in press).[17] Everything works, yet its deepest foundations remain unclear. I can only promote the technique of quanta, you have to do your philosophy. Deep inside, I also no longer believe in the spherical wave. (The anomalous Zeeman effects also have a share in rejecting undulatory theory, by the way.) Do tackle it properly!

Yours truly,

A. Sommerfeld.

15. From Felix Ehrenhaft

Vienna IX, 5 Boltzmann Street, 12 January 1922

Dear Mr. Einstein,

Actually, I am a little cross with you because you haven't written me for so long. What did you say about the splendid etching by Schmutzer?[1] When will you consider keeping your word and coming to Vienna *completely incognito* for a fortnight as our guest? You did promise to do so; the room is ready for you anytime.

Today I would like to ask you for something; I am addressing this letter to your wife so that I am certain that my request will be dealt with *directly* as well. Please send me reprints of your papers since January 1918, perhaps also the reprinting at Teubner's of your major papers on the general theory of relativity.[2] I need them *most urgently* because I have to forward them from Vienna and must have them on hand *before* February 1st. *Please do not postpone this matter*;[3] I was unfortunately unable to write earlier because I was sick.

I take this opportunity to ask you most kindly to send me a personal reference about the private lecturer and titular professor, Dr. Friedrich Kottler.[4] A number of Viennese gentlemen, Thirring[5] and I among them, intend to nominate the above-named ad personam as extraordinary professor of mathematical physics at our university. Considering that under the present precarious financial conditions such a proposal to the Ministry requires strong backing, an opinion from your quarters would be extremely valuable. I recall with pleasure those words you said about Kottler to me and Lampa[6] during your last Viennese stay.

In once again expressing, in the name of my wife[7] as well as my own, the hope of seeing you and perhaps also your spouse at our home for a longer stay, I am, in pleasant anticipation of getting a message from you very soon, with best wishes, yours very truly,

Ehrenhaft

16. To Maurice Solovine[1]

Berlin, 14 January 1922

Dear Solovine,

Much work and an aversion to book-writing make it impossible for me to write what you desire.[2] Soon you will receive my Princeton lectures.[3] That edition has to wait, though, until the same has appeared in America. Conditions for the publisher: 20% of the retail price, thereof you get 5%, I the remainder.

With cordial regards, yours,

A. Einstein.

D[ear] S[olovine], best would be if you wrote Mrs. Untermyer in English so that she sees that you can do so.[4] But you also have to tell her that you can do better in German and French. You also have to say that we spent much time together as young men and studied together.[5] Self-assured forthrightness is necessary everywhere in America, otherwise you don't get anything paid for and are looked down on.[6]

17. From Eberhard Zschimmer[1]

Jena, 59 Reuter Street, 14 January 1922

Highly esteemed Professor,

I must admit that the idea of starting a discussion with you about the known problems did occur to me long ago; but how could I have dared to presume such a thing of you? I consider myself lucky that you are inclined of your own accord to discuss at all seriously matters that I with my stupid layman's understanding deem important—with the vague feeling that from your lofty outpost all of this could only be ludicrous conceptual difficulties that vanish as soon as one has the mathematical wherewithal to be able to draw the conclusions from your theory on one's own. And yet you have now written me such a very promising postcard! Well, rest assured; I shall only make [use] of your kind offer after I have first interviewed your "guardian angel" about how your [mood] is, etc. Besides, it does have to be rewarding for you as well, and that will only be the case once I have brought Prof. Straubel[2] far enough along that the concave mirror for my measurements has descended from the sphere of fantasy into the realm of graspable reality.[3] Straubel is horrendously busy; if you hatch an idea with him, you never know when he will find the time to materialize it. A little "hint" by you, however, would, I believe, have a very accelerating effect. [I am visiting him later on.][4] I assume, as you see, that the glass experiments are of much more importance to you than my stupid philosophical theoretical speculations about the perceptible world. Nevertheless, I would like to permit myself to send you next week my finally finished (short!) manuscript for the *Beiträge z[ur] Phil[osophie] d[es] d[eutschen] Idealismus*, in particular, with the following earnest request: (1) if according to your assessment it is pointless, simply drop it into the mailbox in the enclosed stamped envelope; (2) if you consider the problem per se of merit but find the solution ⟨to⟩ flawed, give me a few pointers in the margin. For this you may take as much time as you please; there is no need to publish the thing so quickly, although it would appeal to me to land a crushing blow on the philosophical pamphlet (*Kant contra Einstein*) by that "virago" [*Mannin*], and specifically in the same academic journal in whose supplements this lamentable product appeared.[5] Alternatively, I could go to the *Naturwissenschaften*; but a philosophical journal does appear to me to be more appropriate.

By the way, may I point out what is surely a typographical error that has remained standing even in the 13th edition of your Vieweg publication: On p. 75

the surface of the sphere $= \pi r^2$ (twice).[6] If Mrs. Ripke-Kühn reads this, she will presumably believe it is a conclusion of the G[eneral] P[rinciple] of R[elativity]. ("Critics" of this Valkyrie genre are capable of anything!)–

Might I now request that you at least leaf through my manuscript once? You will immediately recognize whether it is nonsense and then I would get it back soon; in the alternate case I would then at least have a faint hope that "there's something to it"; and then you are welcome to keep it longer until you are in the mood to make the comments on it. It goes without saying that in a note in the publication I would make due reference to the "checking of the results" demanded of you, similar to your support of Cassirer with your authority in this sense.[7] For, whatever philosophers in general philosophize about in such matters has little standing without your "imprimatur." Then [I] could incorporate the matter into the *Philosophische Briefe*[8] and be of a little service to the theory, or rather, to the people for whom the "new physics" remains a book with seven seals. I often converse with the Jena mathematician Prof. Köbe[9] about the spatial problem; he is considering reworking the mathematical foundations in a "comprehensible" way, initially in a preparatory lecture for the theory. Would you please examine in particular the expression: "invariance of the natural laws" = "invariance of the general analytic form of the natural laws." This term should only express your own ideas.

In great respect, most devotedly,

Eberhard Zschimmer.

18. Expert Opinion on Goldschmidt Patent

[Berlin, after 14 January 1922][1]

Expert Opinion on American Patent No. 1386329[2] Goldschmidt

Mr. Goldschmidt has asked me for my view on whether one of the patents mentioned in the following had a limiting influence on the scope of his own patent as it is defined in the latter's claim [no.] 1. After detailed examination of the material, I arrive at the secure finding that such a limitation on its scope is out of the question.

The achievement of oscillatory motion created through rotation in continuous progressive stepwise motion in the direction of the oscillatory motion, as is explicitly stated by Goldschmidt's patent claim, is mentioned in none of the cited patent subjects.[3] Neither is "means for impressing an unidirectional impulse upon the reciprocatory member, so that it travels or progresses step-by-step in one direction," under consideration anywhere in the patents.[4]

List of indicated patents:[5]

1204245.	Concrete processor. The oscillatory motion does not serve here to generate a progressive motion. A means for the transmission of an impulse onto the oscillating part is missing.
942299 955339	Massage apparatus. The oscillatory motion does not serve to generate progressive motion.
1091533	Car horn. Progressive motion is entirely absent.
1367117 1286617 236697 1280269 1125500 1192502 1249094 1332864 1363495 85721	Sieve shaker. Progressive motion is entirely absent.

A. Einstein.

19. From Richard Courant[1]

Göttingen, 5 Nikolausberger Way, 15 January 1922

Esteemed Mr. Einstein,

The Hilberts and all of us are especially pleased about your intention to come here for the 23rd.[2] As is the way of such things, if a finger is extended, one always does like to grab the whole hand. In this sense, after receiving sanction by Born and Franck about the permissibility of such a request,[3] I would like to ask whether you might want to play something in the afternoon of Hilbert's birthday together with Born as pianist, a very nice little cellist, and my wife? The E flat major piano quartet by Beethoven, that is, the first movement thereof, originally the octet for winds, op. 16, would be very appropriate for Hilbert: it is easy to grasp, very melodious, and not difficult to play, either. You would be entirely free to decide whether to play violin or viola; instruments of adequate quality are also here. I think that Hilbert would be particularly pleased if you wanted to be a part of this little serenade; just the few guests invited to the midday meal will be listening; but everything concerning the timing and the piece could be rearranged according to your wishes. I would be very grateful if you would inform me briefly on a postcard whether you say yes or no; we could then settle the rest here with Born so that you are very little inconvenienced by rehearsals, etc.

Incidentally, Edwin Fischer gave us the idea to ask you to join; he was here recently and told us about his music-making with you.[4]

With most cordial regards, yours very sincerely,

R. Courant.

20. From Michael Polányi[1]

Berlin, 15 January 1922

Esteemed Professor,

I went over the calculations again and discovered an error caused by a misprint in Füchtbauer's paper.[2] There a "2" is indicated where a "λ" should stand. Thus the contradiction is solved and one arrives at the thermodynamically required relation between the decay time and the absorption coefficient.[3] –

It also seems to me that if quantum effects of the kind sought do exist in gases, they cannot in any case become perceptible in the emission process of individual atoms.

In great respect, sincerely yours,

M. Polanyi

21. From Sanehiko Yamamoto[1]

[Tokyo, 15 January 1922]

The negotiations.

Kaizosha has the honor of most courteously inviting Professor Dr. Albert Einstein to Japan and asking him to give some lectures. We are both agreed about definitely satisfying the following commitment:

1. There are planned

a) One scientific lecture, in Tokyo during a period of six days, in particular, about three hours each day; and

b) Six popular talks, once each in Tokyo, Kyoto, Osaka, Fukuoka, Sendai, and Sapporo (for around two and a half hours).

2. If no inevitable obstacles arise, the speaker would embark on the trip at the end of ⟨August⟩ September or the beginning of ⟨September⟩ October[2] 1922. The sojourn in Japan will last about one month.

3. The honorarium (including the voyage as well as accommodation costs) amounts to two thousand (£2,000) English pounds.

Kaizosha transfers to the speaker half of the total together with this certificate through the "Yokohama Specie Bank" in London.[3] The remainder will be presented to him immediately upon his arrival in Japan.

If it is impossible to come to Japan owing to unavoidable difficulties, the advance in the amount of English £1,000 should be repaid to *Kaizosha*.

In great respect, we both undersign:

Berlin, the —   1922

Tokyo, the 15th of Jan. 1922.

S. Yamamoto
(The representative of Kaizosha.)

22. From Sanehiko Yamamoto

Tokyo, 15 January 1922

Esteemed Professor,

We have, pursuant to the enclosed contract,[1] already deposited in the "Yoko-hama Specie Bank," No. 2 Bishopsgate, London, the amount of a thousand (1,000) English pounds in your name, so that you can withdraw the money from the bank either in pounds or in marks at your convenience.

About the matters concerning your voyage, please make arrangements with Mr. Matsubara, embassy counselor at the Japanese embassy in Berlin.[2] We already wrote to him with this request.

Respectfully, sincerely yours,

S. Yamamoto
(for the Kaizosha).

23. "To Allgemeine Elektrizitäts-Gesellschaft Berlin. Remarks on an Expert Opinion Prepared for Mr. Sannig."

[Berlin,] 16 January 1922

[p. 1]   An die allgemeine Elektrizitäts-Gesellschaft Berlin.
                Bemerkung zu einem Herren Sannig gelieferten Gutachten.

I have been made aware by you that my opinion regarding Deutsches Reichspatent 269498[1] could be understood to mean that I thought that a procedure for the cold production of pliable and drawable tungsten wires were not protectable even if the material at the outset is entirely nonductile and changes into the ductile state exclusively through mechanical reworking.

This was not remotely my intention.

Under (4) of my opinion I said:

"The wording for the main claim:

'. . . be repeatedly continually reworked until they are pliable and drawable at a normal temperature'

creates an unclear situation insofar as it does not exclude with sufficient clarity the case that the desired property of the wires be produced otherwise than by repeated mechanical reworking."

My subsequently suggested wording for the claim does not, by the way, get precisely to the core of the matter, according to my present view. The following version for the claim would appear to me entirely to the point: Procedure for the manufacture of cold ductile tungsten wires for incandescent lamps, distinguished in that the cold ductility is produced by mechanical reworking of a previously brittle cold tungsten mass.

Only in this case would the patent be clearly and definitely limited to the case that mechanical reworking (but not other kinds of methods) is used to obtain the ductility.*[2]

[p. 2]   In preparing my opinion and in the correspondence of 18/19 November 1920,[3] I presumed as correct the allegation by Sannig & Co. that their mechanical reworking procedure merely served for the shaping and that the object and consequence was not the conversion of nonductile to ductile material.

* Crossed out text: "In my opinion I had deemed the following wording for the main claim correct:
". . . that pliability and ductility of the wires in a cold state is achieved in that in their manufacture they are subjected to a larger number of reworkings than would be necessary for the mere *shaping* of the wires."

With this I wanted to express that under the circumstances the production of the wires only involved shaping and not an alteration of the material's properties. Thus the procedure is only protectable insofar as ductility not be attainable without this mechanical reworking. Therefore, if, for instance, it were possible to cut out a thin strip from this tungsten mass and already be able to ascertain ductility in it, then any mechanical reworking of such a tungsten mass would be free."

24. From Paul Ehrenfest

Leyden, 17 January 1922

Dear Einstein,

Owing to a somewhat overly hurried trip homeward, your postcard reached me only here.[1] Damn it! If your light experiment really does come out anti-classically[2]—I mean, after not just theoret. but experimental critique—then – – – Well, you know, then you give me the *spooks.*–

Don't laugh at me. I mean it entirely seriously. I'll write to Bohr about it now. I'm very eager to know how he'll react to it.–

Assuming I am capable of reproducing his opinion about these things correctly, I would like to formulate it like this: He is much more ready to give up the energy (and momentum) law (in its classical form) for atomic elementary processes (and therefore just to uphold statist.) than "to shove the blame onto the ether." (He is guided by the totality of his studies on atoms–)

In any event: If your result is true-blue, then—it seems to me—you found something quite enormous.

= : =

This also: You told me at the time that a Spanish professor—(University of Madrid?) is interested in the results that you and I obtained here in Leyden with the concentration of Au ions.[3]–

Thanks to the arrival of a new preparation from Methuen and a preparation that has just been produced here in Leyden proper, the concentration has been brought up to 6.7×10^{-3}.[4] This is already quite a usable figure. So if you pass the address of this noble Spaniard on to me, I can send him the measurement results directly.

Very cordial regards to all of you—but especially to the dear Ilmargotse.[5]

P. Ehrenfest

25. To Max Born and James Franck

[Berlin,] 18 January 1922

Dear Born and dear Franck,

Heavy-heartedly, I do have to cancel after all.[1] But there is no other way. I am so behind with written and other obligations that I cannot in fact afford that escapade into the El Dorado of scholarship. So I shall have to bring my homage to Hilbert in writing.[2] Also tell Courant, who wanted to engage me as a minstrel.[3] Laue is fiercely fighting my experiment, or rather my interpretation of the same.[4] He claims that the undulatory theory does not require any bending of rays at all. He suggested a nice experiment for analyzing possible undulatory ray bending with capillary waves, which exhibit strong dispersion, of course, to replace the theory which is so hard to reach with the required rigor. There was a grand dispute at the colloquium today already.[5] Next time, continuation. Don't be annoyed; delayed does not mean declined.

With cordial regards, also to your wives,[6] yours,

A. Einstein.

Mrs. Born, hearty thanks for the sweet little picture.[7] One evening recently I read out to Laue and Vegard[8] all the verses you had dedicated to us and enthralled them; everyone thought them a sensitive rival to Master Busch.[9] Out of consideration for the little quarrel we had, I send you my special greetings.

26. To David Hilbert

[Berlin,] 18 January 1922

Esteemed Colleague,

I had already quite firmly decided to offer you, in person, my hearty congratulations on accomplishing that period of life.[1] But now it absolutely does not work out because I cannot leave. I can only grasp in a more restricted (and indolent) fashion a mere fraction of your immense life-work, but just enough to divine the framework of your creative mind. Add to that the humor and secure, independent view on all things and—a uniquely hard skull, besides two strong arms to clear the dung out of the faculty stall from time to time.[2] I wish from my heart that you will continue with energy and in good spirits to guide and perfect, as you should, the grand work of art that you have shaped of your life, with the joy and ease with which you have hitherto proceeded.[3] Amen.

Cordial greetings to you and your wife, yours,

A. Einstein.

27. To Arnold Sommerfeld

[Berlin, on or after 18 January 1922][1]

Dear Sommerfeld,

Your cordial letter and your telegram really did please me very much.[2] I would have written you long ago had I not been so much on the treadmill that every minute was devoured out of dire necessity. I had to cancel going to Göttingen at the last minute as well, to my great distress; I would have liked to congratulate Hilbert in person.[3] With great admiration do I observe your step-by-step disentanglement of spectra; how adept you are at conforming these few selection rules to so much material! The light experiment is now finished, also the theoretical parts. Laue[4] had contested, you know, that the bending of light is required by the undulatory theory, and I also had to concede that my proof was faulty. Now, however, I believe I have been able to offer a really precise proof, which will appear as an addendum to the publication.[5] The importance of the subject makes me prefer that the matter be thoroughly gone over as critically as possible.

As regards the *Figaro* article, I would be willing to give you authorization for a denial, if you procured the article for me and I saw that an injustice had been done by it.[6] Students really are in great need. But their political attitude and that of the professors, mainly toward the government, appears to me very unfortunate, indeed foolish. For the men who are now bearing the burden of government are not to blame for the current difficult circumstances but precisely those who are voicing the loudest criticism. In my opinion, for a leveling of the opposing sides it would be propitious if the entire student association held regular meetings at which the adherents of all the parties gave speeches, under strict preservation of certain conventions. In Anglo-Saxon countries this institution has proved to be politically highly worthwhile, also in that young people are prepared for participation in public life.

Thank God that no affair is being made out of that silly article in the *Schaubühne*. But you will surely understand that under such circumstances one does lose one's enthusiasm for such public appearances. You can believe me, my direct personal activities abroad have always contributed toward reengaging old traditions of friendly relations without my ever having sacrificed my convictions. On the other hand, there is no changing the fact that in such cases the poisoners do cancel part of the good effect by distortions and lies. This happens throughout the world, also over here. Besides, I am so remote from politics that it could not separate us, even with such different views as ours. In the end, each of us is convinced of the honest stance of the other, so that alone prevents the possibility of bitterness arising.

What I particularly admire about you is how you raised such a large number of young talents out of the bare earth.[7] This is something entirely unique. You must have the gift of ennobling and activating the minds of your auditors. So do send me the scribble from the *Figaro* sometime at your convenience, not so much for its own sake as for ours.

Best regards to you, your wife, and the Anschütz[8] family, yours,

A. Einstein.

28. "Response to the Expert Opinion of Hans Wolff in the Legal Dispute between Anschütz & Co and Kreiselbau"[1]

18 January [1922][2]

Response to Wolff's Expert Opinion[3] in the Matter of Anschütz versus Kreiselbau[4]

It cannot be my task here to go into all points connected with the ones I would have to comment on about Wolff's expert opinion; instead I confine myself to indicating the most important points of principle.[5]

1. According to my earlier expert opinion,[6] any apparatus that makes use of a horizontal gyrocompass to detect curved flight falls under the protection of the plaintiff's patent.[7] This point of view is not shared by Wolff's expert opinion, in that it makes reference to Rosenbaum's treatise[8] and represents the view that the transfer of the ideas expressed there to airship travel would have been obvious although it does not deny that the application to aircraft offers characteristic technical advantages. This involves an issue that cannot be solved exactly but only intuitively; I am, however, as before, of the conviction that despite Rosenbaum's article the invention of the gyroscopic kymograph for airplanes could very well not have been made or would have required some years' wait to emerge. This issue is perhaps immaterial according to British patent specification 125096 from the year 1916 describing a gyroscopic kymograph for airplanes,[9] which patent specification was not available to me as I was writing my earlier opinion.[10] If this British patent specification should also be recognized in the interpretation of the protective scope of the plaintiff's patent, even though its content was naturally not accessible over here in this country at filing of its patent, the plaintiff's patent's protective scope would have to be narrowed down further than indicated above.

2. This narrower version for possible consideration may be characterized as a "surveyable combination of a gyroscopic kymograph with a device for observing the direction of apparent gravity" (more precisely, its projection on an aircraft plane and lateral plane). That this combination has a certain value and in the customary sense is worth protecting, likewise that it appears for the first time in the Anschütz patent seems hardly doubtful to me. That for Anschütz the device to indicate the direction of apparent gravity is itself a part of the kymograph, while for Drexler it is not, does not seem to me to constitute an essential difference. Even with this[11] interpretation perhaps coming into consideration, Drexler's steering indicator falls within the scope of protection of the plaintiff's patent. I therefore cannot acknowledge as legitimate the decision offered at the bottom of page 9 of Wolff's expert opinion even in case 2 of the narrower scope—unless the combination of gyroscopic kymograph and vertical reference indicator generally is not regarded as warranting protection. Wolff's expert opinion does not draw this possible interpretation of the reach into consideration at all.

sig. A. Einstein.

Berlin, 18 January 1920.[12]

29. "On an Optical Experiment Whose Result Is Incompatible with the Undulatory Theory"

[Berlin, around 19 January 1922][1]

⟨Theoretical part by A. Einstein, experimental part by H. Geiger and W. Bothe.⟩[2]   [p. 1]

Theoretical Part (A. Einstein)

In a notice that recently appeared in these *Berichte*, I took under consideration an optical experiment that promised to deliver interesting ⟨conclusions about the⟩ findings regarding the elementary process of light emission.[3] Messrs. Geiger and Bothe performed this experiment at the Physikalisch Technische Reichsanstalt with the secure result that the diffraction required by the undulatory theory of light emitted by canal-ray particles in dispersive media does *not* occur.[4] The importance of this finding prompts me ⟨in the following⟩ to preface a description of the experiments with a more precise presentation of ⟨the⟩ a few considerations related to it.

In our current ⟨investigations⟩ theories about light, two ⟨hitherto not connectable⟩ theoretical systems run independently alongside each other, both apparently indispensable and yet ⟨irreconcilable with each other⟩ contradictory to each other, the undulatory theory or the electromagnetic field theory, resp., and quantum theory. The undulatory theory has been to this date indispensable to us for the theoretical interpretation of refraction, diffraction, interference, dispersion as well as for an understanding of the connection between optical and electromagnetic phenomena in the narrower sense. It spans an enormous range of phenomena, which we will call the ⟨geometrical⟩ field of "true optics." But it fails for all problems of absorption, emission, and generally for all finer energetic properties of ⟨light⟩ radiation. It is absolutely incapable of explaining Planck's formula, the spectral laws, the laws of the photoelectric effect, photochemical activity, etc. The quantum theory, conversely, proves to be an indispensable guide in the area of conformity with a natural law of energy, but has completely failed until now in "true optics."

[p. 2]   Without a doubt physicists now favor the view that quantum theory covers deeper features of physical reality than the undulatory theory and that in all problems ⟨[on which applic]⟩ accessible to both theories quantum theory has proved to be superior. Since, however, the undulatory theory ⟨was⟩ is capable of describing the phenomena within the range of true optics with exceptional exactitude, without failing in a single case, the belief still prevails today that we shall manage one day to fuse quantum theory and undulatory theory into a single whole without denying the latter exact validity.

Let us now regard the process of light emission from a single gas molecule from the point of view of both theories. According to the undulatory theory, an electron vibrating relative to the molecule generates a system of electromagnetic spherical waves. These spherical waves are concentric if the ⟨particle⟩ emitting molecule as a whole has the velocity zero, eccentric if the emitting ⟨particle⟩ molecule has a velocity relative to the system of coordinates. The color of the emitted ⟨light⟩ radiation then is not a constant function of the direction of emission but a continuous one (Doppler's principle) according to the formula

$$v = v_0\left(1 + \frac{q}{c}\cos\vartheta\right), \qquad \qquad \ldots (1)$$

provided v means the frequency of the emitted radiation, q the molecule's velocity, and ϑ the angle between it and the considered direction of emission.[5] The particle sends out in various directions *coherent* radiation of *various colors*. The distance between opposing planes of the same phase, that is, of wavelength λ, is spatially variable. One could therefore see from the undulatory field, even from a finite part of it, whether it originates from a stationary or a moving molecule—if the undulatory field was perceptible. This ⟨dependence⟩ local variability of λ diminishes, for freely propagating waves, with the distance from the molecule but can be maintained without loss over large distances if one allows the waves to pass through a lens with the particle moving within its focal plane. The wave planes ⟨will⟩ of the same phase are then fanned out behind the lens slightly obliquely from each other. The larger the variability of λ in a direction crosswise from the direction of propagation, the greater the molecule's velocity and the ⟨closer the⟩ smaller the lens's focal distance. So in this case even radiation at a great distance from the emitting molecule has a trait accessible in principle to observation that is characteristic of the state of motion of the emitting particle. Does the radiation emitted from a moving particle really possess this property? One might think that this has already been verified by Stark's observation of the Doppler effect of light emitted by moving   [p. 3] canal-ray particles.[6] Such a conclusion would be unjustified, however. For the existence of the Doppler effect does not prove that the same particle sends out ⟨light⟩ radiation of differing ⟨color⟩ frequency *simultaneously in various directions*, rather only that when a particle sends out any radiation at all in one direction, it then has a ⟨color⟩ frequency in accord with Doppler's principle. This could also occur if in an elementary process of emission the entire radiant energy were emitted in a single direction, e.g., according to Newton's emission theory of light.

Before we go into the question of the demonstrability of the fan-like structure of emitted radiation from moving particles required by the undulatory theory, let us consider the emission process from the point of view of quantum theory. This requires the following:

1) The energy of the molecule is only capable of certain ⟨energy⟩ values $E_1, E_2 \ldots$. (Judged from a system of coordinates relative to the molecule).

2) In the transition from the state with the greater energy E_m to the smaller energy E_n, the difference $E_m - E_n$ is emitted at frequency v, whereby

$$E_m - E_n = hv \quad \ldots (2)$$

3) The emission time is small against the time that, according to the undulatory theory, the emission should take in order to be able to explain the empirically found interference capability of ⟨spectral light⟩ light for great path differences.

On one hand, one knows from Wien's experiments on the light emission of canal rays in high vacuum that the average lingering period in the state of greater energy and the emission time *together* is of order of magnitude 10^{-8} seconds;[7] on the other hand, quantum statistics demands that the transition times be very small compared to the lingering periods in Bohr's "stationary" orbits. The emission times must therefore be smaller than 10^{-10} seconds, which is not reconcilable with an interpretation of the observed interference capability of spectral light in the sense of the undulatory theory. The quantum theory hence comes into conflict with the undulatory theory.

4) The process of emission is, in energetic terms, a *directed* process. The quantum theoretical derivation of Planck's radiation law requires that, upon emission of a quantum, a momentum of the quantity $\frac{h\nu}{c}$ be transferred onto the emitting atom;

[p. 4] this means, according to the law of radiation pressure, that the entire energy $h\nu$ of the quantum emission is given off in one and the same direction. (A. Einstein. Zur Quantentheorie der Strahlung. *Phys. Zeitschr.* 1917. pp. 121 to 128).[8]

These results would necessarily lead to a pure emission theory (corpuscular theory) of light if the interference effects did not persistently pose insurmountable obstacles to an interpretation from this standpoint. In particular, it seems impossible according to any emission theory that one source of light would radiate coherent light in different directions, which is without a doubt the case; the effectiveness of a microscope, for instance, is based on that the light emitted from the object, which is sent to ⟨various⟩ opposing edges of a microscope lens, be subjected to interference.

Although a corpuscular theory of light cannot do justice to some of the phenomena, it does express features of the phenomena of light that ⟨find no place in⟩ one is accustomed to interpreting with the aid of the undulatory theory. E.g., it likewise enables an understanding of Doppler's principle. If a moving ⟨atom⟩ molecule emits a quantum of eigenfrequency ν_0 (seen from the molecule) in ⟨the⟩ a direction that forms the angle ϑ from the direction of motion, then the loss in velocity that the molecule experiences from it is $\frac{h\nu_0}{c} \cdot \frac{\cos\vartheta}{m}$. This velocity loss corresponds to the energy loss $mq \cdot \frac{h\nu_0 \cos\vartheta}{cm} = h\nu_0 \frac{q\cos\vartheta}{c}$. As this energy must change into the radiant energy of the quantum, it becomes, altogether, $h\nu_0 + h\nu_0 \frac{q\cos\vartheta}{c}$ or $h\nu_0\left(1 + \frac{q}{c}\cos\vartheta\right)$, which according to the quantum rule must equal $h\nu$ if ν means the frequency of the quantum from the frame of reference not moving along with it. From this follows equation (1).

One thing is of special importance to us. According to the corpuscular theory, one cannot see from a quantum moving through space whether it originates from a moving or stationary molecule, at least if one imagines the quantum as a point moving at the velocity c that is fully characterized merely by an energy value and perhaps a direction of polarization. The quantum could just as well have originated from a molecule at rest of suitable emission frequency. Conversely, we saw that according to the undulatory theory, the elementary process is endowed with formal

qualities that, in principle, permit one to establish whether the emission comes
from a resting molecule or a moving one.

We now inquire about an experimental criterion for the trait that according to the   [p. 5]
undulatory theory is granted to light originating from a moving particle. Accordingly, the distance between two surfaces of the same phase and therefore the radiation's frequency is a function of position. If we let such "fanned" radiation pass
through a dispersive medium, then at frequency ν the standard propagation velocity
of the equally phased surfaces is also a function of position. From this then follows
that during propagation the surfaces of equal phase must be subject to a rotation,
i.e., the ⟨light rays⟩ wave normals are subject to a continued change of direction,
which must be detectable as a deflection of the light after the wave train leaves the
dispersive medium. Since this derivation has been rightfully criticized by Mr. Laue
as not rigorous enough, I offer another in appendix to the theoretical part, whose
conclusiveness no one should be able to doubt.[9]

We observe the equally phased surface of
waves, which is emitted by the molecule
moving perpendicularly to the optical axis
within the focal plane of the lens L, as it
passes the optical axis. This is a spherical
surface of growing radius up to lens L; after
passing lens L it is a plane that remains per-
pendicular to the optical axis until its entry into the dispersive medium. Let V be

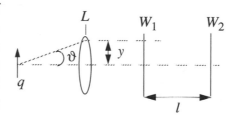

the propagation velocity on the abscissa y.[10] $\left(-\dfrac{1}{V}\dfrac{\partial V}{\partial y}\right)$ is then the deflection angle

of the wavelength normal upward per unit path in the dispersive medium. The
deflection along the entire path in the dispersive medium is l times larger. Through
refraction upon exiting the dispersive medium, this deflection multiplies itself n-
fold $\left(n = \dfrac{c}{V}\right)$,[11] so one obtains for the total deflection A (of the wavelength normal)

$$A = l\frac{dn}{dy}.$$

On the other hand, however,

$$\frac{dn}{dy} = \frac{dn}{d\nu}\frac{d\nu}{dy} = \frac{dn}{d\nu}\frac{\nu q}{c}\frac{\langle\sin\rangle d\vartheta'}{dy} = \frac{dn}{\left(\dfrac{d\nu}{\nu}\right)}\frac{q}{c}\frac{l}{\Delta},\quad [12]$$

where Δ signifies the lens's focal distance.[13] One thus obtains for the deflection
angle the expression already indicated in the first communication:[14]

$$A = \frac{dn}{\left(\frac{d\nu}{\nu}\right)}\frac{q}{c}\frac{l}{\Delta}. \qquad \qquad \dots (3)$$

For $\Delta = 1$ cm $\langle$and$\rangle$, a sulpho-carbonic acid layer of 25-cm length and $\frac{q}{[c]} = \frac{1}{300}$

yields a deflection of the order of magnitude 10^{-2}.

[p. 6]   [This page is missing.][15]

[p. 7]   Let there be an open (or closed) line on the plane with the coordinates ξ, η, s running from a reference point along its measured arc length. Such a line then has solutions to the wave equation of the type:

$$\varphi = \int \frac{A}{\sqrt{r}} e^{j\omega\left(t - \frac{r}{v} + \alpha\right)} ds \qquad \qquad \dots (6)$$

The integral should extend over the curve. r is a function of ξ and η and therefore indirectly of s; A and α are given (real) functions of s. (6) solves the wave equation at distances from the curve that are large against the wavelength. For the following we want to confine ourselves to the case where r is large against the length of the curve and the latter is large against the wavelength of the radiation; $\langle$Then we can set to sufficient approximation:$\rangle$

$$\varphi = \frac{e^{j\omega\left(t - \frac{r_0}{v}\right)}}{\sqrt{r_0}} \langle \int A \rangle \int A e^{-j\omega\left(\frac{\Delta}{v} + \alpha\right)} ds$$

A and α are $\langle$gradually$\rangle$ constantly variable along the curve $\langle$in such a way that$\rangle$ if one depicts the curve length as of the order of magnitude 1. ($s\frac{dA}{ds}$ and $s\frac{d\alpha}{ds}$ are

finite.) Because under these conditions the curve appears to be a small angle from the starting point, the surfaces of equal phase at the starting point are nearly vertical to the connection of an arbitrary point on the curve with the starting point. $\langle$as can easily be gathered.$\rangle$

     Any choice of the curve and of the functions A and α yields, in a part of space that excludes the curve, the solution of a planar problem of deflection (for light sources at rest). We, however, are not interested here in the phenomenon of deflection but in the ray's path, neglecting the deflection. How $\langle$do we find the course that$\rangle$ can this be derived from (6)? For a given starting point, $\frac{A}{\sqrt{r}}$ is a gradually changing function of s; $\omega\left(t - \dfrac{r}{V} - \alpha\right) = H$, however, generally is a rapidly chang-

ing function of s. So e^{-jH} ⟨therefore assumes⟩ oscillates rapidly back and forth with variable s between values of differing sign and equal amount in such a way that the portions of the integrals yielded by finite parts of the curve virtually cancel out. An exception is only made for those locations on the curve for which the starting point being considered, $\dfrac{\partial H}{\partial s}$, vanishes ⟨(the deflection being neglected)⟩. If such exist, then the starting point is illuminated, otherwise it is dark. This condition gives us the path of the ray, neglecting the deflection.

For all the following considerations the curve is henceforth a straight line that extends between $\xi = -b$ and $\xi = +b$ on the abscissa. In this case, one must set for r, developing up to terms of second order in ξ,

$$r = r_0 - \frac{x}{r_0}\xi + \frac{1}{2}\frac{y^2}{r_0^3}\xi^2 , \qquad\qquad \dots (7)$$

where $r_0 = \sqrt{x^2 + y^2}$ is set. Our fundamental condition for the illumination of the starting point then always states that for the starting point and a quantity for ξ between $-b$ and $+b$ the condition

$$\frac{\partial H}{\partial \xi} = 0 \qquad\qquad \dots (8)$$

should be satisfied.

[This page is missing.]   [p. 8]

[This page is missing.]   [p. 9]

Equation (8a) yields   [p. 10]

$$\gamma\left(t - \frac{r_0}{V_0}\right) - \frac{\omega_0}{c}\left(v_0\frac{dn}{d\omega}\gamma - n_0\frac{x}{r_0}\right) = 0 \quad [16]$$

If by the same considerations as in the dispersion-free case one again sets $t - \dfrac{r_0}{V_0} = 0,$[17] then one gets

$$x = r_0^2 \frac{1}{n}\frac{dn}{d\omega}\gamma \qquad\qquad \dots (13)$$

⟨The ray thus moves in a circle. As [emerges] from our earlier considerations⟩ The radiation emitted at time $t = 0$ therefore propagates in a bent line, to be precise, on a circular path. The wavelength normal that according to one of our earlier results has the direction of the radius vector drawn to the starting point r_0 hence undergoes a deflection in the sense of a growing x of magnitude

$$r_0 \frac{1}{n}\frac{dn}{d\omega} \cdot \gamma \,. \qquad \qquad \ldots (14)$$

This result is equivalent to our equation (3). To see this, one only has to take into account that the beam's rotational velocity is $\frac{q}{\Delta}$ outside of the dispersive dielectric; inside it, therefore, it is equal to $\frac{q}{\Delta n}$, which term is set equal to $\gamma\frac{V}{\omega}$. Furthermore, for r_0 one has to set length l of the dispersive layer and introduce for V the refractive index n, for $⟨d\omega⟩\omega$ the frequency ν.

Thus rigorous proof is provided that equation (3), which is not confirmed by experience, is in fact a consequence of the undulatory theory.

30. From Paul Ehrenfest

[Leyden,] 19 January 1922

Dear Einstein,

To facilitate reading this letter:

Borrowing a remark by Gibbs (1886), I believe I can show: even if one grants you the rotation of the wave planes as proportional to the length of the bisulf[ide]-of-carb[on]-tube, there follows (on a purely classical basis) a *negative* result for your experiment on *a classical* foundation.– The point is: *group* of waves![1]

Sheets A, B, C reiterate more clearly what pages 1, 2 present somewhat messily.

Forgive me if I am mistaken. Please reply to me, if only briefly and provisionally.

Don't forget, either, to indicate the address of the Spanish professor, about whom I wrote yesterday on my postcard.[2]

Don't be angry with me if I am wrong; don't be angry with me if I am right.

With warm regards, yours,

P. Ehrenfest

Dear Einstein,

Well, now—you are such a devil of a fellow that you will naturally be correct in the end. But I would still like to grasp *clearly*, with much greater certainty than before, if the classical theory really does demand a bending of the image. I am now full of suspicion about every step of the proof.

In any case, I would like to draw your attention to a very short note by *Gibbs* [*Scientific Papers*, vol. II, page 253 = *Nature* (vol. 33, p. 582, April 1886)].[3] It refers to the *measurement of the velocity of light in a dispersive medium with the aid of the rotating mirror method.*

There light waves, set in fanlike opposition to one another, are also sent back by the rotating mirror; and Lord Rayleigh drew attention to it[4] [*Nature*, vol. 25 (1881), page 52.—He incidentally retracted the *result* he indicated there in favor of Gibbs's better result.], because these waves *rotate* as they traverse the *dispersive* medium (as you, of course, said).

Now Gibbs observes a "group" of such fanned waves and recalls primarily that one must distinguish between the propagation velocity of the waves $V(\lambda)$ and the propagation velocity of the wave "group" $U(\lambda)$. Then he *shows*:

Each individual wave plane does indeed rotate (just as each spoke on a wagon wheel of a moving wagon does); but if one "runs along with the group," i.e., at the velocity $U(\lambda)$, and thus fixes on a "point in the middle of the group," then one sees the following: the consecutive wave planes of the wave fan go through this (running) point, one after another (as their veloc. $V(\lambda) > U(\lambda)$); yet the *orientation of the wave normal* for all these consecutive wave planes is always the same (just like an observer traveling along on the wagon, who is constantly watching the upper rim of the wheel, sees the consecutive spokes passing by in vertical positions).

Why am I telling you all this?– *Not* because I clearly see that these remarks prove anything against your claim that according to the classical theory your experiment has to come out *positive*, but only because it shows how many things one might perhaps have forgotten in these considerations.–

I don't know whether one should fuss at all with group velocities in any way in your derivation (an argument for *yes* immediately crops up)—but assuming *yes*, then Gibbs's result perhaps is important, after all: in that *by running along with the group from the beginning to the end of the bisulf.-of-carb. tube, you can't speak of an "increasing inclination of the wave planes" at all anymore.* For, look at what happens:

The canal-ray particle, as it passes through the slit (or during an even shorter time—but at most during 10^{-8} sec),[5] emits a "fanned wave-*group*." This *group* wanders through the bisulf.-of-carb. tube. *Meanwhile* the wave-fan wanders at the relat. velocity $V - U$ from the tail of the group to the head of the group, *while new waves are constantly forming at the tail and just as many are dying away at the head; and three observers running along at the head, at the middle, and at the tail of the group persistently get to see virtually vertical standing wave planes, even if the group travels for kms through the bisulfide-of-carbon* (thanks to Gibbs's theorem!).

So now, Einstein, I'm pretty sure of myself.[6]

1.) During passage through the slit (or during a fraction of this time), the canal-ray particle transmits a fan-shaped *group* of waves into the apparatus, therefore ultimately into the bisulfide-of-carbon tube. The inclination of the foremost and hindmost planes upon entry into the tube amount to $+\varepsilon$, or $-\varepsilon$, resp.

2.) Follow (with Gibbs) the wandering of this *group* through the bisulf.-of-carbon.—It wanders with "group"-velocity $U(\lambda)$ toward the right.–

3.) Meanwhile the individ. waves are moving *relative to the group* at the veloc.

$$V(\lambda) - U(\lambda)$$
$$\downarrow$$
("wave veloc.")

from left (tail) to right (head of the group) and at the tail end new waves are constantly being *born*, while at the head end just as many are *dying*.

4.) If we follow an individual wave plane, it rotates (in heaven's name) as ⟨much⟩ rapidly as you calculate it (on this point you, Rayleigh, and Gibbs agree)—and if we *!!could!!* follow an individual wave as it wanders through the *whole* tube, it would ultimately rotate by some angle; the tube only has to be long enough.

5.) But we can't do so—because the wave exists only as long as it lies *within* the group.

6.) So we just need to check which wave inclinations can occur *within* the group.—This is easily done with the help of Gibbs's theorem.

7.) For Gibbs proves: If we let a system of coordinates move along in parallel to itself at the group's velocity $U(\lambda)$ toward the right, then—since the waves are moving at the greater veloc. $V(\lambda)$, new waves are constantly consecutively occupying the 0 point of the running coordinate axes. *But all these consecutive waves have exactly the same inclination*[7] [The proof goes such that he demonstrates: In the time required for the $(n + 1)$th wave behind the nth to reach the 0 point, it has rotated just as much as it had been askew to its predecessor.] (at *which* inclination, of course, depends on whether the 0 point of the running coordinate axes is moving at the front, in the middle, or at the rear of the group).

8.) From this theorem by Gibbs therefore follows: No matter how long and far we run along with the group, we always see in it solely the waves with the small inclinations between ±ε, which correspond to ⟨one⟩ the slit breadth (or a part of it)—just the *fictitious* (dead!) waves far *in advance* of the head of the group have the large inclinations.

31. To Paul Ehrenfest

[Berlin, between 19 and 22 January 1922][1]

Dear Ehrenfest,

Everyone is pelting me about my constant doggedness. Laue already had a proper duel with me at the colloquium.[2] But I'm sure of my point. I performed a fine, rigorous calculation that will surely convince you. It's just too long for me to be able to tell you the essence so briefly. In a couple of weeks I'll send you the correction proofs.[3] The error in your consideration[4] lies in the following:

You prove that a wave that enters at *A* after propagating has the same inclination for the wave normal at *B* as at *A*. This would be true—if the wave complex arrived at *B*. This, however, is not what it does; instead it arrives at *C and its wave normal at C has the orientation AC.*The whole problem can only be solved properly if one examines the course of the ray. I started from a strict solution and consider my proof secure— whatever a theoretical physicist may call secure. I am penitently remorseful of formerly committed blunders . . . . In any event I am curious what you'll say about my proof. My love  for it is naturally based for the most part on my having agonized over it. The Spanish fellow scientist interested in the Au-ion affair is called Kuno Kocherthaler, Apartment 425, Madrid, 9 Lealtad Street. I laughed myself to tears about your message regarding him.[5] With Grommer I am trying to address the zero-point energy problem.[6] It's not easy. I also have in mind to determine with Geiger and Bothe the quadratic Doppler effect by means of a few little tricks; but it is difficult.[7] Prod K[amerlingh] Onnes about the superconductivity experiments.[8] Did the young American find fluctuations? I ought to have gone to Paris for New Year's, but could not overcome my weariness of traveling.[9] Kramers recently visited me, an excellent fellow.[10] Are you still thinking about the spectral congress?[11] It really would be fine if Julius were to analyze the solar-center/solar-limb effect photographically/ photometrically by then, so that some clarity could be reached on this.[12] The main fault in his investigations lies in uncritical acceptance of the terrestrial electric arc. In America (Mount Wilson) they were able to eliminate the "pole effect," in the main ("Pfund's electric arc"). It originates from spots of high metallic vapor density. Precision measurements of the wavelengths are undergoing correction.

Affectionate greetings to you and your whole dear company from your

Einstein.

32. From Chaim Weizmann

Oakwood, 16 Addison Crescent, W. 14, 21 January 1922

My dear Professor,

Your lines pleased me very much.[1] I believe it would be best if you would write to Mr. Berliner that he please send you the money. It could then be used for furnishing the Institute of Physical Chemistry, be it with books, apparatus, or the workshop.[2]

I was in Paris & Geneva.[3] In Paris I spoke with a few professors, such as Hadamard and Widal.[4] There is great interest in the university there. In Geneva I saw the gentlemen of the League of Nations and I now do have the hope that our mandate will get ratified for once, at last, at the next session of the League of Nations.[5] That would be great luck. I am thinking of traveling to Palestine on the 25th of Feb., maybe my path will take me to Germany again.

With many cordial greetings from both of us[6] to you and your wife and family, yours truly,

Ch. Weizmann

33. From Charlotte Weigert[1]

Kopenhagen, Westend 4 II, 22 January 1922

[Not selected for translation.]

34. From Heinrich Zangger

[Zurich, after 23 January 1922][1]

[Not selected for translation.]

35. From Koshin Murofuse[1]

471, Omori-Iriyamazu, Tokio-Fu, Japan. [around 26 January 1922][2]

[Not selected for translation.]

36. To Emile Berliner

Berlin, 26 January 1922

Highly esteemed Mr. Berliner,

I am extremely pleased that you are supporting our university in such a magnanimous way.[1] As soon as I was informed by Prof. O. Warburg's wife,[2] I immediately contacted Mr. Weizmann about the best way to use the sum. In both of our names, I ask you please to send me the money in dollars with the reference: "For the university in Jerusalem." This reference is necessary for tax reasons. If you agree, the money will be used for the equipping of the physico-chemical department. We are convinced, in view of the institution's function as a whole and the high procurement costs, that an institute catering purely to the needs of physics is still premature. Therefore, we decided for now just to establish one joint institute for physics and physical chemistry. If you approve of this allocation of your grant, we shall soon purchase with this money, here in Germany, apparatus, books, as well as tools and machines for the workshop and send them to Jerusalem as soon as possible, where the furnishing will begin very soon, once the interior redecoration of a building available for our purpose has been completed.[3]

In expressing my high respect, I am, with kind regards to you and your family, yours.

37. To Paul Ehrenfest

[Berlin,] Thursday. [26 January 1922]

Dear Ehrenfest,

You were entirely right.[1] Today I discovered that there was another mistake in my calculation,[2] upon correction of which one finds that the motion does not influence the trace of the rays and the wave normal at all.[3] It really is an insidious problem, though! Now I want to publish the theory completely nonetheless, so that the matter is finally cleared up.[4]

Cordial regards, yours,

Einstein.

38. To Paul Hausmeister[1]

Berlin, 26 January 1922

Dear Sir,

I join my colleague Paschen[2] in his verdict insofar as I also consider your idea of practical importance. Whether or not in your case the work of electrical decomposition depends on the pressure is difficult to say, because, in the practical conditions under consideration, the decomposition voltage of the water is, in all circumstances, higher than the theoretical value. Under all conditions, the decisive thing is that you do not need any compression system. I advise you to patent the matter and then to contact pertinent firms. The following wording would apply for the patent claim:

Process for the production of gases at high voltage, characterized by the electrolytic disintegration of liquids under high pressure for the purpose of eliminating the need for a separate compression system for the gaseous products of electrolysis.

I recommend you immediately register a general patent of this type and perhaps also protect with a special patent the system for executing the procedure, in case specific arrangements turn out to be necessary or advantageous in the practical implementation.

In great respect.

39. From Paul Ehrenfest

Leyden, 26 January 1922

Dear Einstein,

I naturally thank you very much for your return letter.[1] Well, I wish you and your proof all the best, but I don't believe you at all, zilch; and I fear, I fear – – –. I don't, of course, know *what* you'll be proving in your "correction proof."[2] In any case, note for the record that *I* say the following:

a.) that an *individual* wave plane of the wave-fan turns more and more and more, the farther it travels in the bisulfide-carbon, *I gladly concede without any further proof* (but am also willing to read that proof, if you've proven *this*).

b.) But *although* I concede this, I deny that very slanted waves come out at the end.

c.) For, as one sees by *recalling* the circumstance that the experiment *essentially* operates with wave "*groups*," an *individual* wave never experiences traveling from the left end of the tube up to the right end (see letter).—The waves entering at the

left have all died off during the peregrination (through the disulf.-carbon desert).—
Entirely individual waves arrive at the right that had been *born* in the desert. And
they emerge at the right with the same inclination as the one in which their blessed
parents had entered at the left.

Listen, Einstein, to thy little brother, that thou may prosper!!! (Otherwise Einstein will turn into Whinestein.) And be persuaded that I would be *very* pleased if
you are right, after all. Send me the manuscript or correction proof very promptly.
Regards, yours,

P[aul] E[hrenfest].

40. From Jun Ishiwara[1]

Hota, 26 January 1922

Esteemed Professor,

On 22 Sep. last year I sent you a letter in which I conveyed the wish of Mr. S.
Yamamoto (representative of Kaizosha) to invite you to give some lectures.[2] A
short while ago Mr. K. Murobuse returned from his European trip and informed me
that he had visited you in Berlin and that you gladly granted him your consent to
the invitation.[3] Not only Kaizosha itself, but all of us are heartily pleased about it
and are deeply grateful to you.

I would now like to have the honor of forwarding to you the confirmation you
were seeking for the conference. Thus I enclose the same in duplicate, which
Mr. Yamamoto has already signed personally; please return one of them after
signing it.[4]

I would also like to apologize because the discussed conditions have been somewhat altered in the meantime to take the actual circumstances into account, not as
I had written you last time. Regarding the content of the conference here set forth,
I would like to add, besides the enclosed letter from Mr. Yamamoto,[5] the following remarks:

1. The education minister and the president of the University of Tokyo[6] positively approve of Kaizosha's plans, as they would like to express them here below.
Permission has already been granted for making lecture halls available in the universities at each location of the mentioned lectures.

2. The duration of the lectures may be lengthened or shortened as appropriate.[7]
The popular lectures shall be translated into Japanese; the other scientific ones,
however, do not need to be translated.

3. Any public lectures or speeches (in any other Japanese association) may be
held *only* with the acquiescence of Kaizosha.

4. If your gracious wife would like to come along to Japan, Mr. Yamamoto would be very pleased to carry her total accommodation costs at least for one month.

5. In the unexpected event that you could come here only with difficulty and the return of the prepaid honorarium is to be determined, Mr. Yamamoto does not demand that the entire amount be paid back in cash. Instead, he would be happy if you would only kindly send short popular texts to him for inclusion in the journal *Kaizo* edited by him.

In great respect, heartily wishing you excellent health, I remain yours most sincerely,

Prof. Jun. Ishiwara.

41. To Arnold Sommerfeld

28 January 1922

Dear Sommerfeld,

First, I have to inform you that the experiment on which I had placed such great hopes proves nothing, in that upon rigorous consideration the undulatory theory leads to exactly the same consequences as quantum theory (no deflection).[1] Again somewhat the wiser, and one hope the poorer!

Now, the *Figaro* article.[2] I read everything you had marked. There is no doubt that I sat at the same table as that man at a mutual acquaintance's; Ehrenfest, who was just here, was also there.[3] I recognize the conversation in the article. It is what I said, only cast in a French Bengali light. Why I emigrated, why the Munich school agreed so very little with me, are factually correctly rendered, just the *tone* of my words was different.[4] That in 1914 I made it a condition that I would remain a Swiss citizen is correct,[5] that I acknowledged great admiration for Poincaré, likewise, although I did not describe him as *the* most important mind of our day.[6] My saying that a large proportion of the opponents of my theory are guided by political and anti-Semitic tendencies is correct,[7] likewise the remark about the Manifesto of the 93, and the protest against the same.[8] That I make nothing of all this fuss he also legitimately noted on the basis of a comment made by me in jest.[9] Even the remark about my view re. the future pol. development of Germany, which incidentally can only have a curative effect, applies.[10]

Therefore: The man has no right to reproduce statements made by me. He has furthermore attached a false nuance to some things—whether or not by intent I do not know; but he certainly did not *lie*. There cannot be any question of a genuine denial;[11] at most one could say that it is objectionable to reproduce private conversations in the press without authorization. Best would be, however, to say nothing at all, because one would thus at most stir up the issue again. Paper is forbearing and the newspaper reader forgetful—in a couple of years we shall all be dead and the new generation will fret over and amuse themselves again with new follies. I send you the article back in the same post.

P.S. An honest person should be respected, even if he holds and defends different views from one's own.[12]

42. From Gregory Breit[1]

Leyden, 31 January 1922

[Not selected for translation.]

43. "On the Theory of Light Propagation in Dispersive Media"

[*Einstein 1922f*]

SUBMITTED 2 February 1922
PUBLISHED 27 February 1922

IN: *Preußische Akademie der Wissenschaften* (Berlin). *Physikalisch-mathematische Klasse. Sitzungsberichte* (1922): 18–22.

In a notice that just recently appeared in these *Berichte*,[1] I proposed an optical experiment for which, based on my reflections, the undulatory theory would lead one to expect results that are at odds with the quantum theory. The reasoning was as follows. A canal-ray particle moving within the focal plane of a lens generates light with eccentric surfaces of equal phase that by the lens's diffraction are transformed into nonparallel planes (a "fanned out" system of planes). In such light the frequency, hence also the propagation velocity, is a function of location. If one lets such a

[p. 18]

wave pass through a dispersive medium, then within this medium the propagation velocity of the planes of equal phase is a function of location; therefore, as they propagate within the dispersive medium, the equally phased planes experience a rotation that must manifest itself optically as light deflection.

As Messrs. Ehrenfest and Laue doubted the conclusiveness of this consideration,[2] I examined the propagation of light in dispersive media more closely from the point of view of the undulatory theory and did, in fact, find that this consideration leads to an incorrect result. The reason lies—as Mr. Ehrenfest rightly judged[3]—in that by following the crest of a wave in dispersive media, places lying outside the group of waves under consideration come into reach; although rotated, that plane of the wave crest does not physically exist anymore. Other new ones are formed elsewhere of different orientation in its stead.

Our goal is to find an exact mathematical description of the process taking place within the dispersive medium from the point of view of the undulatory theory. In doing so we can restrict ourselves from the outset to considering two-dimensional processes, i.e., those whose field components are independent of the z-coordinate.[4] We assume that dispersive media act just like nondispersive ones with regard to purely perceptual processes. So if ϕ means a function satisfying the wave equation, e.g., the z-component of the electric field strength,

$$\phi = \frac{A}{\sqrt{r}} e^{j\left[\omega\left(t - \frac{r}{V}\right) + \alpha\right]} \tag{1}$$

[p. 19] is then a solution for the wave equation for all r that are large compared to the wavelength $\frac{2\pi V}{\omega} = \lambda$, ϕ means the excitation at time t at a starting point (x, y), whose distance from a reference point (ξ, η) equals r. A, ω, V, and α signify real constants, where ω and V are linked by a relation due to the optical properties of the medium. Any additive connection of solutions of type (1) is yet another solution because of the linearity of the differential equations.

We now imagine a continuous series of excitations that produce waves of type (1), continuously distributed over a given curve lying in the x-y-plane. The reference points (ξ, η) should be considered given as a function of arc length s measured on the curve. At a sufficient distance from the curve, the integral over the curve

$$\left.\begin{array}{l} \phi = \int \frac{A}{\sqrt{r}} e^{jH} ds \\[2ex] H = \omega\left(t - \frac{r}{V}\right) + \alpha \end{array}\right\} \tag{2}$$

is then likewise a solution of the equations. A, ω, α, and V are regarded as slowly variable on the curve such that their changes by moving along the curve by λ are infinitesimally small. The wave length is very small against the length of the curve and this latter, again, is small against the starting point's distances r to the points on the curve. Calculation of the integral (2) provides a theory for the propagation of light including the Fraunhofer and Fresnel effects in the cylindrical case under consideration here, if ω is set constant. In the case where ω depends on s, one obtains nonstationary solutions, i.e., ones whose ray paths depend on time.

We are not interested here in the deflection problem but in the optical problem, neglecting deflection. We ask: Which points are illuminated at time t and which are not, specifically neglecting the deflection effects? This question is easily answered from solutions in the form of (2). H depends on the choice of the starting point and the point on the curve and generally varies rapidly as the curve point wanders along the curve; then e^{jH} is a rapidly alternating function. That is why only curve locations for which $\frac{\partial H}{\partial s}$ vanishes can contribute substantially to the integral. If such locations exist for the starting point and the point in time under consideration, then the point is "illuminated," otherwise it is "dark."

We now choose as a curve the part of the x-axis between $\xi = -b$ and $\xi = +b$ and regard the solution only for starting points with a positive y. If we are just interested in the axis of the beam, regarding it as infinitely thin, then it evidently suffices to set the illumination condition for the midpoint $\xi = 0$ of the distance. Thus, for [p. 20] the ray's path we obtain the condition

$$\left(\frac{\partial H}{\partial \xi}\right)_{\xi = 0} = 0. \tag{3}$$

Under the geometrical conditions taken into consideration, the wavelength normal evidently has the orientation of the radius vector drawn from the origin of the coordinates to the starting point.

The case of interest to us is a beam in a dispersive medium that changes its radiative direction at a constant angular velocity. We approach this case stepwise by regarding simpler cases.

I. Train of waves[5] of constant direction. We specialize (2) according to the conditions

$$\frac{\partial \omega}{\partial \xi} = 0$$

$$\frac{\partial \alpha}{\partial \xi} = 0.$$

Furthermore, we set here, as in the following, to sufficient accuracy[6]

$$r = r_0 - \frac{x}{r_0}\xi, \qquad\qquad (4)$$

where $r_0 = \sqrt{x^2 + y^2}$ is set. Condition (3) gives

$$x = 0.$$

The propagation of light hence occurs along the y-axis.

II. Train of waves[7] of variable direction in a nondispersive medium. We set

$$\frac{\partial\omega}{\partial\xi} = \gamma$$

$$\frac{\partial\alpha}{\partial\xi} = 0.$$

Then

$$H = (\omega_0 + \gamma\xi)\left(t - \frac{r_0}{V} + \frac{1}{V}\frac{x}{r_0}\xi\right) + \alpha.$$

The velocity V is, in this case, independent of the frequency $\frac{\omega}{2\pi}$. Equation (3) yields

$$\gamma\left(t - \frac{r_0}{V}\right) + \frac{\omega_0}{V}\frac{x}{r_0} = 0. \qquad\qquad (5)$$

[p. 21]   That it really does involve a ray of variable direction one can see from the following. Light that is illuminating the starting point at time t passes the origin at time $t' = t - \frac{r_0}{V}$. The illuminated starting points lie in the direction

$$\frac{x}{r_0} = -\frac{V}{\omega_0}t' .$$

So this direction changes over time t' . The light passing the origin at a certain time t' propagates rectilinearly.

III. Train of waves[8] of variable direction within a dispersive medium. We again set

$$\frac{\partial\omega}{\partial\xi} = \gamma$$

$$\frac{\partial\alpha}{\partial\xi} = 0.$$

Here, however, we must take into account that V is dependent on ω. If we set $n = \frac{c}{V}$, then

$$n = n_0 + \frac{dn}{d\omega}d\omega = n_0 + \frac{dn}{d\omega}\gamma\xi,$$

therefore

$$\frac{1}{V} = \frac{1}{c}\left(n_0 + \frac{dn}{d\omega}\gamma\xi\right)$$

must be set, so[9]

$$H = (\omega_0 + \gamma\xi)\left[t - \frac{1}{c}\left(r_0 - \frac{x}{r_0}\xi\right)\right]\left(n_0 + \frac{dn}{d\omega}\gamma\xi\right) + \alpha.$$

Condition (3) here yields

$$\gamma\left[t - \frac{r_0}{c}\left(n_0 + \omega\frac{dn}{d\omega}\right)\right] + \frac{\omega_0}{c}n_0\frac{x}{r_0} = 0. \tag{6}$$

We now ask: What will happen to the group of waves[10] that passes the plane $y = 0$ at the short interval at $t = 0$? As is known, such a group propagates not at

the velocity $V = \frac{c}{m}$ but at the group velocity $V_g = \dfrac{c}{n + \omega\dfrac{dn}{d\omega}}$. For the starting

points illuminated by this group, the relation

$$t - \frac{r_0}{V_g} = t - \frac{r_0}{c}\left(n + \omega\frac{dn}{d\omega}\right) = 0$$

must be satisfied.[11] Equation (6) thus yields also in this case

$$x = 0. \tag{7}$$

Therefore the group of waves propagates rectilinearly along the y-axis and the   [p. 22]
wavelength standard has the same direction.

Thus is shown that light generated by canal-ray particles moving within dispersive media does not experience deflection—in contradiction to the above elementary consideration. This is also the outcome of the experiment that was performed by Messrs. Geiger and Bothe at the Physikalisch-Technische Reichsanstalt through E. Warburg's[12] kind arrangement. Based on this finding of the theoretical consideration, more profound conclusions about the nature of the elementary process of emission cannot be drawn from this experiment.

It should also be noted that a deflection of light in dispersive media dependent on the state of motion of the emitting molecule would lead to a contradiction with the Second Law of Thermodynamics, as Mr. Laue pointed out to me. Because such a bending is not to be expected by undulatory theory either, however, it is probably not necessary to demonstrate this point more precisely.

I am pleasantly obliged to express my cordial thanks here also to Messrs. Warburg, Geiger, and Bothe.

44. From Hermann Anschütz-Kaempfe

Munich, 6 Leopold St., 3 February 1922

Dear, esteemed Professor Einstein,

You can probably roughly imagine how your letter to Sommerfeld hit, a bomb is just a weak comparison;[1] he gave it to me in complete despair, despair about you and about humanity; I'm afraid my consolation was also bitter and yet perhaps some medicine; my thought was that general human kindness and nationalistic passion had always proved to be opposites and that something could not be right here.

Now the *Figaro* article is surely buried for good. R.I.P.[2]

Last week Prof. Herzfeld[3] presented a review of the canal-ray experiment;[4] he got his science secondhand; the general disappointment was that we will not be able to see and listen to you at all here in Munich. The bloodthirsty article in the *Weltbühne* is, in my opinion, not worth any attention; the students here have long since overcome the postwar psychosis, and in front of you personally, any anti-Semitism definitely stops short, even the most fanatical![5]

This year we want to resettle in Kiel in March already; work, i.e., the floating sphere is calling. We are experiencing good success with the automatic steering control;[6] it is helping make the gyrocompass become accepted.

On Feb. 18th we are hosting the Faculty Evening at home, as we do every winter; if you happened to be in attendance, it would cause much joy. The revolution has prohibited a pullman car, but I would procure you a sleeping-car compartment in Berlin and here so that you have no trouble or exertion at all. And the organ and the music room would also be attractions, particularly if your violin came along.

And then perhaps a little colloquium for the physicists would also spin off of it, if the negative outcome of the light experiment you had initially reported turned out not to be right.[7] That really would be nice. And finally and in the end, doing other people favors really is the best thing in the world.

With best regards to you and your wife from us both, yours,

Anschütz-Kaempfe.

45. From Paul Ehrenfest

[Leyden,] 4 February 1922

Dear Einstein,

Many thanks for your postcard regarding nondeflection of the light ray.[1] I think that the group consider[ation], as I presented it in the letter, shows it most clearly.[2] You wrote me the address of Professor Kocherthaler in Madrid so unclearly that I asked you in my postcard whether I had read correctly "9 Scultur Street".[3] You didn't answer this question—please do so finally.–

As I already informed you, we now managed to raise the concentration of the Au ions up to 6.96×10^{-3}, thanks to the coprocessing of a local preparation of 1.00×10^{-3} and one from Methu[en] 1.08×10^{-3}. I received the report by Villars just now that he manufactured for us a preparation of approx. 0.38×10^{-3},[4] all in our units. But, you see, the concentration is so low that it's perhaps better to wait with it until you come here again and then ask Villars to send it; he'll have a higher concentration by then, all right.

Please *immediately* choose when you'll be coming to Leyden.—Our Easter holidays will be from 8 April incl. until 24 April incl. *Before* Easter, Lorentz is hardly likely to be back yet.[5] In March my wife will possibly be away too.[6] In this short period I cannot arrange *anything* either for "redsh[ift]."[7]–

After Easter the weather ought to be very nice here.–

We could set one part of the eventual *discussions* (e.g., redsh.) *within* the period (vacation!) from 19 April until 24 April.–

Choose quickly and *reliably!!* and answer me immediately. Regards to all of you together, yours,

P. Ehrenfest Leyden Holland[8]

46. From Joan Voûte[1]

Batavia, Weltevreden 11 February 1922 Java

[Not selected for translation.]

47. To Paul Ehrenfest

[Berlin, 12 February 1922]

Dear Ehrenfest,

In the same post I am sending you the correction proof of my notice about the ray path in dispersive media.[1] It's nice how simply the business comes out. Your note about the high concentration of Au ions is very welcome, considering that the state of our examination here makes such a high concentration appear very desirable.[2] The address of our Madrid colleague is: 9 Lealtad (Kuno Kocherthaler), Madrid.[3] It would be preferable if we could shift the congress to the end of September.[4] I was with you for so long just recently.[5] I'm glad that your wife can now also go away for once.[6] I can't grasp her opinion that the quantum difficulty could be cured by means of a fifth dimension. While thinking about whether the primary emission proceeds in a spherical wave, the following experiment occurred to me now. Very obliquely incident cathode rays *K* produce on one of two parallel plates of the same thickness, at *A*, *planar* light. The plates generate interference without phase differences—provided emission proceeds in a spherical wave. Does a similar experiment exist? A positive result is very probable, however.

Warm regards, yours,

Einstein

48. To Hans Albert and Eduard Einstein

[Berlin, 12 February 1922]

My dear boys,

You are shrouded in mysterious silence again, you rascals. I am feeling well, but there's little news. The experiment on which I had placed so much importance proves nothing for and nothing against the undulatory theory, so all the labors of love were actually in vain.[1] Did you send the four photographs that were left over from the 10 + 2 to my friend: Kuno Kocherthaler, 9 Lealtad, Madrid, d[ear] Albert? If not, do it directly, but don't forget. I still don't have a sailboat but am on the lookout.[2] It simply has to be a very sturdy one, because everyone is so anxious that my precious life not be put in danger. I'm very keen on spending our summer vacation in my castle (haha).[3]

Heartfelt greetings to all three of you from your

Papa.

D[ear] Albert, please go to the Zurich Kantonalbank and inquire there about why I supposedly have *4,000 marks* deposited there and then write me about it. I received from there notification to that effect, which I cannot explain. Ask please *when* and *from whom* the money was deposited there.

49. From Hans Albert Einstein

[Zurich, between 12 February and 4 March 1922][1]

Dear Papa,

I really haven't written you for a very long time now; but there are two reasons for it: First, I don't have any news at all; and second, I've got quite a lot to do at school now.[2] The photographs went out.[3] Now we're feeling quite well. Teddy also had the flu. We're going to have to move out by April 1st but I don't quite know where we'll end up; I'll write you when I do.[4] We're excited about this summer, at the castle.[5] Hopefully we can somehow play music together then, as well. I, for inst., played the Tartini sonata in G minor[6] and am very keen on hearing it played nicely for once. I'm also tackling something completely new. I'm learning to dance; imagine that! The modern things are more Negro [*Neger*]—than dancing, but there's no harm in that. You asked about 4,000 marks: they're from Dresdener Bank, as refund for the year 1919.[7] Recently, I've been playing piano quite a lot, I accompany Prof. von Gonzenbach[8] and Miss Hurwitz,[9] for instance, too. The double bass is flourishing as well; we played in two concerts again.[10]

It's a pity that nothing came of that light experiment.[11] What are they doing with the gyroscopes, by the way? Did they finish the suspension mechanism already or not?[12] How did the ice box and its competitor turn out?[13] We're noticing that we're heading for the *Matura* finals now, all right; there's a sense of wrapping up our studies.

Many greetings from,

Adn.

50. From Eduard Einstein

[Zurich, between 12 February and 4 March 1922][1]

Dear Papa,

I'm better again now and can go back to school.[2] I'm very excited about the summer as well. If we can play music then, I can do a bit of accompaniment, too. I can already play a piece by Corelli and a couple of gavottes.

Many greetings from

Teddy.

51. To Madeleine Rolland[1]

Berlin, 15 February 1922

Esteemed Madam,

I surely do not need to tell you how highly I value the cause in which you have placed your service, and how very pleased I am to receive an invitation from you to your meeting in upper Italy.[2] If I am unable to accept this invitation it is not for lack of interest or the courage to profess it, but for the following reason: ever since my theories received such an astonishing degree of popularity, I am being invited from every quarter to engage myself personally for things, each one of which taken on its own thoroughly merits dedication. If I were to give way to this pressure, the rest of my life would be forever lost to tranquil scientific research. I must therefore keep my distance personally from all major social enterprises.

With cordial regards to you and your highly esteemed brother,[3] I am, with best wishes for your Italian meeting, yours very sincerely.

52. From Emil Warburg[1]

Charlottenburg, 25b March Street, 15 February 1922

Esteemed Colleague,

Confined to the house or, resp., to bed by a catarrh, I am unfortunately not in a position to present in person my petition to award the silver Leibniz Medal[2] to Mr. Haenisch.[3] I therefore permit myself to request that the following lines be read out during the pertinent proceedings:

The following considerations of a more general nature seem to me to speak for the award of the medal to a mechanic.

1) The progress of experimental physics depends to a certain degree on the further perfection of measuring instruments, which is achieved through collaborative work between scholar and mechanic. [If,] therefore, mechanics like Mr. Haenisch zealously and successfully devote themselves to such joint efforts, it lies in the interest of science to give them acknowledgment for such expertise, expressed in the form of a reward.

2) The mechanics working in state facilities (now called technicians) are ranked at considerably lower salary levels than the bureaucrats. By this it has been made clear that the status of a mechanic does not yet enjoy the respect it deserves in the state. Science is, after what has been said under (1), interested in raising this status, and for this an academic award to this rank is a suitable means.

I permit myself to direct the petition to the Class to support my application, perhaps also by cosigning the same.

Re. Mr. Haenisch, I can supply more specific details upon request and just remark here that according to the verdict of the prominent expert Prof. Brodhun,[4] the firm Schmidt & Haenisch is [by far] the leader in the area of photometry in Germany and, despite [some] weighty competition, still retains first place in the area of polarimetry.

On this occasion, I would like to add the personal comment to you, dear Colleague, that I cannot, with the best of intentions, involve myself in electing foreign correspondents.[5] To my knowledge, ever since the war, not a single German physicist—with you, of course, excepted—has been nominated as a correspondent to any foreign academy; and as we in Germany have no shortage of competent physicists, one should leave to them the available honors.—Besides, it is doubtful to me whether foreigners benefit from an election. Kamerlingh-Onnes seems to me to stand out for having published in French in the last no. [of the] *Communications*[6] (not just in the Solvay conference *Rapport*, 21 April).[7] Moreover, through the Solvay Institute he maintains ties with extremely anti-German Belgians.—[8]

However, all this is my private opinion; and I do not doubt that the gentlemen have cogent reasons for their nominations.

With best regards, ever yours,

E. Warburg.

53. "Proposal for the Nomination of a Corresponding Member in Physics" [Niels Bohr]

[Berlin, before 16 February 1922] [1]

When someday future generations will describe the history of the advances made in the physics of our era, they will have to associate one of the most significant advances in our knowledge of the nature of atoms with the name of Niels Bohr. It was already known that classical mechanics had failed in regard to the building blocks of matter, and that atoms consisted of positively charged nuclei that are surrounded by a relatively loose layer of atoms. But the empirically almost completely known structure of spectra differed so significantly from what could be expected according to our older theories that no one envisaged the possibility of a convincing theoretical interpretation of the observed regularities. Then, in 1913, Bohr found a quantum theoretical interpretation of the simplest spectra,[2] for which he produced such an abundance of quantitative confirmations in a short period of time that the boldly chosen hypothetical basis of his considerations soon became a secure pillar for atomic physics. Even though less than a decade has elapsed since Bohr's initial discovery, the system of thought which he devised and, for the most part, developed dominates atomic physics and chemistry to such an extent that to the expert all former attempts at interpretation seem to derive from a long-gone era. The theory of Röntgen spectra, of the visible spectra, of the periodical system of elements are all based primarily on Bohr's thought. What strikes one as so wonderful in Bohr as a researcher is a peculiar combination of audacity and cautious deliberation; seldom has a researcher possessed to such an extent the ability for both an intuitive comprehension of hidden things and for acute criticism. In spite of all his knowledge of the particular, his vision is directed unflinchingly toward matters of principle. He is, without doubt, one of the greatest innovators of our time in the field of science.

Niels Bohr is a professor of theoretical physics at the University of Copenhagen. The undersigned recommend him as a corresponding member of our Academy.

A. Einstein.[3]

54. From Paul Ehrenfest

[Leyden,] 16 February 1922

Dear Einstein,

Many thanks for your postcard and printer's proofs.[1]—As soon as I have read the latter in peace, I'll probably write you about it—it appears to me to be very nice indeed.

The following passage in your postcard caused me *much* unease: "It would be preferable if we could shift the congress to the end of September. I was with you for so long just recently."– I begin from the back:

a.) If we correspond in *January* about what you're supposed to do at the *end of April*, it is a terrible computational error to confuse the (quite) correct assessment:

(A) January minus September = "recently"

with the *incorrect* assessment:

(B) end of April minus September = "recently."

b.) There are naturally absolutely no objections to a shifting of the "congress"; *for me* it is even convenient, because I can then prepare everything more calmly. I had just meanwhile, prompted by your penultimate letter about it, written to Lorentz in Pasadena about whether one of the "redshift Americans" might not perhaps be present in Europe around Easter. If Lorentz were to write: yes, then you really can come to Leyden for it for a short time. But presumably Lorentz's answer will be negative, and then the business can very well wait until September.[2]

c.) But it would be very *fatal* for me if you did not come to Leyden this spring.— You know: Nobody demands that you present lectures; not even convening little "conferences" is obligatory. This all can happen one time this way, another time that way, or not at all. But I'll get into real trouble if you simply don't come.

Because I have to write an annual progress report that contains as its essential core the fact of your (twice yearly) presence in Leyden.—Your presence here, even if you "do absolutely nothing," means *for all of us* (especially also for Onnes and me and our older students) *enormously* much. More than you can fathom, because each time, your presence here triggers a long-lasting series of discussions [e.g., if you come in April, then I'll tell you about what I learned at Bohr's[3]—you'll respond—this again provides a lot of inspiration to all of us—and this is just *one* of 5–6 topics that are waiting for you].[4]

So if you *do come* (twice a year) I can bona fide and happily recount all that came out of it in my progress report.—And nobody then mindlessly asks, "What did he do?"—If you *don't* come, however, then it would put me in a pickle. Quite apart from how sorry I would feel not to see you.

Understand me correctly: Neither Lorentz nor Onnes nor anyone else here would permit themselves to say such a thing to you—indeed, I think they are *so very* Dutch that they even understand the art of not just not *saying* it, but even banishing it from their *thoughts*.—And everyone here loves you so much and values the benefits of your presence here so highly (it really is *very* high!—Plain fact.) that one would rather have you here 1/2 the time than not at all. But *I* am allowed to tell you that it would put me in a pickle to have to explain an absence during spring 1922 in my next report, as I already had to do so for spring 1921 (with your America trip).

Oh, please, please, Einstein, come to Leyden, for ex. for the second half of April, let me tell you about what Bohr taught me, have a look at what Onnes is doing, listen to what Lorentz is bringing home from America—[Onnes probably wrote you that

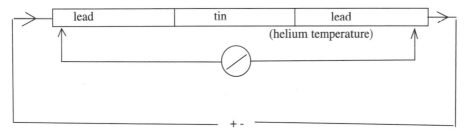

indicates *no* resistance].[5]

In truth, you will not regret it and no one wants more from you! Also, at the *!end of April!* you certainly won't have the feeling that *momentarily* confuses you: "But I was in Leyden just recently!"

Every time that you *are* here, you don't find it that bad at all!

Greetings to all of you, yours,

                                                    P. Ehrenfest

55. From Wolfgang Hallgarten[1]

Heidelberg, 16 February 1922

Esteemed Professor Einstein,

In regard to the friendly letter from your secretary,[2] which came into our possession only now because of the railroad strike,[3] we permit ourselves to inform you in the enclosed of 16 cases of needy students,[4] for whom Prof. Lederer (political economist) in Heidelberg, 28 Kepler Street,[5] and we can vouch as deserving of support. We permit ourselves furthermore to forward to you in the near future a program by our newly founded University League [*Hochschulbund*] (temporary name: Republican Association of Independent Students), which is supposed to span the whole of Germany, is currently already in contact with the Universities of Freiburg, Leipzig, Cologne, Marburg, Münster, Berlin, Tübingen, and Heidelberg, and is supposed to represent the democratic socialist pacifist orientation at these universities. Information about this university league can also be provided to you, aside from gentlemen, Professor Lederer and Priv. Coun. Thoma, Heidelberg, also by Prof. Rosenheim (No. 3 Carmer Street, telephone 8033, Steinplatz, Berlin), who is ready, as needed, to list the gentlemen who are interested in the cause and are willing to provide financial support.[6]

In utmost respect, very devotedly,

The management of the ⟨University League⟩
"Republican Association of Independent Students"
by p[roxy], Hallgarten.
Heidelberg, 182 Haupt St.,
c/o Karl Frank,[7]

56. From Paul Langevin

Paris 5th, 10b Boulevard de Port-Royal, 18 February 1922

My dear friend,

At its last meeting, the assembly of professors at the Collège de France decided, at my proposal, to invite you to give this year's series of Michonis Lectures held by a foreign scholar in Paris each year.[1]

We would be very happy if you would accept. Apart from the pleasure that I personally would have at seeing you here again,[2] I must impress upon you the following reasons of a general sort.

The interest of science wants a reestablishment of relations between German-speaking scholars and us.

You can help in this better than anyone; you would be doing a great service for your colleagues in Germany and in France and, above all, for our common endeavor by accepting.[3]

You will find here the best of welcomes; your work and your personality are equally agreeable here. You will also encounter an audience of students prepared to follow you, in addition to a public at large desirous of getting to know you. My teaching here at the Collège has been concentrating exclusively on the theory of relativity for many years now.[4]

The Michonis Lectures have not been devoted to physics since Mr. Lorentz came in 1912.[5] You alone could renew this link, following the horrible years of the war.

You will hold any number of lectures you please and have them as interrelated as you please. Five or six, for example. Depending on the subject you choose. You will receive for it a remuneration of 5,000 [francs].

It is desirable, so that the audience can follow you, that these lectures be held in French. I think there should be no difficulty in that. If you would like to draft a text in advance, I will translate it for you and you could read at least a part of it if you experience any difficulty.

We would be very happy if you could come before the month of June and if your first visit to Paris were made exclusively for scientific purposes. This will be the best manner of working toward the peace we all want.

As soon as I have received your affirmative reply, the administrator of the Collège de France will send you an official invitation.

Our friends here join me in urging you very enthusiastically and telling you that we shall do our utmost to make your sojourn as agreeable for you as we can. I myself have fond memories of your visit in 1913, not counting the other circumstances in which we saw each other again abroad.[6]

In most affectionate devotion, and do send me a positive response very soon,

P. Langevin.

57. To Paul Ehrenfest

[Berlin,] 20 February [1922]

Dear Ehrenfest,

In that case, of course I'll come for the second half of April, and generally twice a year. You ⟨generally⟩ do know how much I like being there. You did not have to waste so many words on it.[1] I am working with Grommer on quanta in hydrogen. We have already been able to prove that a quantum influence is there; now it still has to be decided whether or not there is zero-point energy. Probably yes according to the findings up to now.[2] There's a nice paper by Gans on magnetism in rapidly oscillating weak fields in the *Annalen* (confirmation of results by Arcadiew).[3]

Warm regards, yours,

Einstein.

58. To Oswald Veblen[1]

Berlin, 20 February 1922

Dear Professor Veblen,

I was very pleased to receive a sign of life from you.[2] If Mr. Tracy Y. Thomas comes to Berlin, I shall be pleased to work with him. However, it will not be easy to find a sufficiently interesting research topic for him. In any event, I will do what I can.

Along with this letter I am sending you the manuscript of the four Princeton lectures on relativity.[3] I did, of course, arrange that they appear in Europe later than in America.[4] I am sending the manuscript to you instead of to Prof. Adam[s][5] because I am not quite certain whether he is in Princeton at the moment. Please give this manuscript to him with kind regards.

With best wishes to you and your wife, I am yours.

59. To Franz Selety[1]

Berlin, 22 February 1922

Dear Colleague,

I grant you permission to publish the material you mentioned.[2] I read your latest version and largely agree with the content. I ask you please to emphasize in your appendix that my remarks should be seen only as passing statements reflecting first impressions, not a carefully weighed argument pertaining to the problem.[3]

In utmost respect.

60. To Juliusz Wolfsohn[1]

Berlin, 22 February 1922

Highly esteemed Mr. Wolfsohn,

A short time ago nobody would have thought possible what we are experiencing today with the Jewish nation. While hitherto, intellectually prominent Jews consciously or unconsciously retreated from Jews as a group for lack of recognition, now our nation is suddenly revitalized before us again and every one of us loses the feeling of isolation that has led so many to behavior that, at first glance, appears to be a lack of character.[2] Your endeavor deserves the thanks of all Jews; and it is a matter of course that I gladly accept the intended dedication.[3] As I myself cannot sight-read music on the piano, I have not yet been able to make your piece known but will do so as soon as possible.

Expressing my warm sympathy, I am sincerely yours.

61. From Theodor von Kármán[1]

Aachen, 22 February 1922

Dear Mr. Einstein,

At the university a chair for mathematics has become vacant and we intend to nominate Max Abraham as first choice.[2] We are of the opinion that someone who knows how to apply mathematical methods in a variety of creative ways is best suited as an instructor of mathematics at a technical university. Because this appointment does constitute somewhat of an exceptional case, I would be very grateful if you would communicate to me your view of Abraham's personality as a scientist and of the matter of his appointment in a letter that I may present to my colleagues here. As this matter *is very urgent*, I request that you please reply as soon as possible.

Sometime ago I wrote you on behalf of Mr. Renner.[3] Thank you very much for receiving Mr. Renner so nicely and for your kind help in that affair. Mr. Renner asked me to excuse him, first, for having altered the wording of the letter some-what, but that this had been definitely necessary as the factual circumstances had changed during the 6 weeks since the first letter had been drafted. Second, Mr. Renner asked me to apologize for having, at the instigation of an employee of the *Manchester Guardian*, demanded that you comment on some issues in the paper. He said that on the spur of the moment, and considering his multifaceted duties in Berlin, he had overlooked the inappropriateness of this request and asks you please not to hold it against him.

I beg your pardon for having to bother you and assure you that I am very grateful for your help in all these matters.

With deep respect, yours very sincerely,

Kármán.

P.S. Mr. Renner also instructed me to ask you whether you would mind if the news of interest to you and G[erhart] H[auptmann][4] regarding the school situation appeared in the newspaper?[5]

62. Review of Wolfgang Pauli,
The Theory of Relativity

[*Einstein 1922e*]

PUBLISHED 24 February 1922

IN: *Die Naturwissenschaften* 10 (1922): 184–185.

[p. 184]   **Pauli, W., Jr., Relativitätstheorie.** Offprint from *Enzyklopädie der mathematischen Wissenschaften*. Leipzig, B. G. Teubner, 1921. IV, pp. 539 to 775. 17 × 25 cm. Price bound 40.– marks; bound 50.– marks.

Whoever studies this mature and broadly conceived work would not believe that the author is a man of twenty-one years of age.[1] One does not know what to admire most: the psychological grasp of the development of ideas, the certainty of mathematical deduction, the profound physical vision, the capacity for clear systematic presentation, the knowledge of the literature, the factual completeness, or the certainty of the critique.

This exhaustive exposition on roughly 230 pages is arranged as follows:

I.   Development of the special theory of relativity with careful account taken of the defining observational data for its foundation.

II.   Mathematical aids for the special and general theories of relativity. The paragraphs on affine tensors and infinitesimal transformations are especially recommended to knowledgeable readers.

[p. 185]   III.   Further development of the special theory of relativity. Exhaustive from the formal as well as physical points of view.

IV.   General theory of relativity (75 pages). Model account of the development of the ideas. Complete presentation of the mathematical methods for solving specific problems. The discussions on the energy equation and the criticism of Weyl's theory are particularly valuable.

Anyone working creatively in the area of relativity should consult Pauli's edition; likewise anyone who wants to familiarize himself accurately with the principal issues.[2]

A. Einstein, Berlin.

63. To Paul Langevin

27 February 1922

Dear friend Langevin,

When I received your kind letter of invitation,[1] I felt great and pure joy; and now, one week later, I hesitantly and sadly take up my pen because I cannot accept the invitation now, as much as I would personally have liked to—even aside from the cordial feelings of friendship I have for you. You know that I am of the view that relations between scholars should not suffer from political causes and that concern for the scientific professional community should go above all other considerations. You also know that I am totally internationally minded and the fact that I am employed by the Prussian Academy of Sciences has had no influence on this mentality.[2] After conscientious reflection, however, I came to the conclusion that at this moment of political tension my visit to Paris would have more adverse than favorable consequences.[3] My colleagues here are still being excluded from all international scientific activities and they are of the opinion that our fellow French professionals are primarily to blame for this.[4] I fully appreciate the deeper causes that led to this attitude. But on the other hand *you*, too, can imagine that these people here, whose sensitivities have been stirred up almost to pathological heights by the events and experiences of the last few years, would perceive a trip by me to Paris at this moment as an act of betrayal and would take such offense that very unpleasant consequences could arise. Even in Paris, unforeseeable complications threaten. I cannot imagine anything finer than being able to chat comfortably with you, Perrin, and Madame Curie again in private and to depict the theory of relativity to your students with a subjective brush.[5] Yet the greater public and—politics— have long since taken possession of my theory and my person and have tried to make both somehow suit their purposes. There would be a considerable number of people watching out for every candid word I utter, to toss it back at newspaper readers, conveniently repackaged.[6] My experiences in this regard in recent times make this danger appear to me to be very great; the end effect is always hatred and animosity instead of reason and goodwill.[7] I would certainly also be interrogated about my political opinions regarding Franco-German relations; as I cannot speak any other way than honestly, my reply would not earn me sympathy either on this side of the Rhine or on the other.

It is true that I did not hesitate to visit North America, England, and Italy in the past few years. However, my American voyage concerned the University of

Jerusalem in the first place; and as far as the other two countries are concerned, the psychological circumstances were incomparably simpler and more auspicious than they are in our case (unfortunately!).[8]

Dear Langevin, it pains me that I cannot oblige, as I like you so much. I also feel the need to thank you and your colleagues at the Collège de France warmly for this generous gesture and the reconciliatory attitude underlying your decision.

With cordial regards and hoping to see you again very soon, yours,

A. Einstein.

64. From Thomas Barclay[1]

17. Rue Pasquier, Paris, VIIIe 3 mars/22

[Not selected for translation.]

65. From Erich Marx

Leipzig, 4 Markgrafen St., 3 March 1922

Esteemed Colleague,

Today I return to your kind reply at the beginning of January.[1] I took a long time thinking about what I should do, since you definitely reject writing an outline of the general theory of relativity for the handbook. I reached the decision to omit the general theory of relativity altogether from the work. After all, it is not absolutely necessary in a handbook on radiology.

But now we still have a manuscript on the special theory of relativity here,[2] and it definitely must be submitted to press now. Wouldn't we perhaps be permitted to send it to you beforehand for a very short addition of a few pages? Lorentz, Zeeman, Laue, Riecke, Debye,[3] and others are contributing to the volume that is now appearing. Surely it would be awkward if the date from before the war appeared under the heading "Special Theory of Relativity," of which I am fortunate to own a manuscript. Please, esteemed Colleague, do us the favor and take a brief look through the manuscript before we send it to press.

Reply envelope is enclosed in case you would like to have the same sent back to you.

With kind regards, yours very sincerely,

Erich Marx.

66. To Erich Marx

[Berlin, after 3 March 1922][1]

Was written in 1912.[2]

I regret, it is far too outdated, impossible to grant permission.

I cannot bring myself to allow the m[anuscri]pt to be published in the present form.

67. To Hans Albert and Eduard Einstein

[Berlin,] 4 March 1922

Dear boys,

From your letter I see that all is well with you, particularly that noble Musica is flourishing. I meanwhile bought a most lovable sailboat with a sail in impeccable condition,[1] so our time together at the so-called castle will be magnificent.[2] We're going to play music on Katzenstein's grand piano, which he's happy to make available to us.[3] We'll have to drive there for ½ an hour or walk for an hour, though. But that won't put us off. You're entirely right about dancing, d[ear] Albert; one has to do something for the fairer sex, whom as experience shows one cannot do without.[4]

The status of the gyroscope affair is this: the electrolyte liquid, of specific weight 2, proved not to be stable enough, after all, so water had to be resorted to, which consequently leads to an enlargement of the suspended part.[5] My electrical arrangement is working quite satisfactorily.[6] The ice box is making progress. We're negotiating a contract with the local firm Borsig right now.[7] The patent issue is still entirely obscure; but we do in any event have the right of joint use—in case the other patent, which is still not published, should come into serious conflict with ours in any way.

It's amusing that you both have such similar handwritings, hardly distinguishable. I'm so thrilled about every sign of life from you, even if you have nothing particular to say; do write short postcards more often. I'm now planning another light experiment. Cathode rays fall onto a small leaf of mica and cause it to glow at the surface. The light comes out in part directly, in part after reflection off the rear surface. The question is

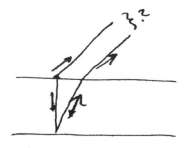

whether these two lights interfere. The experiment is simple, it's just that the leaf shouldn't get too hot.[8]

Warm regards to both of you from your

Papa.

Did you see the son of the pastor from Florence, d[ear] Albert? If not, get to it. Pauli was very sick (pleurisy).[9] It would be nice if both of you wrote my sister a post-card sometime; she would be very pleased. Her address is Dr. M. Winteler. Giuseppe Verdi Street, Fiesole near Florence. What's the situation with the apart-ment? Why do you have to move out? You'd have it incomparably more convenient in Germany.[10]

68. From Edith Einstein[1]

Zurich, 27 Ottiker St., 5 March 1922

Dear Albert,

In attachment I send you my dissertation;[2] there is something very wrong and questionable about this paper, you see, and I turn to you in my distress.

It involves the following:

After you had looked through the paper in the fall,[3] I introduced the general tensor argument following your directions,[4] and then, after Ratnowsky had also checked the paper again and had found everything to be in the finest order,[5] I sub-mitted it to Prof. Schrödinger.[6]

Can you imagine my shock when shortly afterwards Prof. Schrödinger pointed out a few very disturbing weaknesses in the theory that actually place the entire solution to the problem in question! Namely:

1) The coefficients of the h_{xx}, h_{yy}, h_{zz}'s derived from gas kinetics do *not* agree with the coefficients of the tensorially derived pressure component (they even have different signs):

Tensorial: Gas-kinetically:

$$h_{xx} = \frac{1}{2}\sigma\varphi_x^2 + \delta \qquad h_{xx} = \frac{22}{25}\frac{m^2}{\rho R^3 T^2}\varphi_x^2 + NKT$$

$$h_{yy} = \frac{1}{2}\sigma\varphi_x^2 + \delta \qquad h_{yy} = \frac{4}{25}\frac{m^2}{\rho R^3 T^2}\varphi_x^2 + NKT$$

$$h_{zz} = \frac{1}{2}\sigma\varphi_x^2 + \delta \qquad h_{zz} = \frac{4}{25}\frac{m^2}{\rho R^3 T^2}\varphi_x^2 + NKT.$$

2) The pressure derivation does not take into account that N (= the number of molecules in a cm^3) isn't a constant but a function of locality.[7]

Particularly the first objection is, of course, very serious; and because—when the pressure components calculated in this way are compared against some measurement results by Knudsen,[8] you get no conformity between his measured pressure and my calculated one—the theory is untenable in its present form.

Prof. Schrödinger was incredibly nice and gave much attention to the matter. He also gave me instructions on how to posit a new theory and particularly on avoiding the erroneous result (1). He hoped it might be avoidable if ⟨the⟩ α and β, which were constants in the original theory, were regarded as *functions of locality*.[9]

Prof. Schrödinger unfortunately had to go on a trip for 6 weeks and is only coming back in mid-April, i.e., [before?] my examination (beginning of May!).

So now I'm left godforsaken, a few weeks before the examination, with a doctoral thesis that doesn't work! My situation is so dire because I *definitely* must (for financial reasons) take my exam at the beginning of May and during the 7 weeks that still separate me from the exam I just have enough time to prepare myself for the oral exams and it's impossible to find the time to rework this entire wretched radiometer theory!

I tried to set up the new theory now by following the directions I got from Prof. Schrödinger before he left. But who knows if it works!!

And now, Albert, you can guess what I'm asking of you: do take care of this paper of mine. No one but you can help me . . .

Will you come to my rescue in this lamentable situation and check through the "new" theory? I know, Albert, it is a lot of work . . . because you surely have little time, and it's not enough just to give me a couple of hints.

But I know that you're a nice guy and that you'll help me make this poor misfit—which is your child a bit, too—viable again.

Warm regards to Elsa, also to the "girls." All my love, your poor (nebbish) doctoral student

Edith.

69. To Paul Langevin

[Berlin,] 6 March [1922][1]

Dear friend Langevin,

Further reflection and a fortuitous conversation with Rathenau led me to the persuasion that I should have accepted your invitation, despite all the reservations mentioned in my letter.[2] In the endeavor gradually to repair the harm from this war, one should not allow oneself be confused by petty considerations and— *you and your colleagues did not let yourselves be confused, either.*[3] Hence I declare myself willing to come if you have not already chosen another person. But even in case this has already happened, I do feel the need to document by this letter my goodwill and my—courage. If something should come of this matter, I would come in the first half of April. The language, however, will certainly cause me some trouble. But I do prefer to develop the material freely rather than read from something written. The formulas do help a lot and a kindly fellow professional will serve as prompter and pull out the words that get stuck in my throat. Albeit, it would perhaps have been nicer and more productive if we had made it a kind of mini relativity congress in which I would only have had to answer questions; my limited language ability would have disturbed less than it does in a more or less complete exposition of the theory. I can imagine, however, that the foundation statutes might bind you to a particular mode.[4]

Cordial regards, yours,

A. Einstein.

70. From Paul Langevin

Paris, 10b Boulevard de Port-Royal, 8 March 1922

My Dear Friend,

By accepting the invitation of the Collège de France, you made me very happy.[1] You understood, and your colleagues will certainly understand, what this gesture signifies, by which we wish not only to pay deserved homage to you personally and to your ideas but also to reopen the way to improved relations that the uppermost interests of the intellect demand. By accepting, you are doing a great service to the cause of the work reuniting us all.

You will receive an official letter from Mr. Croiset, the administrator of the Collège,[2] whom I am about to see and who was very disappointed about your refusal last week.

Our students and their professors will be on vacation for Easter from the 9th to the 24th of the month of April. Best would be if you could come in the final days of March to stay here for about ten days and have four or five meetings with a day's interval each time so as not to subject you to too much fatigue. By coming on the 28th or 29th of March you would have the necessary time.

I think I can arrange things in the form you wish, as discussions under your direction.[3] If you like, after having confirmed this option, I shall submit to you tomorrow or the day after tomorrow an agenda, taking into account what will most interest those who still encounter some difficulty comprehending you properly. It is understood, of course, that we hope to see you come with Mrs. Einstein, who, I believe, accompanied you to the United States and whom it would be a great pleasure to have here with you.

As the housing crisis continues to rage here and you would find little peace in the very heavily booked-up hotels, I could place at your disposal a small, very quiet apartment situated near the botanical gardens, a ten-minute walk away from the Collège[4] and five minutes away from the place where I currently live.[5] You would be alone there with a housekeeper who speaks German. There is no telephone and the furnishing is very simple but you will perhaps be better off there than in a large caravansary where journalists would come to bother you in person or by phone. Your address could, if you desire, not be known to more than a small number of persons who would have the selfish joy of having you nearby in the academic quarter.

The most urgent thing is that you let me know if you can come before our Easter vacation, from 28 March to 9 April, for example. We could maybe afterwards, if you have time, spend a few days in the countryside together.

'Bye for now, my dear friend, your most affectionately devoted,

P. Langevin.

71. To Paul Langevin

[Berlin, between 8 and 13 March 1922][1]

Dear Friend,

So I shall be coming around the 28th of March.[2] You are a kind, good person to have sought out such an ideal hideaway for me. (The service person does not need to speak German.) Nobody should know about it besides your closest friends[3] and Solovine.[4] I happily accept this proposal. Planck called this decision of mine to go to Paris "heroic," because he thinks that, although beneficial, my visit there will make a thousand printed and unprinted enemies for me. But he doesn't know the thickness of my skin. I find *your* decision was much more heroic than mine and duly appreciate it.

I still have not received either the official invitation or your agenda but am not going to wait for either of them. I will inform you about the exact time of my arrival. I am not going to take my wife along because I think we will be more at ease by ourselves. The simpler and less official everything is, the finer it will be.

Everything else in person. I am as excited as a child about being able to walk the streets of Paris with you again.

Cordial greetings, yours,

A. Einstein.

72. To Bernardo Dessau[1]

Berlin, 9 March 1922

Dear Colleague,

You are going to laugh at receiving an answer to a 1½-year old letter: But nothing else is to blame for this than the postman's generosity.[2]

Going to the Technion in Haifa cannot be advised to an established physicist who is not altogether possessed by idealism. Mere training in lower-level engineering for technical development is what is involved over there.[3] The University of Jerusalem, which will materialize shortly, comes into consideration instead. It is not supposed to be a teaching establishment proper but a type of research institute.[4] You probably know that I was in America last year to raise funds for this institution, which succeeded quite respectably.[5] However, establishing a physics institute has not yet been envisaged, owing to the great related costs. If such an institute is erected, you would certainly fall under serious consideration for it, equally so Mr. Ornstein from Utrecht, whom you surely know as a competent fellow professional.[6] I confidently hope that we Jews of our generation manage to come far enough along to set up modern scientific institutes at our own initiative.

In the hope of making your personal acquaintance, I am, with kind regards, yours,

A. Einstein.

73. From Richard B. Haldane

28 Queen Anne's Gate, Westminster, 9 March 1922

Highly esteemed and dear Professor,

The book—so kindly given to me by you—has arrived.[1] I prize this book very highly as a "care package" from you.

I value it also because it interests me very much.

I read through it, not entirely without comprehension. Weyl's attempt is interesting.[2] I have, however, read the critique of it by Hans Reichenbach.[3]

Important questions concerning the mathematical logic seem to be raised [over] here as well.

I hope you [and] your gracious wife are very well.

In gratitude and most devotedly yours,

Haldane.

74. To the French League of Human Rights

[Berlin, 10 March 1922][1]

Your invitation to the *New Fatherland League*, in which you kindly mentioned me personally, produced the greatest pleasure in our circle.

I would have joined the *League's* delegation if scientific research that is impossible to postpone had not prevented it.[2]

It would have been a pressing need for me, on the occasion of the first interview on French territory between democrats from the two countries, to proclaim that intellectual collaboration between the two peoples is in the interest of humanity as a whole and that our primary duty is to dispel the psychological obstacles to reconciliation.

I hope that in this regard the meeting in Paris will be of great importance.

Einstein.

75. From Lipmann Halpern[1]

Berlin, 10 March 1922

Highly esteemed Professor,

Your research into the infinite spaces, which sets you as one link in the golden chain of Jewish wise men, did not deflect your gaze from Earth. Your love for the Jewish nation, your self-sacrificing engagement for a Palestinian academy,[2] your constantly demonstrated [good]will towards Jewish students, give me the courage and confidence to lay before you a deeply cherished wish of mine.

I am the son of the rabbi from Bialystok;[3] my family had emigrated from Germany in the 17th century and my forefathers had always achieved much in the area of Torah studies; thus they are famed as wise, scholarly men throughout the East. I myself studied ancient Jewish [liter]ature until 18 years of age and believe I acquired some good knowledge in that way. My acquaintance with the work of

Maimonides awoke within me the wish to acquire a European education and make medicine my profession—according to the words in the Talmud: "The beauty of Japheth shall dwell in the tents of Shem."[4]

That is why I came to Germany as a high school graduate in order to study at a university here.

Toward this end I submitted my documents last semester already, but my application was rejected. Now for the coming semester I handed in my documents again at the Prussian Ministry of Science, the Arts, and Public Education and would be very grateful for some support at the ministry for my intentions.[5] Because I heard about the obstacles in Berlin, my application is for admission to the University in Königsberg. A rejection of my application would destroy the course I had planned for my life and would therefore hurt me deeply.

Hear the cry of a young Jew who eagerly wants to study and is denied any access to academia. Your personality is legendary among young Eastern European Jews; and because it is your country that I am entering, it is to you that I address my words of distress.

In deep veneration,

Lipa Halpern.

The application, together with the final-grades report by the preparatory school in Lida, government of Vilna, I conveyed to the Prussian Ministry of S[cience], C[ulture], and P[ublic Education] (4 Unter den Linden) on 7 Mar. 1922.

76. "Theoretical Comments on the Superconductivity of Metals"

[*Einstein 1922k*]

Manuscript completed around 11 March 1922

PUBLISHED after 4 September 1922

IN: *Het Natuurkundig Laboratorium der Rijksuniversiteit te Leiden in de jaren 1904–1922. Gedenkboek aangeboden aan H. Kamerlingh Onnes, directeur van het Laboratorium bij gelegenheid van zijn veertigjarig professoraat op 11 November 1922.* Leiden: IJdo, 1922, pp. 429–435.

The theoretical scientific researcher is not to be envied, because Nature—or more precisely put: experiment—is a merciless and not very kindly judge of his efforts. She never says "yes" to a theory, in the best case merely "perhaps"; but in most cases simply "no."[1] If an experiment agrees with the theory, it means "perhaps"; if it does not agree, then it means "no." Every theory is sure to experience its "no" someday, most theories already do so soon after their formulation. We would here like to cast a glance at the fates of the theories of metallic conductivity and at the revolutionary impact that the discovery of superconductivity must have on our ideas about metallic conductivity.   [p. 429]

After it was realized that negative electricity is embodied in subatomic carriers of a particular mass and charge (electrons),[2] it made sense to assume that the motion of electrons is the basis of metallic conductivity. Furthermore, the circumstance that metals conduct heat far better than nonmetals, likewise the Wiedemann-Franz approximation law[3] that the ratio of electrical and thermal conductivity of pure metals is independent of the substance (at normal temperatures), also led to ascribing thermal conductivity mainly to electrons. These circumstances gave occasion for an electron theory of metals following the model of the kinetic theory of gases (Riecke, Drude, H. A. Lorentz).[4] This theory has assumed that electrons in metals, disregarding collisions they undergo from time to time with the metal's atoms, are freely moving and, like gas molecules, should be endowed with thermal kinetic energy of average magnitude $^3/_2\,kT$.   [p. 430]

This theory was wonderfully successful to the extent that it was able to derive theoretically, with admirable precision, the coefficient of the Wiedemann-Franz law from the ratio of the electron's mechanical and electrical mass.[5] It also explained qualitatively the phenomena of thermoelectricity, the Hall effect, etc. However the theory of electrical conduction may develop in the future, one main support of this theory will surely always be retained, namely: the hypothesis that electrical conductivity is based on the motion of electrons.

Drude's formula for the specific resistance ω of metals is[6]

$$\omega = \frac{2m}{\varepsilon^2}\frac{u}{nl}, \qquad \ldots (1)$$

where m signifies the mass, ε is the electron's charge, u the mean velocity, n the volume density, and l the electrons' free length of path. Unfortunately, three unknown temperature functions u, n, and l enter into the theory, one of which (u), according to the kinetic theory of heat, is supposed to be connected with the absolute temperature by the relation[7]

$$3mu^2 = kT; \qquad \ldots (2)$$

n must be small compared to the mean density of the atoms in order to explain why the electrons do not contribute significantly to the metal's specific heat.

To what extent is the fundamental formula (1) suited to explain the temperature dependence of the specific resistance? Here we run into great difficulties. According to (2), n should be proportional to $\sqrt{T}$. One would not wish to expect any considerable dependence of the length of path l on the temperature. However, one would expect a rapid increase of the number n of electrically dissociated atoms

[p. 431]   with the temperature because the dissociation of a weakly dissociated substance grows rapidly with T. Thus one would think that the resistance of pure metals drops rapidly with rising temperature. This is by no means the case, however, since it is known that the resistance of pure metals increases at high temperatures almost exactly proportionally to T.[8]

To do justice to this striking fact on the basis of equation (1), one would perhaps have to seek refuge in hypotheses: that the number n of free electrons is independent of temperature; the electrons' free length of path l is inversely proportional to the square root of the metal's energy content. With the thus modified formula (1), Kamerlingh Onnes was able to represent with remarkable precision the behavior of metals in nonsuperconductive states.[9] The hypothesis of the dependence of the length of path on the thermal agitation is not very disconcerting; one would have to imagine that the electron is moving in an agitation-free metal as does in empty space, but that the inhomogeneities occasioned by the thermal fluctuations produce electric fields that deflect the electrons. On the other hand, the hypothesis of a temperature independence for n is highly questionable. Also, the assumed law of a dependence between l and the heat content would surely be difficult to justify quantitatively. In any case, though, the success of Kamerlingh Onnes's consideration seems to prove that the thermal agitation of the metal (not of the electrons) is essentially the factor determining the resistance. Only then could one explain the fact that the resistance at higher temperatures satisfies the law

$$\omega = \alpha(T - \vartheta)$$

and not the law

$$\omega = \alpha T,^{(1)}$$

[p. 431]   (1) Compare, e.g., Comm. no. 142a, *Versl. Ak. Amsterdam*, June 1914, fig. 2 for *Sn*, *Cu*, *Cd*, and Suppl. no. 34b, Report Third Int. Congr. Refr. Chicago, fig. 5 for *Hg*.[10]

and that the resistance of nonsuperconductive metals becomes temperature-independent at low temperatures. The curvature of the resistivity curve at low temperatures is thereby indirectly connected to quantum theory. [p. 432]

In fact, according to the above conception, as the temperature drops, the resistance of nonsuperconductive metals ought to approach zero, whereas in reality it approaches a limit other than zero. Kamerlingh Onnes found, however, that this limiting value is strongly influenced by slight impurities.[11] He found out, furthermore, that these trace admixtures cause a vertical parallel shift of the entire resistivity curve, i.e., that they produce an "additive resistance," so that the resistance of the pure, homogeneous metal may very well have the limiting resistance of zero.[12] Let it be said that this extremely remarkable fact obstinately resists explanation by equation (1). For, if the admixture creates special collision opportunities for the electrons, this causes, as is easily demonstrated, a constant contribution of $\frac{1}{l}$. This latter does not change the resistance by a temperature-independent amount but by one proportional to u $\left(\text{or to } \frac{u}{n}, \text{resp.}\right)$; but u cannot by any means be assumed to be temperature-independent, because otherwise the sole great success of the theory, namely, the explanation of the Wiedemann-Franz law, would have to be abandoned. For the same reason, too, it would be difficult to explain theoretically the resistance of impure metals becoming constant at low temperatures.

From this sketch one can see that the thermal theory of electrons fails already for the usual conduction phenomena, let alone for superconductivity. On the other hand, it certainly is conceivable that the Wiedemann-Franz law will result from another kind of theory that relates electrical and thermal conductivity back to an electron mechanism.

The failure of the theory became fully obvious after the discovery of superconductivity in metals.[13] By proving that nonsuperconductive wires with a thin coating of a superconductive material are superconductive, Kamerlingh Onnes convincingly demonstrated that superconductivity certainly cannot be based on electrons in motion from thermal agitation.[14] With time the coating's electrons would have to penetrate into the nonsuperconductor and there lose their mean motion advantage that forms the electric current. Consequently, the system would not be superconductive. [p. 433]

If one wished to explain superconduction by *free* electrons, one would have to conceive of them as agitation-free in such a way that the negative electricity in the superconductor in which a current is flowing has no other motion than the one making up the electric current. Such a notion is made improbable not only by the Rutherford-Bohr theory, according to which there are strong electric fields inside a body, but also from the fact that superconduction is destroyed by moderate magnetic fields. For, the lateral forces produced by the Lorentz force (Hall force)[15] would, of their own accord, balance out electrostatically by charge accumulation on the surfaces, such that no effect of the magnetic field on the electrons would be expected.

It therefore appears that electric conduction must be attributed to the atoms' peripheral electrons, which move around the nuclei at great speed. Indeed, according to Bohr's theory, it scarcely seems conceivable that the energetically orbiting peripheral electrons would lose a considerable portion of their speed, i. e. in the case of mercury vapor with its relatively low-energy liquefaction. This is why, based on the current state of our knowledge, it looks as though free electrons do not exist at all inside metals. Then, metallic conduction would have to consist in atoms exchanging their peripheral electrons. If an atom were to receive an electron from a neighboring atom without roughly simultaneously releasing an electron of its own to another neighboring atom, it would experience a powerful energetic modification, which certainly cannot happen with superconductive currents that main-

[p. 434]   tain themselves without any expenditure of energy. Thus it appears unavoidable that superconductive currents are borne by closed chains of molecules (conduction chains) whose electrons experience constant cyclical exchanges. That is why Kamerlingh Onnes compares the closed currents in superconductors to Ampère's molecular currents.[16]

With our broad ignorance about the quantum mechanics[17] of composite systems, we are far away from being able to condense this vague idea into a theory. We can only attach a few questions to it that could be decided by experiment. It appears unlikely that different kinds of atoms could form conduction chains with one another. Thus the transition from one superconductive metal to another one may never be superconductive.[18] A further idea that suggests itself is that this may be the reason why thus far only metals with relatively low melting points have turned out to be superconductive;[19] for, in such substances, the impurities may not be in a truly dissolved state but rather in the form of small complexes which, in the metal's plastic state, will precipitate out.

Furthermore, there is the possibility that the conduction chains can not carry arbitrarily small currents, but only those of a particular finite magnitude, which would likewise be accessible to experimental verification.[20]

The idea that conduction chains can be destroyed by magnetic fields is likely, indeed almost necessary, as is the fact that a sufficiently strong thermal motion would destroy conduction chains as would sufficiently large $h\nu$-energy quanta of which they are composed. Thus, the transformation of a superconductor into a normal conductor by a temperature increase, and perhaps even the superconductor's sharp temperature limit, may become comprehensible. Electrical conductivity at normal temperature may be based on the incessant thermal formation and destruction of conduction chains.

This conjecturing can only be excused by the momentary predicament of the theory. It is clear that new avenues have to be sought in order to be able to do justice to the facts of superconductivity. It appears probable, but not certain, that conductivity at normal temperatures is based on superconductivity constantly being disturbed by thermal motion.   [p. 435]

This idea is backed by the consideration that the frequency of the electrons' transition to the neighboring atom may be closely related to the orbital frequency of electrons in the isolated atom. One thus arrives at the suspicion that the elementary currents of the individual conduction chains could be of substantial size. If this idea of the quantum dependence of elementary currents were proved valid, it would make sense that such chains could never contain atoms of various kinds.

P.S. The suppositions suggested at the end, which incidentally do not raise any claim to novelty,[1] are partially disproved by an important experiment that Kamerlingh Onnes performed over the last months ago. He showed that at the point of contact between two different superconductors (lead and tin) no measurable Ohmic resistance occurs.[22]

[1] Comp., e.g., F. Haber, *Sitz. ber. Ak. Berlin*, 1919, pg. 506.[21]

77. From Paul Ehrenfest

[Leyden,] 11 March 1922

Dear Einstein,

Thanks very much for the card.[1]

1.° Please inform me when you want to be here.— As orientation: 27 April, first phys[ics] colloqu[ium] again after Easter. Circa 1 May, Onnes back in Leyden. When Lorentz will be back again I don't know but probably unlikely before circa 5 May.[2]–

I would suggest to you: 29 April. If that is too late for you, then 22 or 24 April.

2.° Onnes recently established: Contact area between supercond[ucting] tin and supercond. lead yields no detectable resistance. I.e.:[3]

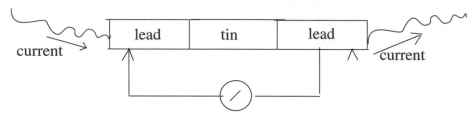

No potential difference detectable.

He will write to you about it.

3.° Please forgive me for still not having corresponded with Prof. Kochert[haler] (Madrid). Permit me to wait with this until you are here because too much is unclear to me. E.g., I don't know whether I should send him the preparation as I have it here or ⟨directly in⟩ must first convert it chemically into a form that he can use directly, which would naturally require a recalculation of the molecular weights.[4]–

In case you don't want me to wait for you about this, do tell me whether I should first convert it chemically or not and whether he should cite it in his book solely under your name or also with mine added (forgive this immodest question—for I naturally know very well that the entire merit goes to you).

Best regards to your d[ear] wife and Ilmargotse.

78. From Heinrich Zangger

[after 11 March 1922][1]

Dear friend Einstein,

Are you already in *Japan*? When you arrive in Tokyo, do have the very interesting physico-chem[ical] experiment by Prof. Hashida shown to you (physiol. institute). He takes continuous measurem[ents] of potential fluctuations that are physic[ally] very interesting. He was also here in Zurich for some years.[2]

You know that Besso is a grandfather; that is surely the best cure against pessimism.[3] The enclosure was written long ago. But what use is mailing, when the mind is in Japan?

This winter I audited Debye's entire course on the structure of matter—very good, just too short, but many connections in all directions and brilliant dimensional considerations.[4] I am working with all spectra as *measurement sensors* in unknown substances—*this* particular med. eye I am developing very well (plate sensitization with mineral oil 1/1000 [5000?] spectral[lines?] is very interesting).[5]

Do you know about *India's critique* of Europe in the book: *God's Conic Sections* [*Die Kegelschnitte Gottes*] (Lange, München, 1920) by Sir Galahad,[6] in which you also play a role (p. 132)—as ideal European, whom Sir Galahad has not discovered yet.[7] These secrets will be solved later—after Japan[8]

(from Zurich) Have a nice trip and enjoy yourself[9]

Zangger

79. To Paul Ehrenfest

[Berlin, between 11 and 13 March 1922][1]

Dear Ehrenfest,

I'm going to be in Paris from 28 March until c. 4 April.[2] Can I come to Leyden then or should I go back to Berlin in order to come 2 weeks later?

It is very fine that K[amerlingh] Onnes did that experiment. Yet another glimmer of hope to understand is dashed.[3] Franck informed me about a very nice point regarding the behavior of free electrons towards atoms. I'll tell you about that then. In other respects I'll be bringing an empty head along, which, after your consoling assurance, shouldn't cause any harm.

Warm regards to you all, yours,

Einstein.

I am probably going to Japan now, after all, with my wife; departure August or September.[4]

80. From Michele Besso

Bern, 42 Ziegler Street, 12 March 1922

Dear Albert,

A little card by Vero[1] may have reported to you that I've become a grandfather.[2] We just returned from seeing the little grandchild (who, for the sake of convenience, was born in the clinic) and my folks are pointing out resemblances. I believe I can see something peering into this world in expectation of impressions and ready to react with determination to them—so that perhaps in 49 years he, too will sit himself down and write to a friend: "It was worth the effort to have taken a look at this poor, rich, strange, and ordinary world."

Who knows, maybe it comes from *his* will to live that I've been thinking for a while already that, since you are currently actually in contact with all sorts of industries, you might perhaps be able to . . . ease my decision to finally let go of the patent application with the enclosed content. The basic idea is that the local intermeshing of textile fibers, which is produced by friction ⟨in the spun fibers⟩ inside each fiber and by the fabric weave, is achieved by chemical treatment and by as localized a fusing of the fibers as possible. Thus with a minimum of work, structures similar to felt are supposed to form, but distortable and washable. The most nearly similar industrial products are 3)[3]hat felts, for which the suction *process* described in the patent description is already being applied; 1) the veil of crossed fibers of artificial silk, which are made to stick together prior to strengthening; 2) and parchmented papers, for which, however, the softening and gluing together of the fibers is not as carefully localized as I plan to have it.[4]

If you should happen to know someone who could be interested in this, I would thank you very much, in the name of my grandchild. Or perhaps not—for who knows, maybe the family soul will need to be fortified again by a frugal diet.

As for mine, it is uncomfortably soft. I was in Trieste for two days and found my mother unusually well. Then the grandchild arrived here with a punctuality that was of mutual benefit to the health of mother & child. I should calm down and finally be able to do a little work for the office again, but every flutter stops me, everything revs me up (or perhaps better said, *down*). The stupid scribbling by poor Guillaume, for example,[5] which ⟨I⟩ makes me ashamed that I can't find the final persuasive word at least for a third-party listener; respectively it actually also concerns Lorentz, purely logically speaking: Namely, local time is introduced without any clear definition. Thus perhaps G[uillaume] thinks he may also play around at will with the Lorentz transf[ormations]. But . . . ⟨not to⟩ find at will, etc. On this occasion I happily chanced upon the properties of the "graphic trajectory" of two-dimensional kinematics, which permits such a clear discussion of the various theories of light and ⟨for⟩ reproduces the Lorentz transformation by the simplest of projective relations (polar line ⟨of the sy⟩ and polar plane). Has this form of representation already been properly employed ⟨and⟩ anywhere? In Weyl it has been sketched correctly, of course,[6] but only for the Lorentz transformation, whereas it is nice to follow ⟨Ritz's⟩ other theories of light on the same picture and it is also nice to ⟨expl⟩ give the derivation of the Lorentz transformation and the Lorentz contraction ⟨with its help, or, resp.⟩ using this diagram as explanation.—

—Vero asked me a few days ago ⟨why⟩ whether the transversal Doppler effect of rapid anode rays has been measured, resp. what experiments exist ⟨at⟩ in this area. I don't know anything more specific about it either.[7]

One thing I'm actually not quite clear about either is ⟨what effect⟩ how the first-order effect fares in the theory of Jupiter's satellites, assuming a velocity against the ether of, e.g., 10 km/sec in the naïvely used theory of a resting ether. Born thinks that the discrepancy would be very large[8]—⟨I still remain doubtful⟩ but I think I recall you saying back then already, that ⟨if one⟩ this would not automatically be the case, ⟨If I imagine that the [light veloc]⟩ I believe because in taking into account the relative velocities of light one would get another position and ⟨form⟩ dimension of the planetary orbit than the one otherwise given.

Intrinsic interest in this issue, at the current state of r[el.] t[h.], is ⟨actually⟩ *very* meager; but it can play a part in *teaching*.–

—In a spherical space I have two touching, infinite ⟨perfectly straight rods⟩, straight rods that therefore eventually meet up with themselves again. After a comparison of their masses has taken place, one of them is brought to a velocity relative to the other along its length. What happens there to the Lorentz contraction? Does it come out somewhat thus: that the masses produced by the "rotating" spherical space are in motion and therefore ⟨however⟩ another spherical space might have to result in a different radius?

Warm regards from all of us, yours,

Michele.

81. To the Prussian Academy of Sciences

Berlin, 13 March 1922

Esteemed Colleagues,

It is now the third time this year that I received an invitation of a public nature from Paris; the first time from the "League of Human Rights,"[1] the second time from the French Philosophical Society,[2] the third time from the Collège de France, asking me to give a few guest lectures.[3] The invitation is issued by the teaching faculty of the Collège de France and was conveyed to me by my friend and fellow colleague in the profession, Langevin. The letter by the latter explicitly points out that this event should serve the restoration of relations between German and French scholars. The relevant passage states: "The interest of science dictates that relations be reestablished between German scholars and ourselves. You can help in this better than anyone else; and by accepting you will do a very great service to your colleagues from Germany and France and above all to our common vision."

I answered my friend's letter initially with a polite refusal, indicating that my main reason grew out of a sense of solidarity with my colleagues here.[4]

I could not rid myself of the feeling, however, that by my refusal I had followed the path of least resistance rather than my true duty. A conversation with Minister Rathenau[5] had the result of turning this feeling into firm conviction. That is why a few days after my first refusal I wrote a second letter to Prof. Langevin[6] in which I retracted my refusal and stated my willingness to accept the invitation if other arrangements had not in the meantime been made regarding the relevant guest lectures. Thus I agreed with Langevin to travel to Paris at the end of the month, in order to deliver the lectures.[7]

In view of the circumstance that the academy is interested in all events that concern international relations, I consider it appropriate to submit the above information to the academy.

With great respect,

A. Einstein.

82. From Paul Ehrenfest

[Leyden,] 13 March 1922

Dear Einstein,

The devil take you—to Leyden!—You drive your contemporaries crazy, not just with $G_{hk}^{st,uv}$ but also with your routine activities.[1]

1.° You are going to Japan in the fall[2]—very rightly so—at least in the sense: *I* naturally would go to Japan, too, if anyone invited me.— But translated from Japanese into Dutch this means: In the fall you are *in any case* not coming to Leyden.—

2.° So I have to thank God Almighty that I grabbed you by your spring coattails when you wrote with such conviction: "Should I come and see you yet again; isn't it better if I come in the fall?"[3]

3.° Now you're piping yet another tune and want to come "through" Leyden instead of "to" Leyden (*on 7 April not a single cat—besides me—will be in Leyden anymore!*)—Oh my, oh my, oh my!—I'd sooner take a mortgage on a soap bubble than on you.

4.° No—Mr. Privy Councillor: You come nicely and like a normal bourgeois to Leyden, not *during* but *after* (our) Easter vacation. Whereby, as a special favor, I leave you the choice whether you would like to come on the 22nd of April, on the 29th of April or on the 6th of May.–

Onnes will not be back here until the first days of May.[4]

Lorentz will be back here around the 12th or the 15th of May.[5]

So I would say that is the *best* time; but I'm scared to death that you'll steam off if I wait even a day longer.

— • —

And if you leave me in the lurch, then I request my dismissal as chronicler of your activities.[6]

!!!! So; and now read this letter to the assembled women of the house so that they will help me out.

With grumbled greetings, yours,

Ehrenfest.

Tschulanowsky will be inquiring of you in the coming days when you'll be coming to Leyden because it's important to *him*.[7]

Addendum

W. J. de Haas is *very* depressed and embittered: A couple of years ago preparations were initiated for a professorship at Utrecht *for Du Bois* with a special laboratory for him (magnetism)—when he suddenly died. These efforts were continued and the faculty in Utrecht negotiated with De Haas. This has been going on now for about 2–3 years. The matter had certainly been dragging on but everyone was sure that it could not last much longer now and that De Haas would soon be called. Suddenly and *totally unexpectedly* De Haas and all of us were able to gather from the state budget that the minister[8] was "abstaining" from the creation of this professorship in connection with the general savings regulations "and is just giving a *lectureship* but will appoint it immediately."– It is now, owing to numerous circumstances, becoming steadily *more probable* (I'm *not* saying "more certain") that Ornstein,[9] albeit definitely not in consultation with the faculty, had guided this "savings wave" in this direction. In any event, it is a *fact*: 1.° That O. has spread himself out like a patch of oil throughout the entire *experimental* operations at Utrecht (Julius, during his illness, was not in a position to direct the laboratory either)[10] 2.° That his pupil Burger resigned at the incandescent lamp works in Eindhoven *already in December* in order to return to Utrecht.[11]—So it just remains to be seen (which nobody doubts) whether *he* is supposed to become the lecturer.

O. is making ever more enemies with his loutishness—which often borders on the unbelievable.—Especially among such people as Fokker.[12] And it could *become necessary* one day that I take a stand against it. I feel the need to let you know about this now already. Previously—less out of neighborly love than simply for the sake of peace—I had always tried to weaken these conflicts by laughter and congenial talking around it. But if O. *really* has acted this way in Utrecht, as is now becoming *probable* (it is not yet certain), then it is gradually becoming a *duty* to erect a dam against this business, which is beginning to become *aggressive*.

If you happen to find the chance to write a nice little letter to the De Haases[13] one of these days, it would be appreciated!

Cordial regards, yours,

Ehrenfest

83. From Paul Winteler

Fiesole V. Verdi 8 [13 March 1922]

[Not selected for translation.]

84. To Thomas Barclay

[Berlin,] 14 March 1922

Dear Sir Thomas Barclay–

With reference to your kind inquiry[1] I can inform you that I am coming to Paris on March 27th for ca. 10 days in order to give a few talks at the Collège de France.[2] Considering my imperfect knowledge of the Fren[ch] language, m[y] duties thus already assumed will be a very great effort for me; so it will be scarcely possible for me to follow yet another invitation to speak at the S[ociété] de Physique. However, it will be a great pleasure for me to discuss with [Mr.] P[ainlevé][3] scientif. questions of mutual interest to us. In hope of seeing you again in Paris, I am, with amic. greetings, sincerely yours

85. To Maurice Solovine

Berlin, 14 March 1922

Dear Solovine,

To a happy reunion in Paris! I'm arriving there on the 27th or 28th of March.[1] Langevin is arranging a hideout for me that he will divulge to you but which I ask you please *to keep strictly secret*; for the days of my Parisian stay will be very exhausting as it is.[2]

I have two objections to the draft contract by Gauthier-Villars:[3]

(1) The paragraph labeled by me as b has to be struck, because I only want to give him the publisher's rights to the French translation, of course. (2) The paragraph labeled a is incomprehensible to me.

We can perhaps settle this affair during my presence in Paris. I'm very much looking forward to our being together, if only my beak were better polished in French.

Kind regards from your

A. Einstein.

86. From Michael Polányi

Dahlem, Berlin, 4–6 Faraday Way, 14 March 1922

Dear Professor,

On an earlier occasion you most kindly agreed that you would intervene on behalf of Dr. E. Bródy.[1] He writes me now that he has some prospect of becoming an assistant in Pasadena (California) in Millikan's new institute,[2] the theoretical heads of which will be Epstein & Tolman.[3] A pertinent suggestion by Professor Born has already gone out to Epstein. Please allow me to request you kindly write to one of the mentioned gentlemen in America. The staff of this institute is being assembled right now and a timely intervention could surely accommodate another man there.

For your kind help I thank you also in my own name and send my humble regards in great respect, yours,

M. Polanyi.

87. To Paul Ehrenfest

[Berlin,] 15 March [1922][1]

Dear Ehrenfest,

If O[rnstein] is doing such terrible things, someone should first appeal to his conscience and, if that bears no fruit, confront him.[2] I'm at your disposal to do anything within my capacity. For inst., I am willing to write him and tell him off.

So now I'm coming to see all of you at the end of April. I really am a poor devil: on 27 March until around 5 April I have to speak in Paris at the Collège de France in—I shudder to say—*French*;[3] I announced a course here for the summer term starting around 1 April;[4] and I have to visit you in May, and do it properly, too, because I was unable to resist the sirens of East Asia.[5] So this is how I'll do it: I'll be teaching *here* just in June and July, but therefore twice a week instead of once. This won't be that bad because after my return, hence still during the winter semester, I can still come to Leyden again. This is possible, since I am putting my duties here on hold for half a year. So, with God's help, everything will turn out all right and you don't need to get agitated. Thank Onnes in my name for having performed the experiment. It would have been nicer still if he had done it with a closed circuit, but *this way* probably suffices as well.[6]

The result of the experiments with sound in $N_2O_4 \Leftrightarrow 2NO_2$ was that the reaction follows the sound vibrations.[7] So the reaction must be very rapid. I then suggested comparing the amplitude of the sympathetic vibrations with the one for normal gases, since a reaction speed not infinitely large acts as an energy dissipator, hence it diminishes the amplitude.

It is actually good that I have so many distractions because otherwise the quantum problem would have driven me to the insane asylum long ago. The fact that light, emitted in diametrically differing directions, can interfere has supposedly been irrefutably demonstrated.[8] How is that supposed to be reconciled with the energetic directedness of elementary processes? How pitifully the theoretical physicist stands before nature and—his students!

88. From Maurice Croiset

[Paris,] 15 March 1922

Sir,

The assembly of professors at the Collège de France has, as Mr. Langevin has already let you know,[1] decided at its last meeting to ask you if it would be possible for you to come to the Collège to deliver some lectures. The favorable response[2] Mr. Langevin has informed me about authorizes me to confer with you about organizing these lectures and to ask you please to let me know your intentions in this regard.

As the Collège is closed every year during the Easter holidays (i. e., this year from April 9 to the 23rd of that same month), the lectures you would be so kind as to deliver would have to take place either before April 9 or after the 23rd. It would be left to you to set their number, depending on the nature of your lecture topic and on your personal preference. I confine myself to suggesting, as a simple indicator, that it could range from between 3 and 5. In accordance with the wish you expressed to Mr. Langevin,[3] they would take place before a restricted audience composed of specially invited experts, mathematicians, physicists, or philosophers.[4] They could, if you deem it appropriate, lead some discussions. The compensation allotted by vote of the assembly to the speaker is 5,000 francs.

I would be obliged if you would kindly indicate to me in your reply the general title under which the issues addressed in these talks are to be grouped and precisely on which date they will commence. We need this for the letters of invitation that we shall have to send out.

It is a my duty to append to this invitation my colleagues' gratitude along with my own. The Collège de France has always had the honor of welcoming those who contribute to progress in science and who open new paths to the endeavors of the human intellect. The universal opinion of the most qualified scholars counts you among these masters of contemporary thought. We should be pleased if for a few days you would occupy one of our chairs.

With high regards,
Administrator of the Collège de France,

Maurice Croiset.

89. To Max Hirschfeld

Berlin, 17 March [1922]

[Not selected for translation.]

90. To Paul Winteler and Maja Winteler-Einstein

Berlin, 17 March 1922

Dear Pauli and dear Sister,

With dismay I gathered from the card[1] [. . .].

I'm going to Paris on 28 Mar. to deliver some lectures at the Collège de France and in the fall to Japan and China. Maybe I'll take Else along.

91. To Arnold Berliner

[Berlin, on or after 17 March 1922][1]

[Not selected for translation.]

92. To Maurice Croiset

18 March 1922

Highly esteemed Colleague,

Mr. Langevin offered a few detailed suggestions about the scheduled presentations, with which I thoroughly agree.[1] It would therefore probably be best if you had the invitations issued in the form ⟨you⟩ he deems right regarding the choice of titles for the individual talks as well as regarding the selection of invited auditors.[2] I feel the need to tell you that I consider this invitation a courageous act of goodwill and reconciliation and that it is also rated highly in this sense by all perspicuous men here with whom I had occasion to talk about it. The fears I mentioned in m[y] first letter appear to be exaggerated.[3]

I anticipate arriving in Paris on 28 Mar. In the happy prospect of soon making your personal acquaintance, I am, with kind regards, yours sincerely.

93. To Gustav and Regina Maier-Friedländer[1]

[Berlin, 18 March 1922][2]

[Not selected for translation.]

94. From Hermann Anschütz-Kaempfe

Kiel, 18 March 1922

Esteemed Professor Einstein,

Your letter from the 10th of this mo. no longer reached me in Munich and followed me here; we have been here for 8 days. Despite very serious worries, I am sitting at my desk; my wife was operated on this evening for appendicitis with quite nasty complications.[1] I just came back from the clinic, where the poor dear is apathetically and yet painfully resting. The prognosis is at least such that a good deal of hope remains.

If the next 3–4 days will only pass without pericarditis, then I could probably breathe easily again.

You are right; vexatious politics spoils so many thoughts and relationships that of themselves are so good and positive; I think, even in the public interest, we shall soon come far enough along to judge bad day-to-day experiences on a higher level. At the moment it upsets me so much that I shut my eyes and ears up against all that ugliness coming from politics. It is probably not quite right but I can't do otherwise.

Aside from the next worrisome days that are full of anxiety and hope, I take pleasure in my work. The new compass is growing like an organism reshaping itself according to its purpose. Platinum and carbon withstand stress in water, likewise rubber and bakelite.[2] And one can work quite well with these materials as building blocks.

However much it would please me if you gave me and the work here a few days, I would like to emphasize that you are free from any kind of obligation; as I already said to you before, I have far too much respect for your work; but it would be nice and it would delight me and my wife if you came. It is possible that Sommerfeld will be returning from his trip to Spain, which he is beginning on the 22nd of this mo., via Hamburg; then he will be visiting us here in ca. 4 weeks. Perhaps you would like to be present then as well?

Then you could certify your electromagnetic experiment with your own eyes.[3] I am currently having a new coil wound that consists only out of 0.5 copper wire and insulation. The last one still had bronze bearings that were disturbing.

And by then I hope all the worries about my wife will have vanished again.

Most cordial regards to you and your wife, yours,

Anschütz-Kaempfe.

P. S. The mailing to Switzerland for April 1st is already on its way.[4]

95. To Michele Besso

20 March [1922]

Dear Michele,

I spoke with Professor Herzog,[1] the director of the K[aiser] W[ilhelm] Inst. for Fiber-Chemistry.[2] He promised assistance, esp. in testing thus manuf. samples.[3] He thought that little could be done with a general patent if it is not known for which specialties the process was of marketable use. For very *loose* fabrics, material costs play a minor role. He said only practically tested items are exploitable.

Do you have facilities for experimenting on a large scale? Through Zangger's intercession, perhaps?[4] I have no contacts with people in this branch.

Cordial greetings,

Albert.

96. To Robert A. Millikan, Paul Epstein, and Richard C. Tolman

Berlin, 20 March 1922

Dear Colleagues,

I learned[1] that a certain possibility exists for my colleague Dr. Bródy to be employed at your institute. As I know Mr. Bródy[2] to be an unusually gifted theoretical physicist, I would not like to fail to recommend him herewith most warmly. I consider this a duty, particularly because as a consequence of a hearing impairment Mr. Bródy comes only secondarily into consideration for teaching positions; so, considering the current oversupply of scientific professionals and the prevailing general impoverishment in this country, he as a foreigner would find it very hard to obtain a position in which his capabilities could be suitably employed.

With cordial regards, I am yours.

97. From Thomas Barclay

Paris, 17 Pasquier Street, 20 March 1922

Dear Professor Einstein,

I am very happy to learn from your letter of the 14th instant[1] that you received the invitation. I showed your letter to Mr. Painlevé who was just eating lunch with me here at home at the very moment of its arrival and who was very happy about this news.

I settled with Mr. Langevin (so as not to have any conflicting engagements) to ask you to give me the pleasure of sharing the midday meal with me on Saturday, April 1st.

I would be happy to know if it would be possible for you to accept my invitation on the set day.

I am, dear Professor Einstein, very devotedly yours,

Thomas Barclay.

P.S. Mrs. Ménard-Dorian is very aggrieved that you will not be staying with her. She has a charming house and she herself is an élite woman who knows how to spend her great fortune without offending the democratic spirit.[2]

98. From Leo Jolowicz[1]

Leipzig, 20 March 1922

Highly esteemed Professor,

Dr. Löwe is out of town.[2] Perhaps you could mark the items in the enclosed catalog that you consider important for "Jerusalem" University; I shall be happy to donate them.[3]

At the same time, I cord[ially] request of you, esteemed Professor, the address or the name of your young collaborator who is supposed to bring your relativity theory up-to-date again.[4] If you do not know his address, then just give us his name; we will then find out the address, all right.

Best compliments & many thanks,[5] yours very sincerely,

Akademische Verlagsgesellschaft m.b.H.

Leo Jolowicz.

99. From Paul Langevin

Paris, 20 March 1922

My dear Friend,

We are all overjoyed at the idea of soon having you here; I am busy organizing the meetings of the Collège for the dates I indicated to you.[1]

I transmit in attachment a letter by Mr. Xavier Léon on the subject of the meeting that he requests of you for the Société Française de Philosophie on 6 April.[2] He would be pleased if you could give him a title for publication to open the dialogue, and a few lines of "theses" as an agenda. If the planned invitations are convenient for you, would you please immediately reserve the following dates:

Saturday, 1st April for lunch at Sir Thomas Barclay's (at noon).

Sunday, 2 April for lunch in Boulogne on the Tour du Monde premises (noon).

Wednesday, 5 April for dinner (7 o'clock) at the home of Mr. Borel.[3]

I think I have reserved Thursday[4] for a program of discussions acceptable to some colleagues and will submit it to you directly.

See you soon, and most affectionately yours,

P. Langevin.

100. From Mileva Einstein-Marić

[Zurich, around 21 March 1922][1]

Dear Albert,

I have a few comments about your last letters to our boys[2] and otherwise also have a few things on my mind; that's why I have to write you these few lines; I hope you won't be cross. Your repeated invitations to the children for the summer vacation heartily please us all, I just have to point out again that I consider that kind of vacation extremely risky for Albert. Right after the vacation he has to take his final exams and should use this vacation thoroughly for studying.[3] Without wanting to criticize Albert, I do have to say that, left to his own devices, he would certainly not work enough and it would be extremely annoying if he did not pass this exam; for us considerably more than for others, for you can imagine how longingly I wait for my older boy, at least, to stand on his own two feet. Surely you understand me well. Albert is good and industrious with subjects that interest him, but he is a bit difficult to win over for subjects he doesn't like and the finals include a variety of the latter type.

Wouldn't it perhaps be possible to arrange ⟨something⟩ a meeting during the spring break? I would very much prefer this also for another reason. A few days ago I received the news that my father died.[4] My mother,[5] who herself is old and weak and is now all alone, asks me to visit her as soon as possible and help her arrange a few things. During the last few years my parents were living in shocking misery. My sister, who has occasionally been suffering from serious mental disorder for some years, turned their lives into hell and abused them;[6] during the war they had lost an important portion of their wealth and, considering the current cost, were anyway unable to put her in an institution. Only now, with the death notice, did I receive the report that Zora is in an institution.—I write you all this only so that you understand how infinitely heavy my heart is. Not just that I myself am completely alone here and have no one to lean on, but those who were closest to me are themselves deeply unhappy. Sometimes I have to think that, despite everything, you would probably be the first to give me a little more personal support in this if only you knew how depressed and sad I am.

If traveling conditions there weren't so difficult, I would really like to take the children with me; I could, of course, also accommodate them well here. But if you could arrange to spend this vacation somewhere with them, they would enjoy it at the same time. I hope you will take my suggestions and please, in any case, send an answer as soon as possible so that I can plan accordingly.

With kind regards,

Mileva

101. From Zhu Jia-hua[1]

Charlottenburg, 29 Kant St. IV, 21 March 1922

Esteemed Professor,

From our legation I learned a few days ago that you have already decided to go on a voyage to East Asia, which naturally interested me extremely. Unfortunately, our envoy,[2] who was stationed here in Berlin just a short time ago, naturally did not have any knowledge about our earlier negotiations, so he had already wired Peking to make inquiries there at the Imperial University. I heard the news about your visit to the legation from the first secretary, the former chargé d'affaires. You, esteemed Professor, will still be able to recall that I personally informed you about the appointment by the University in Peking and that we spoke then about it on different occasions.

To this I add that the Imperial University in Peking actually proposes to have you scheduled there for one year. According to your recent statement in our legation, you now only have 2 weeks left over for Peking. But may I please remind you that you told me at the time that you first absolutely had to travel to America and then China would come next, first and foremost. As I now hear, you have committed yourself to Japan as well. It would interest me very much, however, to know for how long. In any case, I am pleased that you, Professor, decided to first go to Peking, which naturally goes without saying. At all events I welcome your decision very much; and the Chinese government as well as scholarly circles there will very certainly welcome your arrival with complete enthusiasm. They will only be sorry that we shall be hearing lectures by you, Professor, for a period of only 2 weeks. I would otherwise have discussed this with you again as soon as I knew about your visit to the legation. Having waited so long for your decision, however, and not having heard anything in the meantime, I do also still have to wait for the telegram from Peking about how my authorities respond. One thing I can assure you of, esteemed Professor, is that your stay is very welcome there and the authorities will try to make your sojourn there pleasant in every way and ⟨will⟩ also show you much from our culture.

With great respect, very sincerely yours,

Chu Chia-hua.

102. From Paul Winteler

[Firenze, 21 March 1922]

[Not selected for translation.]

103. To Maurice Solovine

Berlin, 22 March 1922

Dear Solo,

I'm arriving on the 28th in the evening on the only available train, otherwise in the morning of the 29th, if I miss the connection in transit.[1] Have already talked my way out of all sorts of things so that we have some time.

Looking forward to seeing you, yours,

A. Einstein.

104. To Joan Voûte

Berlin, 22 March 1922

Esteemed Colleague,

Unfortunately I cannot come along on the expedition as I am obliged to deliver one of the main speeches in September on the occasion of the 100th anniversary of the Society of German Scientists [and Physicians].[1] My being prevented in this way is unimportant, of course, to the extent that I am not an observer and therefore cannot contribute anything directly to the success of the enterprise. I hope on another occasion another opportunity will arise to meet your wish.

Hoping for the success of the expedition, for which so much credit belongs to you, I am, with kind regards, yours.

105. From Paul Langevin

Paris, 10 Vauquelin Street, 22 March 1922

My dear Friend,

I transmit to you the invitation relating to the lunch on 2 April.[1] On the other hand, the Société de Chimie Physique wishes to receive you for dinner on Friday, 7 April, after the last of our meetings at the Collège.

Despite our wishes, it has been impossible to prevent an indiscretion from appearing in the press and your visit has been announced in magazines.[2]—The commentaries have been unanimously sympathetic.[3]

I would like to know the hour of your arrival so that, if feasible, I can meet you ahead of your arrival in Paris and spare you all useless conversations as you are getting off the train. In any case, you will at least find me upon your arrival at Gare du Nord. Contact me by letter if possible or by telegram if you lack the time.

Most affectionately yours,

P. Langevin

106. From Paul Wintcler

[Firenze, 22 March 1922]

[Not selected for translation.]

107. To Paul Ehrenfest

[Berlin,] 23 March 1922

Dear Ehrenfest,

My sister is seriously ill in Florence.[1] As she and her husband are additionally in financial difficulties, please *immediately* send her a thousand Swiss francs.[2] Her address is Florence (Firenze) Sanatorium, 5 Montughi Street.— I am presently reading a major talk by Bohr that makes his entire world of ideas become wonderfully clear.[3] He is a truly ingenious person; what luck that such a man exists at all. I have total confidence in his train of thoughts. The correspondence principle and the way in which he applies it has to be persuasive.[4] Regarding light emission, it cannot—seen from wave theory—be monochromatic if the emitting particle is moving.[5] This is how one sees this:

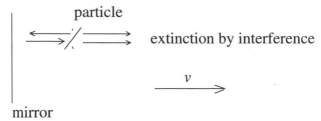

The direct and reflected light extinguish each other in the direction perpendicular to the mirror. This must be the case (because of relativity) even when the entire arrangement is moving in the direction of the arrow. According to the conventional theory this is because there are two Doppler effects: namely, (1) for the emission of the moving particle, (2) for the reflection off the moving mirror. One obviously cannot leave out one of these Doppler effects and retain the other. So if the emission of moving particles is required to be monochromatic, one has to assume that the "guide field" does not change its frequency upon reflection off the moving mirror. The result one would therefore have to reach would be that upon reflection off moving ⟨particles⟩ mirrors the quanta change their "frequency," whereas the guide fields do not. This is impossible. The elementary act of emission therefore cannot be independent of a particle's state of motion in the sense that the emissive field of a moving particle could be substituted for that of a particle at rest of suitable frequency.—

Now I'm completely ripe for the insane asylum. This notwithstanding, warm regards and to a happy reunion, yours,

Einstein.

108. To Paul Langevin

[23 March 1922][1]

Dear friend Langevin,

It took me so long to write because I could not find out when the trains arrive in Paris.[2] If all goes smoothly, I shall arrive on the 28th in the evening (there is only *one* option; I was unable to find out the exact time of arrival). If, however, I am unable to get a connection in Cologne, which supposedly happens frequently, then I shall arrive on the 29th in the morning.

Dear Langevin, I am not as resilient as ten years ago[3] and wish for nothing more dearly than the maximum peace and composure obtainable under the prevailing adverse conditions. Aside from the 4 presentations at your Collège[4] and the evening of debate at the philosophical society,[5] I do not want to attend any public sessions, and certainly not deliver another talk. Second, I beg you *not to accept a single private invitation for me*, not even with colleagues, and generally not schedule anything *a priori*. There will then be ample time for serious scientific discussions. But socializing for its own sake is an ordeal for me; and once one starts with that it is difficult to cease, because one thing leads to the next.[6] Furthermore, I want absolutely nothing to do with journalists. However, I would quite like to say a word to one or the other serious politician, if the opportunity presented itself; maybe something can be done, after all, against the harm that is being borne out into the world from your fine city. Furthermore, I would like to have occasion to speak with scientists about the possibility of restoring international relations in the scientific world, but not just with well-wishers and pacifists. I am objective enough to tolerate anything that is said without getting heavy-handed in any way. All these conversations would best be carried out during short walks but not at meals. This way I also avoid the otherwise inescapable flattery that is the norm at official dinners or such larger social functions generally.

Please do not take my obstinate nature amiss; but otherwise I am not going to be able to endure the strain. Messrs. Barclay and Borel will surely understand my standpoint; I cordially thank them for their invitations. If I refuse all invitations, the two of us can be together much more casually, too. Do please cancel the lunch in Boulogne for me as well.[7] Reason throughout: weak health.

At the philosophical society I would prefer not to present a talk; we could very well limit ourselves to a discussion instead, in such a manner that I reply to opinions stated by others.[8] It suffices that I be informed about the content shortly beforehand. You do not need to send me anything here.

Dear Langevin, I am causing you a good deal of trouble and it is a thankless business, at that. But let us both operate in cheery spirits and try to avoid everything that from an objective point of view is not necessary. Otherwise we and others will just have vain endeavors and pain. The fewer the honors, the greater the informality and enjoyment.

Cordial regards, yours,

Einstein.

109. From Erwin Finlay Freundlich

Potsdam, 24 March 1922

Dear Mr. Einstein,

We forward to you the enclosed letter intended to be directed to the Ministry.[1] We discussed the matter during the last meeting of the board of trustees[2] and ask you please to state your wishes.

The Ministry's explicitly stated wish is the reason behind having a representative of the Ministry join the board of trustees. As the annual subsidies by the Ministry to maintain the building and cover the research expenses are not unsubstantial and even have to be increased considerably, it seemed appropriate to us to take this wish into account. Furthermore, we consider it necessary to expand the board of trustees somewhat in order to connect former financial supporters firmly to the endeavor. Therefore Dr. Jeidels[3] proposed that Director Wassermann[4] of the Deutsche Bank be asked to join the board of directors; and Privy Councillor Müller[5] stated his willingness to approach Mr. Wassermann together with Dr. Schneider.[6] If he declines, Jeidels himself is happy to step in.

Then, aside from Prof. Bosch,[7] Wassermann or Jeidels would belong to the board of trustees as representatives of the founders. Furthermore, we would like to ask Prof. Planck[8] to join the board of trustees, because a certain envy concerning the new installations has persistently emerged among physicists and, as you know, we, for our part, never had the intention of not letting these research facilities be made accessible to others as well.

We would therefore very much welcome it if Planck declared his willingness to join; this, too, Müller was ready to discuss with Planck, if you yourself did not do so, which would surely open up more prospects of an agreement from Planck.

Furthermore, we resolved to ask Dr. Ruge to support us with his counsel as legal advisor.[9] During the past few years I not only had to create the facilities, but also take over the financial settlement of all obligations and in the process saw so many purely legal issues arise that the advice of an attorney is definitely necessary. I cannot in reality assume the responsibility anymore on my own. Dr. Ruge is prepared to give us his counsel as legal advisor with a seat and a voice on the board of trustees.

We hope that you agree with everything in principle and in that eventuality request that you grant us your approval. Only then will Planck and Wassermann be approached.

I forward to you at the same time the printer's proofs of a critical analysis by Reichenbach, in accordance with his wish.[10]

Faithfully,

E. Finlay Freundlich

110. To Thomas Barclay

[Berlin,] 25 March 1922

Esteemed Sir Thomas Barclay–

I thank you cordially for your amic[able] invitation.[1] As I will be very exhausted in Paris by m[y] commitments toward my professional colleagues, I have made it a principle to avoid social events. ⟨That is why I ask you please not to take it amiss if I cannot accept your invitation. At the same time I would not like to fail to tell you even now that it would be a special treat for me to be able to spend a few hours with you in Paris.⟩ In order to be able to uphold this principle without coming into any conflicts, I would like to ask you please to invite at most 2 or 3 other persons besides myself. If preparations made by you make this impossible, then please kindly regard this invitation of yours as declined and give me an opportunity to be able to spend a few hours with you another time.

111. To Zhu Jia-hua

Berlin, 25 March 1922

Dear Sir,

In thanking you very much for your letter of the 21st of this mo.,[1] I take this opportunity to inform you about the status of my projected voyage to East Asia in the fall of this yr., requesting that you please regard this information as confidential; for, I would like to avoid my travel plans becoming generally known in advance because otherwise problems would arise for me in the form of additional invitations, etc. I remember our discussions very clearly but had to desist from traveling to China for the time being because the sojourn envisioned by you could not possibly be brought in line with my other commitments; and moreover, the financial recompense offered me was not large enough to make the journey feasible.

Now the situation has changed insofar as I have received an invitation to Japan with adequate compensation for a planned sojourn of four weeks, namely two weeks in Tokyo and two weeks in other university cities of Japan.[2] This would consequently agree with my also coming to Peking for two weeks. I do not know if the Japanese insist on my visiting Japan before China. However, I had intended to go to Japan first because I think that the winter climate is milder in China than in Japan and the period at my disposal for the Chinese and Japanese visit spans from about the middle of November to the beginning of January. I cannot imagine at all how the sequence of these two visits should be of any importance. Albeit, in having made me the first suitable offer (2000 pound sterling and free accommodations for me and my wife), the Japanese do in a certain sense have a moral right to priority, just as you do by the circumstance of having invited me first.[3]

In the eager hope of being able to reach an agreement with you to your complete satisfaction and of being able to see the cradle of East Asian culture with my own eyes, I am sincerely yours.

112. To Leo Jolowicz

Berlin, 25 March 1922

Esteemed Mr. Jolowitz,

The address of the gentleman I mentioned to you and with whom I already spoke about the matter is:[1]

Dr. J. Grommer, c/o Hamburger, 35 I Dankelmann St., Charlottenburg.[2] It would perhaps be appropriate if you would send him my earlier manuscript.[3]

Your kind offer to donate literature from your publishing house to the library of the University of Jerusalem appears to me to be of such importance that I do not want to assume the decision about which works should be selected, just like that, without first consulting with Prof. Löwe and others.[4] I shall gladly return to this matter, however. I would not like to fail to express to you my deeply felt acknowledgment of this magnanimous gesture toward the cause of Jewish culture.

In utmost respect.

113. To Arthur Nussbaum[1]

Berlin, 26 March 1922

Esteemed Professor,

Yesterday the brother of my best friend, Mr. Besso,[2] was here, for whose company, more spec[ifically], for whose branch in Romania, you are handling a complicated lawsuit. Mr. Besso asked me if I would like to put in a good word with you on their behalf, as the outcome of this trial is of profound importance to this firm founded by my friend's father.[3] I gladly do so because I know how badly struck this firm—directed by excellent men—was by the consequences of the war and herewith permit myself to urge you to represent this lawsuit entrusted to you with special care.

In utmost respect.

114. From Paul Ehrenfest

[Leyden,] 3 o'clock in the aftern[oon], 26 March 1922

Dear Einstein,

Just received your letter.[1]

I request *immediate* notification of your sister's name—you forgot to include it and I don't know it!

— • —

It would be a great pleasure for us if you could make your return trip via Leyden. St. John is coming to Europe in the spring and would be *very gladly!!* prepared to come to Leyden in the *second half of June* for a spectral-shift conference.[2]

Warm greetings to all of you. Naturally, especially also to the dear Ilmargotse,[3] yours,

Ehrenfest.

115. From Hantaro Nagaoka[1]

College of Science, Tokyo Imperial University, 26 March 1922

Highly esteemed Professor,

Mr. Yamamoto, editor of *Kaizo*, informed me that you might arrive in Tokyo at the beginning of October.[2] This is a big surprise; and the Japanese will unanimously welcome you, as your noted name as a world-famous thinker is known everywhere and the principle of relativity is highly reputed. Thanks to translations of various writings about the principle of relativity, popular lectures by Mr. Kuwaki, and books for the general public by Mr. Ishiwara, the Japanese have regarded the principle with great interest. Unfortunately, it is represented in more than one way, particularly by philosophers, all of whom lack mathematical knowledge.[3] A lecture from you personally will clear up this haze and fog and the Japanese will be bathed in the bright light of the genuine principle of relativity.

As I have often heard, German scholars are very annoyed about the conduct of the academicians here during the World War. Perhaps there is some basis for that, as our academy is not on the same level as the European ones; it was often spoken about as the refuge for outdated Japanese scholars.[4] Today, however, the academicians' thoughts are very different; at the last meeting, the president[5] expressed the opinion that the academy should call together a welcoming committee for the founder of the relativity principle. Ideas everywhere ebb and flow, but no academician has any great animosity toward learned Germans; the allusion by one academician to the Royal Society of London probably had political reasons. That is why you do not need to worry about that at all.[6] On behalf of all the members of the Tokyo Academy, it is my duty to pay our respects to you in friendship. I very much admired the advances made in the physical sciences in Germany during the war; I may have studied under Helmholtz, Boltzmann, and Planck;[7] yet the abundance of brilliant research in the areas of relativity and quantum theory during this stimulating period aroused in me such deep respect and admiration that I do not hesitate to devote long hours to them in my talks. You can easily surmise how immensely pleased my students and graduates at the university will be to see you in Japan.

I wrote to Mr. Hioki, the Japanese ambassador in Berlin,[8] to inform him about your oriental travels. He was a school friend of mine and is interested in the progress of science in Japan, so he will be always ready to provide you with every convenience for your trip.

In expressing my utmost respect, I remain most devotedly yours,

H. Nagaoka.

116. To Hermann Anschütz-Kaempfe

[Berlin,] 27 March 1922

Dear Mr. Anschütz,

Your letter, written with such great worries, quite upset me. But I may hope—indeed, I am convinced that everything went well, now that a number of weeks have passed since the serious operation.[1]

Today I write you to send you the enclosed letter,[2] which is not clear to me. I don't recall at all anymore to which patent dispute this matter refers.[3] I wrote back to Mr. M[artienssen][4] that he must arrange a postponement of the court proceedings if he attaches importance to my reconsidering the case. You are obviously going to be able to orient yourself more easily. After I have rechecked the case, which, however, can only happen after my return from Paris (ca. 10 April),[5] let us discuss the affair together. Only then am I going to submit a statement about my earlier opinion. Maybe we can achieve something by direct negotiations with Mr. M. I could perhaps come to Kiel in April for a week, if you consider it right. I hardly believe that he could be somehow right, despite his presumptuous tone.

With heartfelt wishes for your wife's improvement and rapid recovery, I am, with best regards to both of you, yours,

A. Einstein.

117. To Viktor Engelhardt[1]

[Berlin,] 27 March 1922

Esteemed Doctor,

I read through your booklet with the greatest interest. The fondly and vividly written account of individual persons and events is extremely inspiring just as is the purposely subjective presentation of causal connections as well as your thoughts about causality in the areas of science and history.[2]

With thanks for the booklet and for the fine hours it gave me, I am very respectfully yours.

118. To Jun Ishiwara

Berlin, 27 March 1922

Highly esteemed Colleague,

I received your letter of 26 January inst.[1] and the draft contract and agree with everything; I would just like to comment that based on previous experience, I find no listener is capable of following a talk of three hours duration.[2] Nowhere have I yet delivered a talk that lasted longer than 1½ hours. Thus I think that the six scientific presentations should not last much longer than 1½ hours; however the remaining 2½ hours could, e.g., be used for scientific discussions.

Unfortunately I have to postpone the time of my departure by one month because I must deliver a speech on the occasion of the 100-year celebration of the Society of German Scientists in Leipzig on September 21st.[3] I therefore took the liberty of making corrections to the draft contract in this sense in order that our negotiations not get drawn out too much longer.

I received the check at the London bank and thank you sincerely for it.[4] I will keep it until I embark on the voyage. I plan to take my wife along.

Regarding the translation of my writings into Japanese, my booklet *On the Special and the General Theory of Relativity* has, according to information given by my publisher Friedrich Vieweg & Son in Braunschweig, already been translated into Japanese by Mr. Kuwaki.[5] Regarding the original scientific articles as well as two talks of general content, I would be very pleased if you wished to undertake to translate the publications you deem suitable. I would naturally leave to you the selection of publications for translation. As compensation I tend to receive 15% of the sales price per copy. If you undertake translations of this kind, I request that you arrange that the Japanese publisher sign a contract with me on this basis.

In happy anticipation of seeing you next fall and becoming acquainted with your sunny country wrapped here in a mystical shroud of fairy tales, I am with kind regards, yours sincerely.

P.S. Please convey my kind regards to Messrs. Yamamoto and Murobuse.

119. To Hans Reichenbach[1]

[27 March 1922]

Dear Mr. Reichenbach,

At the same time, I send your correction proofs back to you.[2] I agree almost entirely with your critical argumentation, particularly in re. Kantians! and find it exemplarily clear. I just find your opposition to Petzoldt-Cassirer lopsided, without intending to concede the point to Petzoldt.[3] Nor do I understand a sentence on p. 324.[4] Additionally, it would perhaps have been good to discuss Kretschmann (fundamental vacuousness of the invariance postulate), which also really does merit philosophical criticism.[5] Your axiomatic endeavors are very laudable as well.[6]

Cordial regards, yours,

A. Einstein.

120. To Charles Nordmann[1]

[Berlin, before 28 March 1922][2]

[Not selected for translation.]

121. From Wilhelm Mayer-Kaufbeuren[1]

Paris, 28 March 1922.

[Not selected for translation.]

122. To Elsa Einstein

Wednesday morning. [Paris, 29 March 1922][1]

Dear Elsa,

The trip went very smoothly and proceeded according to schedule. In Liège, where we had a 2-hour stopover, I sauntered around for 1½ hours with the man who wanted to travel with me. At the French border Langevin and Nordmann were there to pick me up (5 hours away from Paris).[2] That is touching hospitality! Arrived in Paris, a police officer informed us in the train that a crowd of journalists was waiting for me. But we crossed over countless prohibited tracks through a small exit out of the train station and escaped completely unnoticed to my abode, a modest, nice room on the 5th story of a building on Humboldt Street.[3] But all of you shouldn't write me here so that my hiding place remains undiscovered. The men didn't find my written interview practical; they are writing another. The business will go well, all right. Besides these two men and the maid I haven't seen anyone yet but am writing just after getting up. This letter costs 17 marks postage; in consideration of this I'm not going to write very often; but I'm sending this first one off right away so that all of you know that after arriving in Paris I still exist, happy and in good spirits.

Warm regards, yours,

Albert.

Be careful with Dr. M.: Love him but (otherwise) don't let him get away with anything.

123. To Elsa Einstein

Friday evening. [Paris, 31 March 1922]

Dear Else,

Today I gave my first lecture—my only one, as all the rest are discussions.[1] The whole of next week is garnished with commitments. Life in this small apartment is splendid. A friend of Langevin's, Mr. Malfitano,[2] ceded it to me and is staying with Langevin. I am being splendidly regaled, as never before in my entire life. A kindly Alsatian woman cooks for me and Langevin, who comes here daily to eat with me; Solovine,[3] who is helping me very much as the strain is huge, is usually also here. But my being here is very useful and I am getting to know the most interesting people. Tomorrow I'll be visiting Barclay.[4] There's no real socializing. All scrapped. My reception among fellow professionals is extremely cordial. It's midnight, and I'm too tired to report more. God knows when I'll have some time again!

Warm regards to all of you,

Albert.

I hope you're keeping your friend[5] at arm's length and you're feeling well.

On Tuesday evening I'll be with Aunt Mathilde.[6]

124. From Peter Debye

Zurich 7, 36 Gloria Street, 31 March 1922

Dear Mr. Einstein,

About my paper, Nernst published a remark on the attractive power of polarization and the Van der Waals forces[1] in his book.[2] He rejects the matter on the justification that I arrived at forces that are independent of temperature, which was supposedly unacceptable.[3] In a notice I composed, after entirely perchance seeing Nernst's remark in his book, I summarized the foregoing development of the subject and contended that Nernst's remark was completely erroneous. The notice has not yet been printed; I initially sent it to Nernst in manuscript so that he could express his opinion on it. Today, now, I received an answer from him, from which I gather that he has not understood me at all. In particular, it is evidently unclear to him that whenever attractive forces exist between molecules, in the equation of state an increasing term for the attraction must appear with decreasing temperature; therefore the grounds for his rejection cannot hold.

I would not bother you at all with this matter if Nernst had not himself named you as a witness. He writes: "Perhaps the following fact also interests you. A few days ago I asked Einstein whether he had read my remarks about your theory. He immediately answered that he knew these remarks very well and subscribed to every word of it. Perhaps you will doubly reconsider whether you really want to pick an entirely unmotivated fight."

I cannot imagine that you would want to back Nernst's assertion about a purported temperature independence of the attraction term following from the theory.

In the notice, I wrote that such temperature dependence occurs according to the same law and the same principles from which one concludes that the density of a heavy gas was greater below than above. I really do not know how I could be more explicit.

I believe it would help matters if you would define your position more precisely.

I am afraid that you are now going to have very many things to do in Paris and that that surely interests you more than the little tiff that this letter relates. Nevertheless, I do believe that this business is of some interest, too, if only due to Nernst's position; and I hope that you can comply with my request of relaying your standpoint.

With best regards, yours very sincerely,

P. Debye.

125. From Beatrice Jahn-Rusconi Besso[1]

Firenze, 31 March 1922

[Not selected for translation.]

126. To Paul Langevin

[Paris, 1 April 1922][1]

Dear friend Langevin,

My wife wrote to you and surely expressed 10^8 wishes;[2] they may only be attended to from the vantage point of lyrical poetry. I am sending you a letter from me to Pictet with the request that an entry ticket to the discussion be enclosed for him.[3]

It is good that you do not have to go to the photographer but it is a shame how much time you must be spending on organizational work. I have a bad conscience that I am not helping matters.

Cordial regards and *auf Wiedersehen* today at noon. Yours,

Einstein.

Please, also one invitation to Dr. Moscovici, Arcueil-Cachan (Seine). (If possible; he is a friend of Solovine's and 0.5 versed.

127. From Hermann Anschütz-Kaempfe

Kiel, 2 April 1922

Dear, esteemed Professor Einstein,

I owe the address at which you are reachable to those at your home in Berlin, where I telephoned.[1]

First the good news, that my wife is feeling much better again, that she was transported back from the clinic yesterday, and is now looking forward to a complete recovery.[2]

With this letter I am pursuing you all the way to Paris because we, i.e. the firm, is appealing to you for help as a consequence of Mr. Martienssen's impudence. There is supposed to be a hearing at the regional court on April 11th to decide on the confiscation of Martienssen's apparatus.[3] It essentially revolves around an opinion prepared by you years ago that is now, of course, being contested by Martienssen and a fellow expert he had called to his assistance, a certain Dr. Zahn.[4] In particular, on grounds that are so flimsy that one has to wonder that he found any physicist to back him.

Since you were so kind as to write me that you would be returning to Berlin on the 10th of this mo. and would then poss. be prepared to come over here, I inquire whether it would not be possible for you to arrive here in Kiel on April 10th or, of course, earlier, taking the route from Cologne via Hamburg. You would then have a shorter train ride overall and to us it would be extremely valuable if you could express your views about your expert opinion in person on the 11th. The opponents are counting on your not being reachable at the moment[5] and want to use this time to launch an offensive; I assume that it will fail miserably in your presence.

If you arrive here during the day of the 10th, or if it doesn't work otherwise, during the evening, there will still be plenty of time for you to orient yourself about the situation; I think that 1–2 hours will perfectly suffice.

Now there just remains for me to ask you please to send a short cable to us, i.e., to the firm, addressing the telegram to: *Anschützco Kiel*, about if and when we may receive you here. If it is simply not possible, then we shall just have to see that we quickly find a helper, who can prove to the court that we have a good right to do so.

And your old room in the house is ready and waiting for you and would be heartily pleased, along with us, to be able to welcome you as a guest once again.

With very cordial regards also from my wife, yours,

Anschütz.

128. From Emile Berliner

Washington, D.C., 2 April 1922

Highly esteemed Professor,

After having written you on the 24th of February, I herewith have the honor of handing over to you the promised ten thousand dollars for the Jewish University in Palestine.[1]

I do so with particular pleasure, as it has been some 18 years since I initially grasped the opportunity of supporting this idea, in a club in Philadelphia, because I recognized it was something very promising for the renown of the Jewish nation. And now that you decided to back this highly significant plan personally, a very special guarantee for its success is given.

I hope this letter reaches you and your highly esteemed wife in the best of health and am sending you respectful greetings in the name of all of us,

Emile Berliner.

129. From Ludwig Hopf[1] and Theodor von Kármán

Aachen Polytech., Aerodyn. Institut, 3 April 1922

Dear Mr. Einstein,

The papers report that you are staying in Paris;[2] a glance at the railway map reveals that the return route from Paris goes through Aachen. The undersigned therefore dare to cherish the hope that you might perhaps feel like spending a day in Aachen on your way homeward. We would mark this day red on the calendar and return the favor by promising: (1) no lecture; (2) not a single word more than you yourself would want about relativity; (3) not the faintest notice in the newspaper.

It would be so nice; so please do us this favor!

Most cordial regards! Yours,

L. Hopf, Th. Kármán.

130. From Paul Block[1]

Paris VIII, de Ponthieu Street, 4 April 1922

Esteemed Professor,

The kindness of the administrator of the Collège de France made it possible for me to listen in on your first talk on March 31st.[2] Because this report, which I telephoned back to Berlin on the same evening and appeared on April 1st in the morning edition of the *Berliner Tageblatt*, is the only one that became known in Germany, you may perhaps be interested in reading it.[3] Please excuse the mistakes; acoustical errors from two-way transmission by phone—Paris-Frankfurt, Frankfurt-Paris—aren't always avoidable.

I presume that my first letter was handed to you by Professor Langevin and that you could not find the time to reply to me. Permit me to inform you now by letter what I would have liked to tell you personally.

There are about 200 Germans here in Paris right now—civil servants and members of the Reparation Committee[4] and of the Embassy. Besides their own professional duties, they all have the often quite difficult task of bringing those French people they encounter around to a better opinion of Germany. All these fellow countrymen of yours were deeply pleased about the honors bestowed on you, venerated Professor. They would be happy if during his stay in Paris Albert Einstein also devoted half an hour to the Germans abroad, in order to tell them something about his theory. On the far too flattering notion that it would be my due as the most senior German journalist working in Paris to see you, I was given the mission to speak with you about such an entirely unceremonious event. Had you been so inclined, the German ambassador, whose letter was surely handed to you by the tracks,[5] would without a doubt have made one of the rooms at the embassy available for this purpose at my request.

This, esteemed Professor, I wished to say to you. With me, you would have been safe from journalistic pestering because I do not do such reportage and otherwise admire most highly the tactful reserve with which you avoid any political statements.

In utmost esteem, respectfully,

Paul Block,
Correspondent of the *Berliner Tageblatt*.

131. "The Theory of Relativity." Discussion Remarks at a Meeting of the Société française de Philosophie

[*Einstein et al. 1922*]

DATED 6 April 1922
PUBLISHED July 1922

IN: *Société française de Philosophie. Bulletin* 22 (1922): 91–113.

Mr. Einstein. — I just have a word to say about Mr. Hadamard's remarks. Mr. [p. 97] Hadamard said that a physical theory must first of all be logical, then agree with the experimental findings. I do not believe that this is sufficient and, in any case, this is not evident *a priori*. To say that a theory is logical means that it is built of symbols that are linked among themselves by means of certain rules; and to say that theory is in conformity with experience means that one possesses rules of correspondence between these symbols and the findings. Relativity arose out of experimental necessity; this theory is logical in the sense that one can give it deductive form, but clear rules that make its elements correspond to reality still need to be known; thus there are three postulates, not two, as Mr. Hadamard thought.
    [. . .]
Mr. Einstein. — Which tensor is that?
    [. . .]
Mr. Einstein. — Geometry is an arbitrary conception; one is always free to [p. 98] adopt the one one wants, in particular, a Euclidean geometry; but Euclidean concepts do not have any physical meaning and cannot serve us physicists. Moreover, the relation between the real continuum and imagined geometrical space is not uni- [p. 99] vocal and one cannot say that one manner of speaking is preferable to another.
    [. . .]
Mr. Einstein. — One can always choose the representation one wishes if one believes it is more convenient than another for the task one proposes to undertake; but it has no objective meaning.
    [. . .]

[p. 101]   **Mr. Einstein.** — Regarding Kant's philosophy, I believe that each philosopher has his own Kant; and so I could not respond to what you just said because the few indications you gave do not suffice for me to know how you interpret Kant. I, for my part, do not believe that my theory harmonizes on all points with Kant's thought as I see it.

[p. 102]   What appears most important to me in Kant's philosophy is its mention of *a priori* concepts to construct science. Thus one can contrast two points of view: Kant's apriorism, in which certain concepts preexist in our consciousness, and Poincaré's conventionalism. These two points of view concur about the point that science needs arbitrary concepts to build upon; whereas I can say nothing about knowing whether these concepts are *a priori* givens or are arbitrary conventions.

     [. . .]

[p. 107]   **Mr. Einstein.** — So the question can be posed thus: Is a philosopher's time the same as a physicist's? A philosopher's time, I believe, is both a psychological and a physical time; consequently physical time may be derived from the time in the conscience. Individuals have a primitive notion of the simultaneity of perception; so they can confer among themselves and decide on something they perceive; this would be a first stage toward objective reality. However, there are objective events independent of individuals and so the advance was made from the simultaneity of perceptions to that of the events themselves. And, in fact, for a long time this simultaneity did not lead to any contradictions because of the great velocity at which light propagates. So the concept of simultaneity was able to bypass the perceptions to the objects. From there it was not far removed to deduce a temporal order among events; and instinct did so. Yet nothing in our conscience permits us to conclude the simultaneity of events, because these are no more than mental constructions, logical entities. Consequently, there is no such thing as a philosopher's time; there is only a psychological time differing from a physicist's time.

     [. . .]

[p. 111]   **Mr. Einstein.** — In the four-dimensional continuum definitely not all directions are equivalent.

     On the other hand, from the logical point of view there does not seem to be much relation between the theory of relativity and Mach's theory. For Mach there are two points to distinguish: on one hand, there are things that we cannot budge: these are the immediate facts of experience; on the other hand, these are concepts that we can, on the contrary, modify. Mach's system studies the relations existing between the facts of experience; the ensemble of these relations, for Mach, is science. This [p. 112] is a false standpoint here; all in all, what Mach had made was a catalog, not a system. As good as Mach was as a mechanician, he was a deplorable philosopher. His shortsightedness about science led him to reject the existence of atoms. It is probable that if Mach were still living today, he would have changed his mind. I do, however, insist on saying that I am in complete agreement with Mach regarding one point—that concepts can change.

132. From Oswald Veblen

[Princeton,] 6 April 1922

Dear Professor Einstein,

Thanks very much for your letter about Mr. Thomas.[1] He expects to leave for Germany early in June and to stay until September when he will return to this country for a few months and then come back once more to Germany. I hope that you will find him a satisfactory student. He will at least be thoroughly familiar with your work on Relativity.

Your manuscript arrived considerably later than your letter, but I finally received it and turned it over to Professor Adams.[2] The Princeton University Press is taking the steps necessary for its publication.

With best greetings to yourself and Mrs. Einstein, I am, sincerely yours,

Oswald Veblen.

Translator's note: Original written in English.

133. From Paul Winteler

Florence, Sanat. via Montughi 5, 6 April 1922

[Not selected for translation.]

134. To Elsa Einstein

[Paris,] 9 [8][1] April 1922

Dear Else,

All went brilliantly well. Yesterday was the last discussion session and yesterday evening, a festive dinner with all my fellow colleagues.[2] You can hardly imagine with what sympathy I am being met here.[3] Even in political respects I only encountered calm observations on issues and goodwill toward intercommunications, incomparably better than I had expected. Tomorrow I'm off by car to the battlefield ruins.[4] From there I'm driving onwards to Kiel whither Anschütz has summoned me about an impending important hearing that is going to be decided on Tuesday.[5] If the Germans only knew what services I performed for them here by this visit. But they are too small-minded to grasp it. Yesterday I visited Rothschild, who handed me a not-empty leather bag in farewell. A very astute man. Wife and daughter were there.[6] Wife and daughter of patriarchal simplicity in dress and essence. Toward the end of the week I hope—at last—to arrive home to you all, in order to enjoy my well-earned rest. Mrs. Deng gave me a heavy, sweet package for you.[7] I'll bring you the accompanying letter and hand the packet on to Mrs. Langevin.[8] Instead, you have to buy 200 marks worth of sweets for yourself on my orders, which I herewith categorically assign you to do.

Warm kisses now to all of you from your

Albert.

I have to go to Painlevé's to eat—a splendid person![9]

135. From Chenzu Wei[1]

Berlin, 8 April 1922

Esteemed Professor,

I have the honor herewith of informing you that I have just been asked by the president of the Imperial University in Peking[2] to tell you that the university would most joyfully welcome it if you could possibly arrange to give lectures for a period of time at the Peking Imperial University. About the terms, Mr. President informs me that the university is very willing to carry the costs of hotel accommodations and expenses for the duration of your teaching activities in Peking and additionally to pay a monthly honorarium of 1,000 Chinese dollars.[3] I permit myself to ask you to kindly send word to me about whether our academic circles will have the honor of receiving you in my home country.

In extending my great respect, I have the honor to be most devotedly yours,

Wei Chenzu,
Chinese Envoy

136. From Gustave Le Bon[1]

29, Rue Vignon, Paris, 9 April 1922

[Not selected for translation.]

137. From Georg Maschke[1]

Wannsee, 31 Kleine See St., 9 April 1922

Esteemed Professor,

I congratulate you cordially on your great success in Paris. My wife, daughter, and I have been following your accomplishments up to now with great interest and genuine pleasure.

I would have written you long ago but the unpleasant message I received from Mr. Wankmüller about the ugly manner in which your relationship with the company was solved held me back.[2] Mr. Wankmüller, who like me was abroad for a longer period, surely shares little of the blame in this affair because his managing director acted, as on other occasions, entirely independently. (That gentleman has, of course, been sacked.) I urged Mr. Wankmüller various times orally and in writing to apologize to you in person. This still does not seem to have happened and therefore I wanted at least to inform you about it. An excuse for Mr. Wankmüller might be that he has been burdened by very severe worries about retaining his firm during the past two years. I had left his company so that he could get a younger investor.–

Ultimately the blame lies with me, of course, for having acquainted you with this company and thus ask you please to forgive me. I meant well and really could not have predicted such a thing. The war psychosis and its consequences altered many things and many persons and so let us, with your permission, also attribute this to it.

I would be very pleased to receive a few lines from you and am, with best regards to you and your wife from my family and me, yours very sincerely,

G. Maschke

138. From Paul Oppenheim[1]

Frankfurt/Main, 9 April 1922

Dear Professor,

It doesn't take much tactfulness to notice how right I was to postpone my assistance in the manuscript affair upon your inquiry by telephone until other ways out had failed: I had the feeling, and still have it, that our friendship might be strained by it, without being of service either to someone or to science.[2] That is why I ask you to make the decision on your own, whereby you might still find a way to take the known, somewhat divergent point of mine, a little bit into account.

But I broached the manuscript issue *only as a symptom*, you know. For me, the main thing was to repair the broken relations between the two parties, out of friendship and a love of science. Together we performed psychoanalysis by subjecting the causes, as far as they may have been unwitting, to conscious scrutiny and thereby eliminated them so that henceforth another attitude emerges: You wanted henceforth to regard some things as the opposites to hot-blooded emotion and temperament, or however else you wish to call it, that you had previously only evaluated logically and condemned. To my exceedingly great joy you promised me, in ethical greatness equating that of your science, to at least resume *scientific* relations, in the interest of science.[3] I have the confidence in you that you will keep this promise, particularly because, as I believed to sense, when it was made, a little bit of friendship spoke along in its favor, which although perhaps not deserving the spirit of it, surely would warrant some loyalty.

I do not need to tell you how pleased I would be soon to hear that you had spoken with each other; I am convinced that you will find it satisfying.

In this hope I am, with many cordial regards across the board from both of us,[4] your truly devoted

Oppenheim.

29 April 1922

I can only send off the foregoing today because I did not receive a reply from Freundlich, who had been away, earlier. It changes none of my arguments but rather prompts me to the following, hopefully not importunate suggestion, which at the moment appears to me to be the best solution: After my refusal, you have to decide. Wouldn't you want to wait with it for a while? Such decisions should *not* be made in the grip of emotions. For the interim you could deposit the manuscript in a safe place that is at your personal disposition at all times (e.g., in the manuscripts collection of the State Library[5] or ultimately also in your home). I naturally promise you that I won't mention a word to you about the whole affair unless you directly force me to. All in all: Wait before executing the alternative. Maybe you will want to thank me for this small modification. Again, most cordially yours truly,

Oppenheim.

139. To Lucien Chavan[1]

[Paris, 10 April 1922]

After ten productively busy days in Paris, I send you friendly greetings; your old

A. Einstein

140. From Paul Langevin

Paris, 10 April 1922

My dear Friend,

I think this letter, along with its enclosure, will reach you at the moment of your return to Berlin, to tell you one more time how happy I was to see you again and to work with you toward repairing, within the limits of our powers, the immense damage caused by the war.

I would like to know whether you returned without undue fatigue. Rest assured that your visit, despite the effort it demanded of you, was an excellent thing and that it produced here the best impression, far exceeding what I would have dared to hope.

I had the embarrassment of not finding the slip of paper on which you had written the address to which I should send your money.—[1] Would you please give it to me again so that I can acquit myself of this obligation as soon as possible?

Please present my respectful regards to Mrs. Einstein and believe me to be your very affectionately devoted servant,

P. Langevin.

141. To Ilse Einstein

Kiel, Tuesday [11 April 1922]

Dear Ilse,

So, I made it here safely today at 2 o'clock in the morning and received your correspondence.[1] The court hearing is only taking place tomorrow.[2] I still have to stay here on Thursday because of an experiment.[3] So I'm not coming until

Friday 9h 17 in the evening ([Lehrter sta[tion]). About the packages of candy from Mrs. Deng[4] and Mr. Höchstetter, both of which I gave away as gifts in Paris, you don't have to shed a single tear; I herewith assign you to take 400 marks and, after consulting with Margot and Mother, to buy all kinds of candy. Protest by Mother doesn't count; for this is a matter of replacing gifts given over there. I'll write to Debye.[5] So, I'll be happy to see you all again. I'm very eager to go home again. Don't tell anyone, except Schottki,[6] that I'm coming back so that I can snooze for a few days. I'm very glad now that I was in Paris, because I was able to do a really good deed. Warm regards to all three of you, yours,

Albert.

142. From Hans Albert Einstein

[Zurich, 12 April 1922]

Dear Papa,

As busy as your life now is, as I gather from the postcard and newspaper, mine is as quiet. Imagine: Vacation, and all alone at home. (Teddy is in Rheinfelden.)[1] This is fine relaxation after such an exhausting quarter as the last one was.[2] You gave me such smart advice about the "fair sex."[3] I just have to comment that this fair "s." isn't always "all that fair," I realized that on that occasion!

Music really is better; you can [choose] it yourself and don't have to wait and see what happens to come crawling along. With Mr. Gonzenbach I played 2 Tartini sonatas and a Bach sonatina and the Beethoven concerto (of Kiel); that was really fine.[4] Nothing beats a beautiful booklet. But I also did something you never did: I accompanied a clarinet: Mozart and Brahms; nice too, but difficult!

How are you, by the way? I guess you don't have any time for yourself again!

Well, greetings from

Adn

to you and to Anschütz!

143. From Peter Debye

Zurich, 14 April 1922

Dear Einstein,

I just received your letter[1] and hurry to reply.

To start with the second paragraph of your letter, I first note that you share my view that for statistical reasons polarization forces also produce a growing attraction as the temperature drops. Nernst rejects my arguments in his textbook "if only because Debye's assumptions lead to an attraction independent of the temperature" (quoted from memory, therefore accurate only in meaning).[2] It is thus evident that you do *not* "subscribe to it word for word," as Nernst writes in his letter.

In the first paragraph of your letter, you now present a new reason why my considerations should be rejected. You think that although everything is in order theoretically and the molecular forces I suppose are in fact present, I am incorrect insofar "as until now, owing to a deformation of the molecule, the attraction effect merely constitutes a practically vanishing fraction." This, too, I must reject, however. It happens that I can justify this position irrefutably and, in fact, without it being necessary to resort to the numerical calculations. Indeed: For all monatomic gases *the polarization attraction is the sole one* known up to now that could come into consideration. If you wanted to take the orientational attractions here as well, according to the classical calculation followed by Van der Waals, junior, and by Keesom,[3] then, at the same time, you are also implicitly reckoning that the individual particles possess a rotational energy measurable by the equipartition theorem, i.e., you assume the value 5 instead of 3 for the specific heat. For you, I surely do not need to elaborate further on this remark! I would just still like to point out that (1) Zwicky demonstrated in the *Phys[ikalische] Zeitschr[ift]* that the theoretical temperature dependence of polarization forces also sets the practical course of the attraction for the noble gases,[4] and (2) that Keesom meanwhile completed his calculations for hydrogen by taking the polarization forces into account.[5] Regarding the order of magnitude, one can specify more precisely thus: Polarization attraction is of the same order of magnitude as the total attraction observed in the noble gases.

I would very much like to receive a response from you.

With best regards, yours,

P. Debye.

144. From Paul G. Tomlinson[1]

Princeton, New Jersey, 14 April 1922

My dear Professor Einstein,

We have received the proof sheets of your book and the translation is now complete.[2] It has been suggested by one of the members of the Mathematics Department here at Princeton[3] that possibly the book would have more appeal to the general reader if certain passages in it which are rather technical could be explained and somewhat simplified. The suggested method of doing this is either by a prefatory article or an explanatory appendix. Before proceeding, however, we wish to have your approval, and I should appreciate it greatly if you would write me whether you have any objection to this or not. We feel that it might help the sale of the book.

Yours very truly,

Paul G. Tomlinson,
Manager.

Translator's note: Original written in English.

145. To Georg Maschke

15 April 1922

Dear Mr. Maschke,

Don't you worry;[1] as I in actual fact had no business with the factory for the longest time, the whole affair completely slipped my mind. I cannot imagine what an employee of the factory could possibly have done against me and regret very much if he was hurt so keenly by this affair. If he has not perpetrated anything else, I would like to put in a good word for him on this occasion.

In the hope of soon meeting you again sometime, I am, with kind regards to you and your family, yours.

146. To Paul Oppenheim

Berlin, 15 April 1922

Dear Mr. Oppenheim!

Returned from Paris I hasten to rid myself of the repulsive manuscript affair.[1] I herewith propose the following to you: You arrange for the sale of the manuscript; the Jewish University in Jerusalem[2] gets one half of the proceeds, the other half you dispose of in the manner that your conscience prescribes. I do not want any report about it but just comment that I want to have none of the proceeds for myself and my family. If after consulting with your friend[3] you agree to this proposal, then please notify me so that I can send you the manuscript. If you do not agree, I am not going to accept another proposal from you but shall entrust the sale to a reliable person and donate the entire proceeds according to my own choice of charitable causes.

With amicable regards to you and your wife.[4]

P.S. I do have the urge to tell you again [that] Freundlich has, in fact, absolutely no right to the manuscript and that I insist his entire conduct was dishonest and odious.[5]

147. From Jacques Hadamard[1]

Paris, 2[5] Humboldt Street, 16 April 1922

My dear Colleague,

The project for which I submitted the text to you has just appeared in the *Cahiers de la Ligue des Droits de l'Homme*.[1] The text that just appeared is exactly the one I submitted to you: consequently, just as I said to you, it should not be considered definitive. I would, of course, nevertheless be very happy to have your opinion and that of Mr. von Gerlach on it.[2]

I enclose with this letter a notice (I do not know whether you have taken note of others) showing you that you have definitely attained Parisian glory.

In the hope that your stay in Paris leaves you nothing but good memories, which will be repeated, believe me yours most devotedly,

J. Hadamard.

148. To Heinrich J. Goldschmidt

Berlin, 17 April 1922

Dear Mr. Goldschmidt,

From a cousin of Dr. Paul Hertz,[1] Mrs. Francis Sklarek,[2] I hear that a certain possibility exists that Dr. Hertz be assigned translation work through you. I would not like to fail to heartily recommend to you this man, who fell into difficult material circumstances as a consequence of the war. Mr. Hertz is not only a talented theoretical physicist but also a man of rare general cultivation and an outstandingly gifted critic. You can be confident that work completed by him will meet the highest standards.

With most amicable regards to you, your wife, and your son,[3] yours.

149. From Paul Ehrenfest

[Leyden,] 17 April 1922

Dear Einstein,

You'll understand why I *opened* the letter.[1] That I *read* it afterwards as well, you'll pardon.

Perhaps in *answering* the letter the following may be useful to you:

1.° Because my attention as a student in Göttingen was drawn by *Klein* to Hamilton's original papers, I learned to *admire* them *exceedingly* and always regretted (just like Klein!) that there is no published collection available particularly of his papers on optical mechanics.[2]

2.° I don't know about his quaternion papers, but those on optical mechanics are very grand. Their genesis is this: As an astronomer, H. was naturally interested in geometrical optics (like Gauss, Kepler, etc.). The computations were always governed by the concept: "path of *light rays*," so the question arises with him: What can be gained if, in the sense of the *wave theory* of light (but neglecting the terms representing diffraction), the rays are interpreted as the normal trajectories of *wave surfaces* and hence characterizes an optical instrument as an apparatus for transforming wave surfaces (always excluding diffraction)?– It is only in a discussion of homogeneous media that he pursues the thought further: Instead of the *regular* differential eq. governing the course of the light *rays*—comes the *partial* differential eq. of *1st order* (neglecting diffraction), which governs the propagation of the *wave surfaces. If it is integrated, there follows by differentiation the path of the (orthogonal) rays of light.*

(By imagining the integrals of the wave surface as arranged in series expansions, he obtains a complete "theory of the errors" of an instrument: spher[ical] aberration and the like).

Now, however, he also notices this: If one regards optics as *emanative*, therefore

$$\text{refractive index } n = \frac{v_{\text{eman. in glass}}}{C_{\text{eman. in vacuum}}} \text{ then}$$

$$\underbrace{\int n\,ds}_{\textit{minim[um]} \text{ per Fermat}} = \int n\frac{ds}{dt}dt = \underbrace{\frac{1}{C}\int v^2 dt.}_{\text{"action integral"}}$$

On the other hand, $\int n\,ds$ is constant for all light rays that lead from one *point of light* to one and the same *light-wave surface. So he sees:*

light rays

p. of light

Surface: $S = \int mv^2 dt = $ const.

What is, in *undulatory* optics, a *wave surface* is, in emanative optics, a "surface with an action integral of constant value."

His *optical* inquiries thereby take a *mechanical* turn:

Instead of integrating over the *usual* differential equations governing the motion of an (emanative) point of light in the field of force of a given—inhomogeneous medium, one integrates over the *partial* differential equation of 1st order which determines the "surfaces of constant action" and then looks for their orthogonal trajectories.

He soon realizes that this can be taken over from the emanative point of light to an arbitrary (conservative) mechan. system.

Thus Hamilton made his mechanical discoveries, which Jacobi then perfected in the "Jacobi-Hamiltonian" integration method (this later merges with Cauchy's methods for the integration of part. diff. eqs. of 1st ord. and is completely absorbed in the integration method by *Lie*.[3]

!!!! On the side,[4] he discovered conic refraction from the inquiry into how his emanative-undulatory analysis should be translated from isotropic media to anisotropic ones. (Here the "quadratic" artificial energy $\int mv^2 dt$ is substituted by a square root from a fourth-order polynomial.)

— . —

To Hamilton's discoveries—or more accurately, to disparate individual results that diffuse, are furthermore added:

1.° The development of line geometry [*Liniengeometrie*], which later took on colossal proportions, thanks to Kummer, Möbius, Plücker, Klein, Lie.[5]

2.° Bruhn's theory of "eikonal" optical instruments.[6]

— . —

I see just now that in the *Jahresbericht der deutschen Mathematiker-Vereinigung*, vol. *30* (1921), pag. 69, there is a habilitation presentation by *G. Prange* (Halle) about Hamilton.[7]

Hamilton was a really great fellow. It is a shame and *very* unfortunate that his papers haven't been compiled.– (Maxwell published optical studies as a follow-up to Hamilton.)[8]

— . —

E. H. Synge himself is, God knows what.[9]— He published in the *Philosoph. Magaz.* in March 1922 (pg. 528) "A Definition of Simultaneity and the Aether" (3 pages), a machinery with which *absolute* synchronicity is establishable.[10] Droste was unable to fight his way through it.[11] To me "it smells of F. Adler."[12]—So *some* caution is in order.

Another Synge (J. L. Synge) also planted some other sort of relativity cactus in *Nature* (27 Oct. 1921).[13]

Every day I fearfully check in the papers whether you hadn't let yourself be interviewed about your Parisian impressions. I hope that the absence of such just proves that you are "sensible" and not, perhaps, that you're sick.

Heartfelt greetings to all of you and to a happy reunion.—Lorentz is coming back in mid-May.[14]—Yours,

P. E.

150. To Peter Debye

Berlin, 18 April 1922

Dear Debye,

Don't get so excited.[1] You do know Nernst and his temperament. I only asserted that in the empirical equations of state, which are currently being viewed as the summary of experience, there is no term in the attraction term that does not vanish for $T = \infty$. The question of how much your polarization forces count against the ones based on mere orientation ultimately comes down to whether or not this behavior expressed in the formulas for $T = \infty$ corresponds to reality.

I cannot acknowledge as correct your argument that the polarization attraction is absolutely the only theoretically possible one for the noble gases. Thus one could also wish to prove that a monatomic body could not be paramagnetic. *There can, however, very well be a statistic for orientation, without, therefore, there having to be rotational degrees of freedom in the sense of molecular dynamics* (e.g., Bohr's monatomic hydrogen; the Ag-atom according to the Stern-Gerlach experiments also).[2]

Best regards, yours.

151. To Charles-Eugène Guye[1]

Berlin, 18 April 1922

Dear Colleague,

First of all, my heartfelt condolences on the heavy loss you and science suffered. Langevin told me in Paris that your brother became the tragic victim of an erroneous medical theory.[2]

As regards the presentation in Geneva,[3] I have to seek your kind forbearance, as I am so very occupied that I cannot think of taking a special trip to Geneva. Nor can it be assumed that I would be coming to Switzerland anyway in the near future, as the lethal valuta conditions always force me to bring my boys to Germany.

In our seminar we took a detailed look at your paper on the law of electron motion and were unanimously of the opinion that your confirmation of the theory is the most precise of them all.[4]

Please don't be indignant that I went to Paris before I had been to Geneva; the trip was necessary in the interests of international relations; without that I would not have gone. The scientific literature provides nicely for the promulgation of relativity theory.

In great respect, yours.

152. To Romain Rolland

[Berlin,] 19 April 1922

Highly esteemed Romain Rolland,

Only today do I manage to thank you for your kind letter.[1] I am glad that my stay in Paris went so harmoniously and I have the happy conviction of having contributed toward bringing minds a little closer together. It particularly pleases me that I did not see any trace of triumphant pride or bravado, only people filled with a sense of responsibility. I see it as the main difficulty here as well as in Paris that such a rigid conviction exists about the causal relations and about "the guilt," making it very hard to go beyond it. Personal contacts between people from both camps are consequently difficult and yet would definitely be necessary for a restoration of relations to gradually eliminate the mistrust on both sides.

My schedule in Paris was so tight that I unfortunately had to forgo visiting you; but I will look out for a future opportunity, to make up for it. Your open letter to the *Clarté*, or Barbusse, resp., has my full backing.[2] In Paris I had a chance to speak along these same lines with a few representatives of this group.

In cordial admiration, yours,

A. Einstein.

153. To Paul Block

Berlin, 20 April 1922

Highly esteemed Mr. Block,

I feel the urge to thank you again for your letter of the 4th of this mo., no less also for the friendly invitation it contains.[1] By your report about my first lecture you contributed toward attaining the goal that my French colleagues and I myself had had in mind: namely, a relaxation of the terrible tension between the French and German intelligentsia. Only shared, purely objective interests could promote this aim. That is why I strictly followed the principle of avoiding all meetings of a political or social nature and to use the entire time at my disposal to create or renew ties with French colleagues in my field. I am convinced that in this indirect way I could also do more toward clearing the political atmosphere than in any other way. Please be so kind as to convey this to your collaborators for the German cause in a form that seems appropriate to you and thank them kindly again in my name for their invitation.

In my opinion it would be advantageous if in the German press due acknowledgment were given to the teachers of France's highest scientific school, who offered a similar example of reconciliation and courage by taking the first step toward restoring friendly relations between Germany's and France's scholars.

Kind regards.

154. To Maurice Solovine

20 April 1922

Dear Solovine,

Cordial thanks for sending me the things I had forgotten in Paris.[1] Those were unforgettable days, but darned exhausting; I still feel them in my nerves. I haven't seen anyone here yet but, they tell me, the newspapers acted very well, all right; so the whole enterprise fulfilled its purpose.[2] The corrections aren't finished yet. But you will get them.[3] The Anschütz hearing went well; it is good that I was there.[4] I am mailing you a letter for Baron Rothschild that I ask you please to pass on.[5] Hopefully, someday, we shall spend some time together more in the old Bernese style.[6]

Cordial regards from your

A. Einstein.

155. From Peter Debye

Zurich 7, 36 Gloria Street, 20 April 1922

Dear Einstein,

It's not a matter of getting agitated.[1] I just wonder, and very much so, at how calmly you accept the totally false quotation of you that Nernst fabricated, according to your own report. In this way you are supporting him in his effort to intimidate me, as he is evidently trying to do by having you say: "You know these remarks very well and subscribe to every word of it"! To which is appended the good advice that I should surely doubly reconsider whether I wanted to pick an entirely unmotivated fight. By the way, I enclose herewith a transcription of Nernst's statement so that you are acquainted with it. I cannot imagine how to arrive at half-way reasonable agreement with Nernst if even you prefer, despite everything, to stay on the sidelines and to hide his mistakes even from himself under a cloak of love.

Now you write: you only asserted that in the empirical equations of state, which are currently being viewed as the expression of experience, there is no term in the attraction term that does not vanish for $T = \infty$." Not even this mild version conforms with the facts. In my previous letter I already referred to Zwicky and the more recent calculations by Keesom.[2] But now I have to think that a witness from earlier times may be preferable to you. Such a witness exists in Kamerlingh-Onnes himself. All his empirical formulas (not a single one excepted) to describe the second virial coefficient contain the term in question not vanishing for $T = \infty$. You can probably most speedily be convinced by the enc[yclopedia] art. by Kamerlingh-Onnes and Keesom (*Enc. d. mathem. Wissenschaften*, vol. 5, [part] 1, issue 5, p. 730).[3]

You do not want to accept my argument about the noble gases, pointing out instead a possible future, but as yet not extant, statistic of orientation. I agree with you that such a statistic is conceivable but, on the other hand, of course, it is also clear that it is not needed at all, at least for the time being. The existence of polarization forces is just as secure as the fact of electrostriction. The order of magnitude of these forces fits excellently with the observed attractions in the noble gases. Even their temperature development is described, as far as current observations go. What forces us in this situation to first want to anticipate a future, still very arbitrarily, formulizable theory for what the existing one today has long since achieved? This, when we are already certain that even in that future theory the polarization forces will continue to play their old role!

Best regards, yours,

P. Debye

1 addendum:[4]

"Even in the useful equations of state by *D. Berthelot* and lately by *Wohl*, the volume correction and the pressure correction occurs, similarly in *van der Waals*.[5] Although the former poses hardly any basic difficulties of comprehension, we are kept entirely in the dark about the effect of the molecular forces, regarding both their nature and the law governing their forces. Nor can I detect any essential advancement in the experiment recently done by *Debye* (*Physik. Zeitschr.* 21, 178 (1920)),[6] if only for the reason that by attributing the attractive forces to a kind of electrostatic influence of the individual molecules, the author arrives at forces that are independent of temperature, consequently at *van der Waals's* formula. But a useful theory, even if only for weakly compressed gases, would necessarily have had to lead back to *D. Berthelot's* formula.[7] One can demonstrate, e.g., that the molecular forces in gaseous hydrogen act c. 50 times more weakly in the vicinity of 1000° than near boiling point; for this *Debye's* theory offers no explanation.[8]

156. From Maja Winteler-Einstein

Quinto pr. Firenze 5 via Strozzi, 20-IV-22.

[Not selected for translation.]

157. To Paul Ehrenfest

[Berlin,] 21 April 1922

Dear Ehrenfest,

Many thanks for the English letter and your statement about it.[1] Of course a Hamilton edition merits being published. Now Ilse showed me your postcard with the comment that St. John wants to come in June.[2] Well, we have a problem:

Should I come now or [rather] in June? Twice is impossible because I have announced a course here. As I would have to leave on the 27th, I request an immediate decision.[3] The young Brillouin would also like to come to Leyden, a fine fellow.[4] All the French took a great liking to you in Brussels.[5] They were *very nice* to me in Paris, particularly Painlevé![6]

Warm regards from your

Einstein.

Answer instantly!

158. From Romain Rolland

Paris (XIV), 3 Boissonade Street, Friday, 21 April 1922

Dear A. Einstein,

I thank you cordially for your nice letter.—[1] Yes, I am convinced, as you are, that your coming to Paris will have done much toward a rapprochement among intellectuals.

Duhamel[2] spoke to me about the evening he spent with you, which left him with such fine recollections. I regret not having played my piano part in your mathematical and literary concert.

I am leaving Paris at the end of the month and am settling in *Switzerland, in Villeneuve (Vaud), Villa Olga* (near the Byron Hotel). I am going there to seek tranquility and isolation to write a long novel[3] and other works that I have in my head: for in Paris I am constantly disturbed by the comings and goings; one puts oneself out so totally for others and nothing remains for oneself. Thus rediscovering oneself, defending it, and constantly renewing it is still the best way to be useful to others.

But I am not simply withdrawing myself from my Parisian friends. Less so than ever, now, when we are at the point of founding a major French periodical of independent and truly international thought—outside of any political or social party and even resolutely leaving politics aside. We would like it to be a rallying center for literary, scientific, artistic, and philosophical thought, hence the universally human element. Among the organizers of this *Revue* are such leading liberally minded French writers as Duhamel, Vildrac, Arcos, Bazalgette, Jules Romain,[4] etc. We intend to apply to major personalities from other countries; and we would be proud if you would occasionally think of our periodical, which will probably be launched in Paris in October.[5] The directors will be writing you soon about the rest; I recommend them to you in advance for your kind consideration.

Do not forget my address if you come to Switzerland, either. I would very much enjoy seeing you again.

Believe me to be, dear A. Einstein, your affectionate and devoted admirer,

Romain Rolland.

159. From Paul Colin

Brussels, 22 April 1922

[Not selected for translation.]

160. From Paul Ehrenfest

[Leyden,] 22 April 1922

Dear Einstein,

That's fine!! Your postcard arrived today.[1]—I immediately sent you a telegram: Come. 29 April academy.: i.e., come if possible *so that* you can attend the academy meeting on April 29th.—Your presence could perhaps be of benefit to W. J. de Haas (confidentially!)[2]—The business with St. John[3] isn't certain enough to take upon oneself the risk of a "prolonged Einstein"—No: better one Einstein today than a dozen Einsteins "day after tomorrow."

I would be very pleased if Brillouin came[4]—the only difficulty is that I cannot host *him* now. (You, anytime—because you can be bedded and boarded in any stable.) As soon as you're here I'll write him immediately.

Please write me *right away* when you are arriving where.—For us, and very specially for the children, it is a great pleasure to be able to fetch you!

You would have known from the Indian whooping that the children let out when I read out your postcard how deeply happy all of us are that you are coming.

You are going to encounter here—at least for one whole evening (but perhaps also at the academy) the British physical chemist McDonnan.[5] He will be presenting a talk each in Utrecht, Amsterdam, Leyden, and Groningen.

If Sommerfeld's IIIrd edition has already appeared, please bring it along.[6]

After the academy meeting in Amsterdam you'd also be seeing Hertz, Fokker, W. J. de Haas, Holst,[7] and many other physicists.–

You would have to depart from Leyden on the 29th at *10^{00} in the morning* for Amsterdam in order to come to the academy.

I have to close now for the mailing. Very many regards to your wife and the dear girls! Yours,

Ehrenfest.

That the Fren[ch] physicists reacted about me with sympathy pleases me *very* much, because I liked them very much, too.—I'm rapidly liking myself less and less, to the point of "jumping out of my skin."

161. To Maja Winteler-Einstein

[Berlin,] 23 April 1922

Dear Maja!

You've really put yourselves out[1] a lot recently . . .[2]

. . . I have to go to Holland this week already . . .[3]

I spent some very fine hours with Solovine.[4] He has remained very youthful and fresh and is just as anxious to learn and intellectually ambitious as he ever was. He knows many of the Parisian professors well and has a nice social life. Even financially he's doing acceptably well, but just poorly enough that he is not in danger of getting married, which surely wouldn't suit him.

162. From Sebastian Kornprobst[1]

Berlin,[2] 23 April 1922

Esteemed Mr. Einstein,

After many a long, long year I received from Mr. Rosenthal the pleasant news that he had been in your esteemed company and had spoken with you about long since bygone times.

My wife twice had the pleasure of visiting and speaking with your esteemed mother Mrs. Einstein[3] in Berlin. She also wanted to visit us sometime but then unfortunately the war got in-between with its inconveniences and so we haven't heard anything more since c. 1916.

Years ago we read in the *E[lektro-]T[echnische] Z[eitschrift]*[4] that your uncle, Mr. Jakob Einstein, had died.[5] I would gladly have fulfilled my duty and paid him my last respects; however, I was informed about it too late [and] Vienna was also too far away given my circumstances.

Yes, indeed, we all are simply growing old now and have to make room for the youngsters. That is why my hour will soon strike too . . .

My wife and I still feel relatively well for these times; Wasti[6] is married and has been living in Spain since c. 15 years. He presently resides in Madrid. Our daughter Annita is likewise married since 3 years and lives in Hamburg. Both are doing well.

It would give us much pleasure if we could see and greet you again sometime as well, after so many years. I read about your lectures in America, England, and Paris with great interest and wish you the best of success.

Mr. Gehring, who had also been in the employ of your father Mr. Einstein in Munich for many years, lives here too and is working for Maffei Schwarzkopf.[7]

In closing, I send you and your esteemed family my best regards and sign in the hope of seeing you again soon, as yours very sincerely,

S. Kornprobst and spouse.

163. To Paul Ehrenfest

[Berlin, 24 April 1922]

Dear Ehrenfest,

Thanks for the telegram.[1] I'm arriving Saturday, 11h56 in Amsterdam, which I hope is in time for the session.[2] For I don't remember precisely at what time it starts. That was a good thought of yours.

Cordial regards and looking forward to seeing you, yours,

Einstein.

Then we'll write the piece about Hamilton's works together.[3]

164. To Sebastian Kornprobst

24 April 1922

Dear Mr. Kornprobst,–

It was a great pleasure for me to hear news about you, recently indirectly and now even directly, after such a long time and to hear that you and yours are doing well.[1] My mother died already 2 years ago in B[er]l[i]n; just recently also Mrs. Ida Einstein in Italy.[2] Yesterday Robert E[instein], formerly Bubi, who lives in

Rome, visited us.[3] Unfortunately, I now have to leave town for a few weeks;[4] but then please do come and see me once on a Sunday after letting me know in advance by teleph., so that we can renew our old camaraderie.

With best regards to you and your wife, yours.

Do you still remember that wonderfully pretty little dynamo-engine you made for me back then?

165. To Otto Soehring[1]

Berlin, 24 April 1922

Highly esteemed Legation Counselor,

I thank you very much for the information you kindly gave me on repeated occasions. In October I have to travel to Japan and China and intend to return roughly over the course of February.[2] If this date seems suitable to my Spanish colleagues, I very certainly would be able to give the desired lectures somewhere around that time.[3]

I am not going to be able to go to the Dutch Indies at the time of the solar eclipse because, after repeated urgings, I agreed to speak before the Scientists' Convention in Leipzig, which I cannot cancel without exciting some irritation.[4]

With utmost respect, sincerely yours,

A. Einstein.

166. From Paul Block

Paris, 3 de Ponthieu Street, 24 April 1922

Esteemed Professor,

Sincere thanks for your letter. Your message to our German countrymen in Paris has been delivered.[1] The ambassador, a very intelligent and highminded man,[2] was earnestly disappointed that you did not come and see him. Tickets to your first lecture were made available to the German embassy by Sir Barclay,[3] not by the Collège de France—that was somewhat bitter and the gentlemen deemed it more tactful not to impose themselves.[4] But that is past now. I already expressed words of thanks to the Collège de France in the article "The Hidden Einstein" [*Der verborgene Einstein*], which you may have read (*B[erliner] T[ageblatt]*, 12 April, morning edition).[5] More I cannot do without mentioning at the same time that not all professors at the Collège de France are like your friends. Read the pamphlet by Professor Georges Blondel, "The Disappointments of Peace and the German Peril"[6] [*Les mécomptes de la paix et le péril allemand*] and you will understand me.

I have been working here for two years as best I can toward an alleviation of the tension. It was only for this purpose that I gave in to my friend Theodor Wolff's[7] wish, left my home and my books in Berlin, endured the semiannual separation from my wife (Rosa Bertens),[8] and took up the effort and struggle again on this hot soil, instead of my comfortable employment in Berlin. I do not want to complain personally. Most of my friends remained loyal to me; and the opponents at least know that I tell the truth and am well intentioned. But the press and the chamber are hopeless at the moment.[9] Here in Paris I learned to love my homeland more honestly, despite all its faults.

Nevertheless, one should not lose hope. Your appearance in Paris was a very fine thing. A glimpse into the future; perhaps an avenue into the future. I wish I could live to see it.

With obliging regards and in great respect, yours,

Paul Block.

167. From Paul Langevin

Paris, 25 April 1922

My dear Friend,

I was happy to know that your return proceeded without incident and that you are going to be able to take a well-deserved rest in the knowledge of having performed, from all points of view, useful service. The too short days you spent here will remain with me as happy days. Like you, I regret that we did not have the time to chat at leisure and, that this pleasure must be postponed until your next visit, which I hope will in fact be soon.

I just expedited to Mr. Kocherthaler[1] a check representing the amount you left with me. Certain measures were required to obtain the necessary authorization and for this reason the remittance took place a little late. I asked our correspondent to acknowledge receipt to you directly.

Please find in attachment a word by Mr. d'Ocagne,[2] member of the institute and a distinguished mathematician. He followed our discussions and knows their topic well. See if you can gratify him.

I personally delivered to Mr. Finaly[3] the letter you had instructed me to relay to him. I added my own thanks for the delicate manner in which he had proceeded.

My work on mechanics is appearing imminently. I shall send you a copy of the manuscript as soon as it is finished.[4]

My children and their mother[5] have not yet returned from their vacation. While I wait, I take my meals with Malfitano,[6] who sends his best regards, just as does Selma.[7]

I am completely upset about the bad work that my representatives are doing in Genoa and would nonetheless hope that nothing too serious comes of it.[8] All of this is truly tiresome.

In anticipation of better days, we have our interest in our work to console us and also the precious support of friendships that allow us to mutually support each other.

Your very affectionately devoted

P. Langevin.

168. From Maurice Solovine

Paris, 27 April 1922

Dear Einstein,

I was extremely pleased to learn from your letter that you arrived safely in Berlin.[1] Your efforts in Paris were, in fact, extraordinary. But if one takes into consideration the great result you achieved, you will admit that it was worth the effort to come here. The standing of your theories here is entirely different now from before; and regarding personal impressions, people consider themselves very lucky to have made your ⟨personal⟩ acquaintance. Painlevé told me last week, when I was speaking with him about you, "I do hope that we are now friends, Einstein and I."[2] Everyone wishes to see you here again.–

Now that the papers in Germany also acted decently, you can really be glad to have been an assuaging and beneficial influence, particularly at a time when hatred is working so disastrously and destructively.[3] It is also very nice of you to give me the prospect of our spending wonderful hours together again one day.–

I mailed the letter to Rothschild from the post office.[4] His address is:

Baron Edmond de Rothschild
41, Faubourg Saint Honoré
Paris (8).

Now I, for my part, would like to ask you kindly for the address of Dr. Beck in Chicago. Mrs. Untermeyer wrote me such a very dry letter in which she said that she could do nothing for me.[5] That is why I would like to try with Dr. Beck. As he is academically active, it may be expected that he will receive my inquiry differently.–

You surely remember, dear Einstein, that I told you here about a volume in which scientists of the first order would present their discoveries in brief (about 25 to 30 pages in -8 [octavo]) so as to make them more easily accessible to scholars and the educated general public.[6] In the first issue I would now like to reprint your fundamental paper of 1905.[7] Does the permission for translation depend on you, on Ambrosius Barth, or on Teubner?[8] After finishing off the translation of your lectures delivered in America, I am going to get started on that excellent article of 1916.[9] If the permission for translation of these latter papers does not depend entirely on you but on the publisher, I ask you please to write him that he should not assign it to anyone else.

I would also be very grateful if you would draw up a list sometime noting the names of physicists and chemists who could produce something for the above-mentioned collection. I particularly want to ask Planck to describe his quantum theory in condensed form.

Enclosed please find the contract that you forgot.[10] You will write your name on it and keep it for yourself.

In most cordial friendship,

M. Solovine.

My kindest regards to Miss Ilse Einstein and your esteemed wife.

169. To Emile Borel

Berlin, 28 April 1922

Highly esteemed Colleague,

I read your article in the *Ciencia* as well as the one on space and time with great interest.[1] Furthermore, I spoke with an experienced colleague here about the feasibility of an organization for intellectual workers in Germany.[2] He told me that such an organization would certainly be very necessary but that it would have to contend with strong resistance by the currently almost omnipotent industrial enterprises. The organization could be initiated with some prospect of success only if its legitimacy could also be motivated by non-political reasons. This would be the case if some guarantees could be given that a representative of this corporation was able to collaborate officially with the relevant department of the League of Nations. So

I ask you please to discuss this with suitable individuals (e.g., perhaps with Mr. Barclay)[3] and keep me be informed about the result of these discussions in whichever form appears most suitable to you.

I feel compelled to assure you once again that I am very glad to have gotten to know you and our other Parisian colleagues in the profession better and to have encountered such a friendly reception from you.

With amicable regards, I am yours very sincerely.

170. To Hans Delbrück[1]

Berlin, 28 April 1922

Dear Colleague,

This evening I have to leave for a few weeks[2] but would not like to do so without first briefly reporting about a conversation I had occasion to have in Paris with Prof. Aulard.[3] Aulard gave me the impression of a man honestly concerned about truth and an improvement in Franco-German relations. He is, however, like all the others are—if only subconsciously—under the heavy influence of the prevailing view there regarding the evaluation of political events. He seemed to harbor the bias against you that you do not regard him as an individual but as a spokesman of a political clique; but he was very willing to be persuaded by me of the inaccuracy of this preconceived opinion. In this matter, as everywhere, accumulated distrust can only be overcome by personal communication. I do believe, though, that your debate with him did have a beneficial effect and it would be hoped that there will be a more private continuation of your discussions. Only thus can mutual trust be regained, which in my opinion is an indispensable precondition for a gradual restoration also in political relations.

With kind regards, yours very sincerely,

A. Einstein.

171. To Jacques Hadamard

Berlin, 28 April 1922

Dear Colleague,

It seems to me necessary to tell you that the notebooks of the League of Human Rights still have not arrived.[1] I am very satisfied with my stay in Paris, happy to have made the acquaintance of Parisian mathematicians and physicists, and hoping to have contributed to the restoration of friendly exchanges between French and German scholars.

Please, dear Colleague, accept the expression of my sincerest sympathy,

A. Einstein.

172. To Moritz Schlick[1]

Berlin, 28 April 1922

Dear Mr. Schlick,

Mr. Koehler, who as a philosopher and psychologist has now been called away from Göttingen to Berlin,[2] asks me[3] to put in a word for Dr. Max Wertheimer,[4] taking into account that the latter could eventually be appointed to Göttingen or Kiel.[5] I am the more happy to act on this invitation as I personally know Mr. Wertheimer well and value him very highly as a person.[6] The focus of Wertheimer's interests lies in the field of psychology, where he has been mainly creatively employed. Epistemologically he is less suited than Reichenbach to the extent that he is much less acquainted with the exact sciences than the latter.[7] In any event he is no follower of the petrified philosophy of words (Kant Society) but an active person who can think and experience for himself and in this sense would also have a liberating effect on young people. I have a slight impression that psychology is being neglected somewhat in Germany compared to epistemology.

These lines are not an attempt to influence you in any way, rather just to point out an option to you that may not have occurred to you in view of your own field of expertise.

This letter needs no reply, of course.

With cordial regards, yours,

A. Einstein.

173. To Mario Viscardini[1]

Berlin, 28 April 1922

Highly esteemed Sir,

With reference to your letter of 21 Mar. this yr.[2] and the enclosed article, I inform you of the following: The hypothesis expressed in the article that light in empty space has constant velocity c with respect to the light source, not the system of coordinates, was first discussed extensively by the Swiss physicist W. Ritz[3] and had been seriously considered by me before I postulated the special theory of relativity.[4] I rejected this hypothesis at the time because it leads to great theoretical difficulties (e.g., an explanation for the shadow cast by a screen moving relative to the light source). The most convincing refutation of this hypothesis, however, was given by the Dutch astronomer De Sitter, who drew attention to the fact that the light of one component of a binary star would have to be emitted at a temporally modified velocity, which, however, absolutely does not agree with the results of observations.[5]

In great respect.

174. To Elsa Einstein

[Amsterdam,] Saturday [29 April 1922][1]

Dear Elsa,

After a good trip I arrived successfully in Amsterdam, where Ehrenfest[2] was waiting for me at the station. In Bentheim[3] they said—no second class. So your plan backfired, because it was more expensive than otherwise. We went to the academy.[4] In the afternoon I gave a brief talk at the phys. society.[5] In the evening I visited Zeeman.[6] Now I am sitting in Amsterd. with Ehrenfest in the train and am writing you before the departure. I slept well last night. Amsterdam breathes contentment and abundance as always. I'm already looking forward to Leyden. Now it's Ehrenfest's turn.

Warm greetings to all of you. Yours,

Albert.[7]

Undersigned confirms proper receipt in undamaged condition (umbrella included) of the above-specified shipment.

P. Ehrenfest,
importer.

175. From Max Born

Göttingen, 30 April 1922

Dear Einstein,

Laue was here a short while ago; we had a very good time with him.[1] He told us that you are traveling to Holland.[2] I hope this letter reaches you nevertheless.

First of all, I have to request your help again, and to be specific, concerning Bródy.[3] When I spoke to you in Berlin at Christmas, you told me that it was perhaps possible to get him a position in Kaunas [*Kowno*]. I recently talked to I. Schur[4] about this in Berlin (you were in Paris right then); he has all kinds of contacts in Kaunas. He wanted to take care of the matter. It is now becoming urgent that something happen. My wife,[5] who is taking care of Bródy's family (he had his wife and their young child come over here a while ago), tells me about the misery they are living in. I give him something out of our private fund (around 2,000.– marks per month), but that's very little for a family.[6] We help otherwise as best we can. But the man really must get out of this degrading situation. I regard him *very* highly as a physicist; if he were energetic and in better material circumstances, he would surely achieve very much. He now has a nice paper forthcoming in the *Phys. Z.*;[7] he is also collaborating with me on thermal expansion. Hilbert appreciates him very much,[8] particularly because he gives excellent talks at the seminar. I could certify a habilitation thesis of his here without a problem, if I wanted to; I just consider it pointless because, as a Hungarian Jew, and with his thoroughly Eastern mannerisms, he wouldn't receive an appointment anyway. With Paul Hertz, who is also living from hand to mouth, I have enough worries and responsibility that I can't do anything about as it is.

Couldn't you find a modest position for Bródy in Holland? Or anywhere else in the world? I applied for a stipend for him from the ⟨Helmholtz Soc⟩ Emergency Association [*Notgemeinschaft*] but didn't get an answer yet.[10] Couldn't you put in a good word for him there? Or is there any other way?

Now to other things. I'm continually writing away at my encyclopedia article on lattice theory.[11] I hope it will be finished in May. It's a quite tedious job. There unfortunately turned out to be an error in my recently published theory of the equations of state for crystals;[12] I had assumed that Grüneisen's law of the proportionality between energy and expansion was not exactly valid, rather that at low temperatures the former is proportional to T^4, the latter to T^2. But that was nonsense. It was based on a howler. It's a bit depressing that such a thing could happen to someone of my venerable age. If you find the error yourself, though, it isn't quite as bad; and I console myself that it is such a tricky business; besides, Pauli[13] and Bródy both pored over the paper and didn't find the error.

Pauli is gone, unfortunately: to Hamburg with Lenz.[14] Not long ago we started some research together,[15] a continuation of the one published with Bródy on the quantization of nonharmonic oscillators.[16] The approximation procedure developed there can be applied to all systems in which the "unperturbed" system is conditionally periodic and the perturbation function can be developed in powers of a parameter. Even the case of a degenerate unperturbed system fits in with it and leads right to Bohr's method of "secular perturbations." Indeed, now we really do understand some of Bohr's arguments. We also began to calculate orthohelium (2 coplanar electrons) and were immediately able to confirm Bohr's old postulate that the inner electron revolves rapidly along an ellipse whose large axis is constantly pointing to the more slowly orbiting outer electron.

 Pauli took the paper with him to Hamburg and wants to finish it up over there because I don't have any time, owing to the encycl. article. The con[founded] semester is just starting again, which is a real disruption to contemplative work.

Franck has the entire institute teeming with doctoral students and all of them, tanked up with his ideas, are working on fine theses.[17] Hilbert is in Switzerland and is only coming back in 8 days.

My family is well despite the constantly horrendous weather; none of us has a cold right now. My wife sends all of you warm regards. Greet our Dutch and Berlin colleagues for me.

Yours,

Born.

176. From Paul Painlevé

Paris, 18 rue Séguier, Paris VIᵉ, 30 April 1922[1]

[Not selected for translation.]

177. To Chenzu Wei

Berlin, 3 May 1922

Highly esteemed Sir,

With reference to your kind letter of 8 Apr. of this yr.,[1] I have the honor of informing you that I am gladly prepared to give a few lectures this winter, within a period of two weeks, at the Imperial University in Peking. I must note, however, that I see myself compelled to make other proposals respecting the envisioned compensation. I regard this step necessary because—as much as I would have liked to do this differently—by accepting your terms I would be placing other countries too much at a disadvantage that had offered me incomparably greater compensation, some of them, such as a few universities in the United States of America, having already paid it out.[2] Under these circumstances, I take the liberty of submitting to you the following proposal regarding my compensation:

1) 1,000 American dollars

2) payment of the travel costs Tokyo–Peking, Peking–Hong Kong, as well as the hotel costs in Peking, both for me and my wife.

In the hope that you will understand my way of dealing with this and will approve of it, I am most respectfully and sincerely yours

178. To Elsa Einstein

Thursday [Leyden, 4 May 1922][1]

Dear Else,

My conscience is pretty rotten. For I haven't written to you since Saturday. I'm feeling very well and it's a true rest here, thanks to Ehrenfest Cerberus. The trick simply is not to let one thing follow another, instead you have a feeling of free will. That's what's missing in Berlin. There is much thinking and working going on in the process and music making as well. Mrs. Ehrenfest[2] conveys her kind thanks for the purse. It took some persuasion to put the brooch on Tatya,[3] but I managed. I'm not going to extend my stay longer than 14 days, returning rather around Sunday in 8 days. I'm still going to see Lorentz.[4] Zeemann will be performing an experiment I thought up;[5] I'm very pleased about that. I hope all is well with you, despite my not having received any sign of life from you yet. This evening I'm going to be listening to a historian about the occultist wave that is now going around the world.[6] Tomorrow, a physical chemist is coming from England to give a lecture.[7] Then, I plan to invite the children[8] to the seaside. We already played some music with the young Onnes.[9] Today I wrote to the Englishman about the Hamilton edition.[10] I just sent about half a dozen frogs to Kuno,[11] contrary to your view, because I'm not going to be coming here for a long while. We mostly have fine weather; it's windy but not cold.

I'm very enthusiastic about the sailboat.[12] Did you have a look at it already with my friend K[atzenstein]?[13] Ehrenfest has been depressed these last few days but perked up again. I'm good medicine for him and he is dear company for me.

Heartfelt greetings to all from your

Albert.

179. To Hans Albert Einstein

Leyden, 5 May 1922

Dear Albert,

First of all, contrary to my habit, I wish you a happy birthday in advance, because I happen to be thinking of it.[1] I'm already very excited about the vacation. As regards the help, I'll try to get Anna to come.[2] If she can't, we might have to try to take care of ourselves with the help of the inn. The boat still has to be varnished and then we're ready to go.[3] We'll have fun ferrying about. Reassure Mama about your studying; you'll manage the final exams, all right; and it doesn't have to be brilliant, either.[4] Write me as precisely as possible and soon when both of you can come so that I can make the arrangements. Send the Anschütz photographs that arrived about 4 weeks ago to: Kuno Kocherthaler (private), Lealtad Street, Madrid, same as before.[5] It's a pity that we couldn't meet during the Easter holidays. But the visit to Paris and Leyden was absolutely necessary. We can play music very well at Katzenstein's, whose house is not a far ride away at all, just ¾ of an hour.[6] But maybe we'll find a piano out there, too. I hope Tete is well, despite your having put him up in Rheinfelden.[7] Write me about that.

Kisses to both of you from your

Papa.

Kind regards to Mama, who is, I hope, back again.[8] Anschütz invites both of you this fall to his vacation home, which he is setting up by Lake Constance for Munich University-ites.[9] There are technical problems to solve there too.

180. From Edgar Zilsel[1]

Vienna, 18 Währingerstr. 71, 5 May 1922

[Not selected for translation.]

181. Two Aphorisms

[Leyden, 8 May 1922][1]

"Nature with its sublime regularity seems to the scientist a conscious being of superior reason."

"No one shall be proud of his knowledge; for others have to work in his stead so that he can study."

A. Einstein, Leyden 1922.

182. To Paul Painlevé

Leyden, 8 May 1922

Dear Mr. Painlevé,

I unfortunately received your letter late; it was forwarded to me.[1] I send you the requested autographs nonetheless in the belief that they have been made use of.[2] The conversations with you were among the most exquisite I experienced in Paris; your intensity and objectivity pleased me very much. All in all, I think back on that sojourn in happy gratitude.

Cordial regards from yours truly,

A. Einstein.

183. From Henri Barbusse

Miramar per Théoule (Alpes Maritimes), 8 May 1922

My dear Maestro,

No one realizes better than I how little leisure you must have at the moment, having become the center of a great movement of scientific renewal and revolution. I nevertheless permit myself, relying on the kind benevolence of your earlier attitude,[1] to ask you whether you would not consent to writing a few lines about your sojourn in Paris for our periodical *Clarté*.

Excuse me for abusing your time like this, dedicated as it is to such important and valuable work. But on the other hand, I think that it is my duty to profit by your importance and prestige to further the cause that my friends and I are defending throughout the world: that of rectifying a sister truth to scientific truth, which, like it, is being menaced and oppressed by routine and preconceived ideas. *Clarté* rises above partisanship and develops the simple and evident ideas that poor humanity is taking so much time to understand and will take even more time to adopt but will nevertheless ultimately realize, because it is a matter of life or death for mankind. It is completely natural for persons who take pride in being the modest defenders of a considerable cause to try to find support for their work from those who in another domain knew how to look upward and ahead, who knew how to see what others were unable to discern at all for centuries. Your name is now universally admired. You are placed definitively among those who have discovered a novel and arduously attainable form of truth. Everything you say resounds and echoes; and that is why some impressions signed by you and published by us will consolidate our propaganda of organized fraternity and positive and constructive wisdom; and we hope that despite your innumerable and pressing duties you would be so kind as to grant us this pleasure and honor.

Believe, my dear Maestro, in my admiration and devotion.

Henri Barbusse.

184. Paul Ehrenfest to Niels Bohr

Leyden, 8 May 1922

[Not selected for translation.]

185. From David Hilbert

[Göttingen, 9 May 1922][1]

I know Mr. Roos personally and can only support Nelson's wish.[2] I thank you most heartily for the greetings you sent me for my 60th birthday; it was as much of an honor for me as it was a pleasure.[3] I was extremely delighted. We haven't seen each other in such a long time![4] When will we have another chance for that?

Best regards from your

D. Hilbert.

186. To Elsa Einstein

[Leyden,] Wednesday [10 May 1922]

Dear Else,

Yesterday I received your disheartening letter. I feel quite sorry for you in all your misery. Don't worry about the sailboat,[1] but nurse yourself and then misbehave more cautiously. I am arriving at the Zoo [Station] Sunday at 7^h 05 in the *morning*.[2] It is my only option to travel. Aren't I punctual with my return? I've been here exactly 14 days.[3] Tomorrow we'll be seeing Lorentz.[4] I was not able to teach a course but did manage to get some other things done.

Warm regards to all of you from your

Albert.

187. From Edward H. Synge

Knockroe, Dundrum, Co. Dublin Ireland. 10. V. 22

Dear Dr Einstein,

I wish to thank you for your kindness in sending me an appreciation of Hamilton's work.[1] I am sure it will be very useful to us in getting people to take an interest in the project.

When you mentioned that the *genesis* of Hamilton's ideas was hard to understand it occurred to me that it might interest you to know, what appears clear from Graves' Life of Hamilton,[2] that a philosophic viewpoint which is very unusual among scientists seems to have been responsible here. Hamilton was a Neo-Platonist, and one gathers that he expected to find in Nature the exemplification of laws which satisfied his sense of intellectual beauty. In fact he regarded such satisfaction of one's intellect as the "raison d'être" of these laws, and went so far as to question whether the laws of Nature were not specially *created* to develope the intellects of men "and angels"!! However odd such a standpoint may seem, it appears to be a powerful agent on heuristic grounds, since it makes explicit an instinctive feeling which I believe is shared in by many mathematicians, that the most abstract researches will be found adequately exemplified somewhere in Nature. For instance it would lead one to look for some physical exemplification of Riemann's surfaces, and from the way in which "quanta" occur here one might perhaps hope for some bridge between your quanta and gravitational theories in this direction.

I am sanguine enough to believe that a great deal in the way of general viewpoints may still be derived from a first hand study of Hamilton's work, and I especially hope that mathematicians may realise that Hamilton's system of quaternions was conceived by him as a *Space Time* symbolism, and that the possibilities of a general development of a theory of functions of quaternions (perhaps on a *sheeted* Space-Time) may be explored, with a view to finding *physical* exemplifications and seeing whether there may not be something in a guess which Hamilton has put on record "that a function of a quaternion *might* be the law of the universe."

It seems not a little curious that a standpoint which is so fundamentally different from yours should have brought Hamilton to results so clearly related to your theories.

Again thanking for your courtesy I remain Yours very truly

E. H. Synge.

Translator's note: Original written in English.

188. To Paul Langevin

[Leyden,] 12±ε May 1922[1]

Dear Langevin,

Here in quiet Leyden we often think amicably of you, your $x^{[2]}$ [$yzh\vee g_{\mu\nu}$] brethren,

A. Einstein and P. Ehrenfest.

189. From Emile Borel

Rome, 13 May 1922

[Not selected for translation.]

190. To Max Born

Berlin, [on or after 14 May 1922][1]

Dear Born,

It is immensely difficult now to find a job for theoreticians.[2] Holland is suffering from overproduction. The fact that something could be done for Epstein[3] is based on the extraordinary importance of his achievements. There are some excellent theoreticians over there (e.g., Fokker) in modest preparatory-school teaching positions.[4] Some months ago I wrote to Millikan and Epstein in Pasadena on Bródy's behalf[5] but haven't received any reply yet. I'll talk with Laue, who, if I'm not mistaken, exerts some influence on the *Notgemeinschaft*.[6] I became acquainted with your perturbation method through Becker's habilitation thesis[7] and enjoyed it.

I made a monumental blunder myself a while ago (experiment about light emission with canal rays).[8] But be consoled. Only death can keep one safe from blundering. Bohr's papers fill me with great admiration for the steady instinct that governs them. It is very nice that you are working on helium. The experiment by Stern and Gerlach is the most interesting at the present time, though.[9] The atoms' orientations without collisions cannot be explained by radiation (according to current methods of considering the problem). By rights, an orientation ought to persist longer than 100 years. Ehrenfest and I did a brief calculation of it.[10] Rubens considers the experimental finding absolutely secure.[11]

Do soon dispose of the money for purchasing the X-ray apparatus! Why is it taking so long?[12]

Cordial regards to all of you from your

Einstein.

191. From Paul Ehrenfest

[Leyden,] 16 May 1922

Dear Einstein,

All of us miss you very, very much!![1]–

— . —

I had to settle a few things for you after your departure: 1. One Public Universit. Rotterdam—You had left and would return only late next spring 1923. 2. Enclosed letter from Prof. J. van Baren.[2] I answered: You had left, would *not* be going to Java. But would forward letter to you as you might possibly satisfy his request indirectly. (I was thinking via Freundlich[3] or the like.)—*He deserves this help* because he did much work on these things. 3. Dismissal of a (very unappealing) Zionist-lectures organizer. ("Einstein is in Holland exclusively for specific advice on physics.") 4. One newspaper photographer.

— . —

I hope very much that now you'll finally get to know my dear, dear Joffe.[4] He is a *very* fine physicist and person. You'll enjoy him in *every* respect. Together with my wife, you, and Bohr, he is my closest friend; and maybe no man has shown more love for *me* than he.– The danger that the two of you will while away your time together with "newspaper-feuilleton" chit-chat is enormous.

— . —

— . —

I'm afraid that I'm not going to be able to finish up our note about the Stern-Gerlach experiment by the next academy meeting.[5]– *Breit* propounded the following hypothesis:[6] Silver atoms are always *in some sort* of magnetic field (even in the melting-pot) and there are already oriented toward the *local* magnetic field.— While vaporizing and flying up to the strong magnetic field, they then undergo only adiabatic changes (no radiation).— *This way* seems nonsense to me. But the following *modification* of Breit's assumption could perhaps be right:

At the instant of vaporization!! the atoms adjust themselves *unadiabatically* to the *magnetic field at A* (terminal field + dispersive field of the magnets), namely almost 50% positively 50% negatively.– During the flight from A to B both groups adjust themselves *adiabatically* to the magnet's field *orientation.*—

This suffices for Stern and Gerlach.

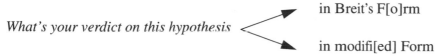

What's your verdict on this hypothesis

in Breit's F[o]rm

in modifi[ed] Form

I request a speedy, clear response.– And how might I formulate this hypothesis in our note?

— . —

— . —

In my next letter I'm going to tell you about a hugely funny hypothesis by Breit and Borelius[7] on the character of superconductivity that can be checked by a very simple experiment. If they are right, it really is hilarious.

— . —

Best regards from each of us in turn! Yours,

P. Ehrenfest.

Convey my sincere wishes to your wife and the Il[se]-Margot [*Il-mar-go-tse*].

192. From Eric Drummond[1]

Geneva, 17 May 1922

Esteemed Professor,

I have the honor of informing you that the First Assembly of the League of Nations adopted a resolution inviting the Council to present at the Second Assembly a report on the possible utility of creating a technical organization for coordinating the work of intellectuals. Subsequent to the report by Mr. Léon Bourgeois[2] on the question, the Second Assembly adopted the following resolution during its session on 21 September 1921:

"The Assembly approves the projected resolution presented in the name of the Council by Mr. Léon Bourgeois regarding the nomination by the Council of a Committee responsible for studying international issues of intellectual cooperation; this Committee will be composed of at most twelve members and will include women."[3]

In compliance with this resolution, the Council of the League of Nations commissioned[4] me with inviting you to be so kind as to be part of the Committee for Intellectual Cooperation.

Similar invitations are likewise being sent out to:

Messrs. BANERJEE, D. N., professor of political economy at the University of Calcutta.[5]

BERGSON, H., professor of philosophy at the Collège de France. Member [of] the Académie Française.[6]

Ms. BONNEVIE, professor of zoology at the University of Christia[nia]. Delegate of the League of Nations Assembly.[7]

Mr. DE CASTRO, A., director of the faculty of medicine and of the University of Rio de Janeiro.[8]

Mme. CURIE SKLODOWSKA, professor of physics at the University of Paris, honorary professor of the University of Warsaw, member of the Academy of Medicine in Paris and of the Scientific Society in Warsaw.[9]

Messrs. DESTRÉE, former minister of the sciences and the fine arts. Member of the Royal Academy of Archeology of Belgium.[10]

MURRAY, G. A., professor of Greek philology at the University of Oxford, member of the Council of the British Academy, and delegate of the League of Nations Assembly.[11]

DE REYNOLD, G., professor of French literature at the University of Bern.[12]

RUFFINI, F., professor of ecclesiastical law at the University of Turin. Former minister of public education, president of the Union of Associations of the League of Nations. Vice-president of the Royal Academy of Turin.[13]

DE TORRES QUEVEDO, L., director of the Laboratory of Electromechanics in Madrid, member of the Council for Advanced Studies.[14]

The nomination of the 12th member of the Committee has been reserved for the moment.[15]

The Council expressed the opinion that the Committee's 1st session could take place on the 1st of August 1922.

You will find as an attachment the report by Mr. Léon Bourgeois envisioned by the Assembly's resolution, the report presented to the Assembly by Professor

Gilbert Murray, and the report by Mr. Hanotaux, adopted by the Council on 14 January 1922.[16] I would be happy to furnish you with any additional information you may want.

I permit myself to express in the name of the Council of the League of Nations the sincere hope that it will be possible for you to accept this invitation.

Very respectfully,

Secretary General.

193. To Paul Ehrenfest

[Berlin,] 18 May 1922

Dear Ehrenfest,

I already took care of Joffe,[1] although he's away from Berlin at the moment. I want to invite him alone and use the time well to initiate contacts with him. I'll make sure that the request by the Dutch geologist is fulfilled.[2] Thank you for dealing with the things that arrived after me. The Zionists are shameless and pushy; I have difficulty always assuming the proper stance each time, considering that I am, of course, sympathetic to the cause.

Breit's hypothesis is not acceptable because the strong electric molecular interactions surely far outweigh the weak magnetic fields and produce statistical disorder.[3] If you say the atoms adjust unadiabatically to the magnetic field *at the instant of vaporization*, that's a hypothesis without any sort of legitimacy. For, as long as the interaction with other atoms persists, it causes disorder. Once the interaction has ceased, though, the orientation problem is the same as at any other time. Unadiabatic adjustment means doing without theory but adds other difficulties.

How should the temperature influence $e^{-\frac{E}{KT}}$ be accounted for in the adjustment based on paramagnetism? (It should occur even in this case without collisions, shouldn't it?) (How are you faring, esteemed southern orbit?) I wouldn't say anything about this eventuality in the note.[4] Regarding the problem with the factor 2, I was thinking that *two* transitions are surely involved in the repositioning, but that the a priori probability of the mean state vanishes with a vanishing magnetic field, hence that the absorption probability is much greater for this state. This question doesn't affect us, by the way, thank God.

I'm very curious about the Breit-Borelius hypothesis, especially if, as you say, it's testable.

Haber is convinced of the conclusiveness of the Stern-Gerlach photographs. Cordial greetings to all of you, y[our],

Einsteins.

The Dutch did so much for the expedition.[5] I cannot join them on the expedition. Do you believe that anything is served by my stopping over in Batavia on my return trip (which isn't possible without a considerable loss of time)?

194. Recommendation for Paul Hertz

[Berlin,] 18 May 1922

I have known Mr. P. Hertz for many years through his scientific papers and in person.[1] He is, without a doubt, one of the most discerning theoretical physicists in Germany, with great expertise and remarkable originality. His high degree of general culture should be particularly pointed out, which allows him to appear excellently suited for literary work of a broad-ranging scope. I value him as highly as an intellectual as I do his personality and am convinced that he will complete in a satisfactory manner any mission that he feels confident to take on.[2]

A. Einstein.

195. To Gustave Le Bon

Berlin, 19 May 1922

Sir,

Your note[1] interested me very much. It is truly remarkable that by means of the equivalence of mass and energy you arrived at consequences in conformity with those of the theory of relativity. It would interest me to become acquainted with the method you used. I would appreciate being informed about this point. If you would like to do me this favor, I ask you please to take into account that I have difficulty reading your handwriting.

Most sincerely yours,

A. Einstein.

196. To Felix Rosenblüth[1]

Berlin, 19 May 1922

Esteemed Doctor Rosenblüth,

I already wrote to Mrs. Hausmann[2] along the lines you wished.[3] Regarding the restriction on the use of the available funds, I see no serious difficulty, considering that the plan is anyway primarily to foster the biological sciences. I think that we shall be able to use the funds in the most useful way without having to arrange for a change to the will. Besides, let us contact Prof. Weizmann about the matter as soon as his mind is freed from a settlement of the mandate question.[4]

With amicable regards.

197. To Oskar Heimann[1]

Berlin, 20 May 1922

Dear Mr. Heimann,

When the boat was picked up at Naglow Wharf,[2] two deficiencies not in conformity with our purchase agreement were found, completely disregarding the necessity of a few minor repairs, namely:

1) The delivered tarpaulin does not fit on the boat and is therefore not usable as such.

2) The batons for the sail were missing, so the wharf had to make new ones.

Accordingly I request of you:

1) To hand over to me the fitted tarpaulin belonging to the boat in exchange for the one delivered.

2) To please settle that part of the wharf's bill concerning the batons.

Requesting a reply soon, I am, very respectfully.

198. To Hantaro Nagaoka

Berlin, 20 May 1922

Highly esteemed Colleague,

I thank you most cordially for your kind and appreciative words.[1] It will be a great pleasure for me to enter into closer relations with Japanese scholars, who in such a short time became productive collaborators in the scientific endeavors of the West, without betraying their own national traditions. I believe that in these times of political unrest it is the primary duty of all scholars to keep science and relations between researchers free from any political influences. I have the impression that in this regard the conduct of Japanese scholars may be viewed as exemplary; and I cherish the hope that among Western scholars, too, the genuinely scientific spirit focused on factuality will soon win the day. The friendly attitude of your academy about which you wrote me in your letter honestly pleases me.[2]

In expressing the hope that I can make my most humble contribution toward solving the problems concerning my field of work within your country, I am, in utmost respect, yours truly.

199. From Robert A. Millikan

Brussels,[1] 22 May 1922

Sir:

I have the honor to inform you that you were elected a foreign associate of the National Academy of Sciences at its annual meeting held in Washington April 24 to 26, 1922. The Academy thus desires to express its high appreciation of your services to science, and trusts you will signify your acceptance of this election.

In due course a diploma signed by the officers of the Academy will be sent to you.[2]

Assuring you of my personal pleasure in counting you as one of our members, I am very respectfully yours,

Robert Andrews Millikan
Foreign Secretary.

Translator's note: Original written in English.

200. To Paul Ehrenfest

[Berlin,] 23 May 1922[1]

Dear Ehrenfest,

At first, your shock hypothesis made a great deal of sense to me and I had already even seriously thought about this interpretation myself. But now I have reached a different view, which, however, I cannot establish rigorously. I'll tell you about it purely schematically.

Each atom is always directionally quantized—no matter where it is. In transit from one field into another (even from an electrical one into a magnetic one), each quantum state changes adiabatically into another state but into one that again obeys the spatial quantum condition. So each atom in the solid body is already direction-ally quantized or bonded with other atoms in such a way that upon vaporizing within a field it is directionally quantized. There aren't supposed to be any degen-erated states in the true sense.

Consequently, from the outset the vaporized beam is completely directionally quantized, hence also in a (strong) field. The statistical distribution among the quantum states is determined by the distribution in the solid body and corresponds to its temperature, taking into account that only in the solid state can sufficiently frequent jumping between quantum states occur.

Our calculation[2] is valid for the transition between quantum states through radiation. It has no practical significance, however, because the experiment cannot give any information about its validity or invalidity. If it were possible to magnetize the vaporizing body far beyond the proportionality limit, the one quantum state in the vapor beam would degenerate at the expense of the other.

Of course I do know that objections to this consideration exist. If such a silver vapor is in a weak magnetic field and prior to a collision two atoms are directionally quantized, then, as I see it, after the collision they should also be that way. Because the magnetic field is arbitrarily weak compared to the forces exerted during the col-lision between the atoms, it is hard to believe that the nonquantization doesn't get into some disarray in the process. At this point I can just say: The dir[ectional] q[uantization] of an individual mol[ecule] during collis[ion] is defined by the strong forces, but the transition to freedom takes place as if it happened infinitely slowly. [. . .] the adiabatic processes, which do *not* happen infinitely slowly.

Now, I don't know whether we should publish our notice yet.[3] In any event, not while a controversy exists between us. If we cannot clarify it enough for the two of *us* to reach agreement, then it's better if we let it be.

Yesterday I spent the entire evening with that magnificent Joffe,[4] throughout which we kept strictly to the agenda but did not manage to finish up nonetheless. His analyses on the conductivity of crystals interested me very much.[5] What's going on with Breit's hypothesis about superconductivity? That interests me.[6] Joffe definitely promised to write to you yesterday. He smiled so nicely as he recalled his sins.

Cord. regards to all of you, yours,

Einstein.

201. To Robert A. Millikan

Berlin, 25 May 1922

Highly esteemed Sir,

Please convey my most cordial thanks to the Washington Academy for my nomination as foreign associate of the Academy in Washington,[1] which I had the honor of visiting last year.[2] I naturally accept the election with thanks, welcoming it not just for myself personally but also as a positive sign of the gradual restoration of international collaboration in the area of science.

In great respect, yours sincerely.

202. To Robert A. Millikan

Berlin, 25 May 1922

Dear Mr. Millikan,

I thank you for relaying the good news[1] and seize this opportunity to express my admiration for the excellent and multifaceted work with which you have recently enriched physics.

With my best wishes for your efforts I am, with kind regards, yours,

A. Einstein.

203. To Erwin Finlay Freundlich

Berlin, 26 May 1922

[Not selected for translation.]

204. From Max Planck

Grunewald, 26 May 1922

Dear Colleague,

Just let me say a word here, after some reflection, about what I think of the plan for a conference of physicists in Berlin. I absolutely do view it as appropriate, in the sense of fulfilling a duty of courtesy as well as in the interest of our German science, to return various honorable invitations abroad that you received with a corresponding counterinvitation to foreign physicists. On the other hand, however, we should carefully avoid anything that may be conceived as going beyond the scope of a simple return invitation. I wrote to Nernst[1] in the same sense and asked him to arrange for a meeting of the Academy's physicists in one of the upcoming ⟨c[ongres]s⟩ days. Then we can discuss the various details.

Cordial greetings, yours,

Planck.

205. From Marie Curie-Skłodowska

[Paris,] 27 May [19]22

Dear Sir,

You were, as was I, invited to participate in the International Committee on Intellectual Cooperation of the League of Nations.[1] I would like to know whether you accepted this invitation. I, for my part, believe that acceptance by you, just as by me, would without a doubt be necessary, if we firmly hope to do a real service. This is also the opinion of our mutual friends. But I don't yet gather what the committee's forms of action are nor the work it would have to accomplish. I would be very pleased to know what you think about this subject. My only feeling is that the League of Nations, although still imperfect, is one hope for the future.

Most sincerely yours,

M. Curie.

206. From Uzumi Doi[1]

27 May 1922

[See the documentary edition for the original English.]

207. To Marie Curie-Skłodowska

Berlin, 30 May 1922

Esteemed Madame Curie,

Although it wasn't at all clear to me what the constituent committee will be able to achieve,[1] after brief reflection I did accept.[2] The intent of international rapprochement surely lies behind this endeavor; whether we can gain some influence then depends on whether we go about the business properly. It would sincerely please me if you likewise have submitted your consent, especially since I know that agreement exists between us on all such questions.

With warm regards to you and our dear friends, yours.

208. To Eric Drummond

Berlin, 30 May 1922

Highly esteemed Sir,

I acknowledge receipt of your letter of the 17th of this mo.[1] and state that I gladly accept the nomination to the "Committee for Intellectual Cooperation" [*Commission pour la Coopération Intellectuelle*]. Although I must confess that I have no clear picture of the character of the work to be conducted by the committee, I do feel it a duty to answer this call, because no one in these times should refuse to collaborate in endeavors to attain international union.

In great respect.

209. From Hermann Weyl

Zurich, 52 Bolley St., 31 May 1922

Dear Colleague,

After the students had such success with their invitation to Langevin (Langevin conquered all hearts here; his talks about the founding of dynamics on the principles of energy and relativity, and his kind personality, his inspirational liveliness were enchanting),[1] after this success, the students' courage swelled and they want to have you come and see us again and give a few lectures for your old fellow countrymen. They took the university convention as an excuse to apply to you with this request. Maybe you are tired of travel and will categorically decline. Then I don't want to pressure you further. But all of us would naturally be very happy to have you among us again sometime; so do consider sometime whether you couldn't do us this favor! You won't need to fuss with the authorities because you aren't going to have to deal with them at all.[2]

I am writing in the hubbub of a move. Cordial regards, also from my wife, yours,

Herm. Weyl.

210. From Maja Winteler-Einstein

Florence-Colonnata, Via Strozzi 5, 31 May 1922.

[Not selected for translation.]

211. To Friedrich Heilbron[1]

Berlin, 1 June 1922

Esteemed Ministerial Department Director,

With courteous reference to your letter of the 5th of last mo.,[2] I have the honor of informing you of the following. Only after repeated urgings on the part of my colleague Planck and long hesitation did I accept the talk at the Scientists' Convention, which prevents me from leaving on the solar eclipse voyage to Batavia.[3] If I were to retract, great annoyance would result, especially considering that a certain latent tension already exists even so between me and some leading German physicists, which clearly manifested itself in certain events at the Scientists' Convention in [Bad] Nauheim.[4] In the interest of good relations with my local colleagues, I must therefore avoid a cancellation of my talk. In view of the most commendable way in which the Dutch made the German solar eclipse expedition possible and promoted it in every way,[5] I find it hard to decline; but under the prevailing circumstances, unfortunately, no other choice remains to me.

In expression of my exceptional respect,

A. Einstein.

212. To Humboldt-Film-Gesellschaft

Berlin, 1 June 1922

Dear Sir,

Recently I have been receiving a number of letters that are all based on the misunderstanding that I had somehow participated in the film being shown by you about relativity theory.[1] The reason is that the film was named "Einstein Film" instead of "Relativity Film." I would herewith like to urge you to do me the favor of choosing a title for your film that excludes such mistakes in future.[2]

In great respect.

213. On the "Einstein Film"

[*Einstein 1922g*]

PUBLISHED 2 June 1922

IN: *Berliner Tageblatt*, Evening Edition, p. 2.

Professor Einstein and the Einstein-Film. Professor Einstein asks us to receive the following lines: "Through remarks by friends and many inquiries I have been made aware that I might be collaborating or otherwise participating in the film on relativity which is now showing. I therefore feel obliged to deny this explicitly. Since I believe that this error is chiefly due to the title "Einstein-Film," I have asked the film company in question to choose for the film an suitable objective title.

214. From Leopold Koppel[1]

Berlin NW. 7, 6 Pariser Place, 2 June 1922

Esteemed Professor,

With reference to our conversation today by telephone, I forward to you in the enclosed a statement of approval prepared in duplicate concerning the transfer of Mr. Jacob Koch's syndicate share to the Argo corporation in Glarus.[2]

Please *provide your signature on both copies* and forward one, using the enclosed envelope, to my firm, the other to Advocate Winteler[3] in Lucerne. Please have Advocate Winteler likewise provide his signature and ask him to send the copy signed by him and by you[4] to Director Oscar Curti, Zurich, 75 Susenberg Street.[5]

Respectfully, with best regards,[6]

Leopold Koppel.

215. From Chaim Weizmann

2 June 1922

Dear Professor,

Kindly forgive me for writing you only now: I was away on a trip to Paris and Geneva—in the latter location on the hope that the League of Nations would finally ratify the Palestine mandate but, as you know, the business has been postponed again to the 15th of July.[1] In the meantime all the shady characters the world over are working against us. Rich Jewish lackeys, dark fanatical Jewish obscurantists in combination with the Vatican, with Arab assassins, English imperialist-anti-Semite reactionaries—in short, all the dogs are howling. Never in my life did I feel so alone—and so sure of myself and confident as well! But the attitude of the Jews is disgraceful. It's a crying shame, a sky-high scandal.[2]

Regarding the 10,000 dollars for the university, I would be grateful if you would leave them in your account for the time being.[3] Our American friends are coming here in July;[4] they will then tell us exactly what they plan to do; we shall then also create a formal committee that will manage the funds as well. Of course, those 10,000 dollars should be spent exclusively on physics, according to the donor's wish.

What plans do you have for the summer? Maybe you will be in Switzerland in July. I have to be in Geneva again on the 15th.

Most cordial regards to you and your wife and children, your truly devoted

Ch. Weizmann

My wife[5] also sends warm wishes.

216. To Friedrich Vieweg

Berlin, 3 June 1922

Dear Sir,

For the new edition of *Relativity* the following important corrections are necessary:[1]

p. 75, first paragraph, 6th line from the top: "If the universe is Euclidean, then $F = 4\pi r^2$; if the universe is spherical, then F is always less than $4\pi r^2$ [2]

p. 81, equation (11) $x'^2 + y'^2 + z'^2 - c^2 t^2 = \sigma(x^2 + y^2 + z^2 - c^2 t^2)$

p. 90, 3rd paragraph, 5th line from the top: "on the so-called cyanogen bands, *likewise Perot on the basis of his own observations*, considered the existence of the effect almost beyond doubt; other investigators, particularly *W. H. Julius and* St. John, have been led to the opposite opinion based on their measurements, *or are not convinced of the conclusiveness of the empirical data used*."

Now, regarding the inquiry by Prof. Chalas, Constantinople,[3] Prof. Einstein thinks that you should reply to this inquiry to the effect that you are not in a position to grant authorization under conditions other than the usual ones, even taking into account the circumstances prevailing in Greece.[4] There is absolutely no reason to grant authorization and in addition pay compensation that cannot be anticipated at all.

Very respectfully,

The Secretary.

217. From Aurel Stodola[1]

Zurich, 5 June 1922

[Not selected for translation.]

218. To Hellmut von Gerlach[1]

Berlin, 6 June 1922

Esteemed Mr. von Gerlach,

I gather that on our visit to France[2] neither Painlevé nor any other prominent representative of French scholarship will be present.[3] As my speech on Sunday is supposed to be one by a nonpolitician devoted especially to welcoming scholars, I am not going to rise to speak at the meeting on Sunday. This omission will be less awkward as I will take every opportunity to attend to our guests. I am particularly looking forward to the opportunity of spending a few hours in your home on Sunday.

With amicable regards.

219. To Hermann Weyl

[Berlin,] 6 June 1922

Dear Mr. Weyl,

Tell the students that I, as an old Zurich boy, was very pleased about their invitation, and heartily so.[1] But I surely need a little tranquillity; and whatever science I could relate—with all due respect—all the sparrows are noisily chattering from the rooftops, so I'm reluctant to open my mouth as well. Please don't misunderstand me if I fail to answer their friendly summons; and don't say: "To Paris he went, but not to us." Declining the Parisian invitation would have been a betrayal of the internationalist ideal that needs attention now more than ever.[2] But for my fellow countrymen it is not a matter of "reparations." They always preserve their temper, equanimity, and tolerance. I am glad that Langevin was there to see you. I can't tell you what he means to me and how much I like him.

Right now I'm trying to understand your work on the mathematically preferential position of the quadratic form.[3] I'm not getting any further on the physics. I don't believe in your link between the electric field and the path's curvature.

Eddington's arguments strike me as Mie's theory does: It is a pretty frame but one absolutely cannot see how it has to be filled in.[4] Have you thought Kaluza's approach through? To me it feels closest to reality; but it does not yield a singularity-free electron either.[5] Permitting singularities does not seem the correct approach. I think—to really make some headway—one would have to find another general principle eavesdropped from nature.

Cordial regards to you and your wife from your

A. Einstein.

220. Response to Ernest Bovet's Question to Paul Langevin

[*Einstein 1922j*]

DATED 7 June 1922
PUBLISHED 1 September 1922
IN: *Wissen und Leben* 15 (1 September 1922): 902.

Dear Sir,[1]

Your "Question to Mr. Langevin"[2] stimulates me to a reply. Regarding the general questions of interest to you, relativity theory does not change the situation at all. This is because it is no more than an improvement and modification of the basis of the worldview of physical causality, without any alteration to the fundamental aspects. It is a kind of logical system for describing spatio-temporal events in which the mental noumena (the will, emotion, etc.) have no immediate place. To avoid a collision between the different sorts of "realities" that physics and psychology deal with, Spinoza or, resp., Fechner invented the doctrine of psychophysical parallelism which, quite frankly, fully satisfies me.[3] Physics is one possible way among many equally legitimate ones to arrange experiences into a certain order. The foundations of this system are freely chosen by us, and in particular, according to the aspect that best accommodates the given known facts with a minimum of hypotheses. Thus "belief" is not what is involved but free choice according to the viewpoint of logical completeness and adaptability to experience, which is expressed so well in the quoted statements by Henri Poincaré.

The question "What's the use of . . .?"[4]—if it really should have a clear sense—always only means something with an extension that expresses *for whom*, or better yet, for the fulfillment of whose wish should the relevant matter serve?[5] Beyond this truism I cannot go.[6]

A. Einstein.

221. From Hans Delbrück

Grunewald, 7 June 1922

Esteemed Colleague,

Thank you very much for communicating the impressions you gained in Paris regarding Prof. Aulard.[1] If he really is honestly seeking the truth, which may well be the case, he evidently does not have the strength of character to think his thoughts through to the end nor a sufficient sense of responsibility to go over carefully what he allows to be printed. If you have been following my Open Letters to him, the latest in the *Berliner Tageblatt*, 25 May and 28 May, you will have observed that he simply leaves questions I positively posed to him unanswered; and in the perfidious and foolish attack on Ambassador von Schoen, whom he accuses of having purposefully falsified his last coded telegram, he himself committed a gross falsehood.[2] Mr. von Schoen points this out to him in an essay that will be appearing in the *Deutsches Revue*.[3] It is hard to put oneself into the mind-set of the French today. After Aulard evaded the debate with me, Professor Basch[4] presented himself in his stead, not publicly but within a closed setting. This too has now been retracted again. Can such behavior be explained otherwise than that an awareness already exists that, alongside the Russians, the president of the French Republic was the true instigator of the war,[5] and that there is a lack of courage to admit this and let it be proved to them?

I am having the publisher send you the latest issue of the *Deutsche Nation*, in which I tackle the scandalous article by Mr. von Gerlach about the Munich proceedings.[6] It is a true misfortune that the Franco-German meeting should take place in parliament under Gerlach's presidency.[7] Thereby it is discredited from the outset. If Gerlach had a trace of real political instinct, he would have used this opportunity now also, for the salvation of humanity, to join the great united front to solve the [war-]guilt issue. But as totally biased as he is about aspects of domestic policy and full of hatred of the old regime, he did not envision this great opportunity and thus cut off for a long time to come any possibility of successful propaganda for the pacifist movement in Germany. Luckily, as Lehmann-Russbüldt[8] informed me, it had been resolved anyway not to touch the guilt issue on Sunday. Hopefully we really shall succeed in preventing transgressions of this limit.

In utmost respect, yours,[9]

Hans Delbrück.

222. From Henry S. Hatfield[1]

Berlin, Kurfürstenstr. 124, 7 June 1922

[Not selected for translation.]

223. From Gustave Le Bon

29 Rue Vignon Paris, 7. VI. 22

[Not selected for translation.]

224. From Heinrich Zangger

[Zurich, between 8 and 18 June 1922][1]

[Not selected for translation.]

225. From Hermann Anschütz-Kaempfe

Kiel, 9 June 1922

Dear, esteemed Professor Einstein,

Of course the money will be disbursed on time to the new address you indicated.[1]

I have had good and less-good experiences with the new apparatus, i.e., the blow sphere, but none of them entirely amiss. The bakelite did not work out; it evidently does not tolerate the acid I have to add to the water; however, I now have the sphere covered with a layer of hardened rubber and have been getting quite good results. At the present time I am trying to enrich the hard rubber intensely enough with carbon powder through vulcanization to make it conductive at the current contact points. I don't know yet whether it works, but it would be very elegant.[2]

Through a higher saturation of iron I recently improved the blow ring-magnet by 20%; the differential curve got steeper as well. Now I am trying to reduce the distance between the poles; I hope further improvement will come out of that. I managed to make a very nice relay via electro-inductive repulsion, too.

And otherwise: materials testing, you know, and the usual struggles attached to technological developments; but I am content and definitely count on good success.

We repeated the heat-rotation experiment without copper, with negative result; I next want to repeat it with a new coil that fits better, i.e., smaller diameter with reference to the distance between the poles and a higher winding, and oil at 250° as the heat source.[3]

My wife and I are still hoping that it will be very hot in Berlin in July, provided you will be there around that time; then the temperature will perhaps remind you that it is always somewhat cooler and breezier over here than elsewhere and that your little room up there on the IInd story is waiting for you.

I hope your wife is better again; both of us are feeling extremely well since the removal of the appendix.[4]

Many warm wishes between households, yours ever truly,

Anschütz.

226. "Second Supplementary Expert Opinion in the Matter of the Gesellschaft für nautische Instrumente vs. Anschütz & Co."

[Berlin, between 9 June and 10 July 1922][1]

In addendum to my opinion of 16 Dec. 1918[2] alongside the first addendum of 12 Apr. 1922, I state below my opinion on the new material presented to the Appeals Court for the Kiel Region as follows:

I. G[erman] R[eich] P[atent] 211 634, supplement to G. R. P. 174 111[3] as extensively discussed in the appeal grounds of 9 Jun., does not come into consideration for a decision on the present question because it does not exhibit any effective installations to eliminate or reduce the rolling error; as, in all the installations described in the patent, horizontal accelerations affect unimpeded oscillations of the direction-finding system around the axis of its gyroscope, which oscillations are known to cause the rolling error. The contrary assertions of the appeal bases of June 9th are incorrect.

II. The expert opinion by Prof. Lorenz, Danzig,[4] based on the alleged main property of the plaintiff's patent that all gyroscopes are supposedly served to generate the torque at the same time as raising the oscillation time around the N/S axis (Lorenz opinion, page 2, 2nd paragraph to page 3 middle; page 3, 5th line from the bottom to page 4, 1st line; page 4, last line to page 5, 10th line). To contradict this assumption, I refer to the version of Anschütz patent 241 637,[5] in which the case of the distinction between the directional and stabilizing gyros is clearly and purposefully expressed, namely, page 1, lines 47–51; page 2, lines 5–17; claim 1, lines 95–99.

The conclusions drawn from the erroneous assumption are no more probative than the general observations about the inventive work by the creator of the later design, as the present case merely revolves around the issue of dependence, not the patentable redundancy.

III. The opinion by Prof. Meldau[6] uses an argument that has already been generously acknowledged in the 1st judicial instance and which in my opinion is, among all the newly presented material, the only one that has real connection to the disputed issue. I therefore again summarize my interpretation of this point in the following:

a) The realization that the rolling error depends on the oscillation time of the direction-finding system around the gyro axis was first revealed in the plaintiff's patent; and moreover for the first time ways to eliminate or sufficiently reduce the rolling error were indicated.

b) The examined apparatus also uses gyro effects to prevent those oscillations.

c) The gyroscopes for increasing the oscillation period in the examined apparatus are not directly borne by the frames carrying the directional gyros, but—what is technically thoroughly equivalent—by a frame that, with reference to the degree of freedom under consideration, is rigidly connected to the other frame.

In my opinion, an infringement on the protected scope of Anschütz patent 241 637 would already exist if the relevant design only applied to point (a). The existing coincidence also respecting points (b) and (c) mean, however, that the technical effectiveness of the compasses by the Gesellschaft für Nautische Instrumente is exactly the same as of the plaintiff's patent 241 637.

A. Einstein.

227. From George Jaffé[1]

Leipzig, Ferd. Rhode Street 26[III], 10 June 1922.

[Not selected for translation.]

228. Address to the German-French Peace Meeting

[*Einstein 1922h*]

PRESENTED 11 June 1922[1]
PUBLISHED 1922

IN: *Die Brücke über den Abgrund. Für die Verständigung zwischen Deutschland und Frankreich. Bericht über den Besuch der „Französischen Liga für Menschenrechte" in Berlin und im Ruhrgebiet.* Otto Lehmann-Russbüldt, ed. Berlin: Bund Neues Vaterland, 1922, pp. 13–14.[2]

[Not selected for translation.]

229. From Aarau Cantonal School Class of 1897

[Aarau,] 11 June 1922
[Not selected for translation.]

230. From Wilhelm Westphal[1]

Zehlendorf, 12 June 1922

Esteemed Professor,

A while ago I established contacts with local representatives of the Russian Committee for the Improvement of the Situation of Scholars, particularly with Mr. Maxim Gorki,[2] who is currently here. These gentlemen have a plan that I find very reasonable and would like to interest you in as well. The intention is to give more substantial form to the already existing informal Organization in Support of Russian Scholars. Specifically, a hostel for Russian scholars is supposed to be created, in which they can find lodging and counsel, etc., during their sojourns in Berlin: a kind of intellectual center for Russian scholars in Berlin. In moral support of this and other endeavors later to be further specified (but entirely apolitical!!), a committee composed essentially of Germans would have to be formed to advise the Russians and also to legitimize them here in Germany. I declared my willingness to join this as yet unborn committee. I would consider it an exceptional promotion of the cause if you would also take an interest in it. By no means would any substantial burden of work be involved. Of essence is a warm heart for the great need of our Russian colleagues. The thought is to draw the local organization into closer relations with the "House of Scholars" in Petersburg,[3] from where the funds will primarily originate. As far as I have been able to see, it is not expected that much will be allocated to this.

I am convinced that this enterprise is worth furtherance in every respect; and I think that you would earn the greatest gratitude if you espoused this cause a little as well. Perhaps I can discuss this with you again on Wednesday shortly before the colloquium.[4] I shall be at the institute around 5 o'clock. If it agrees with you, I would like to arrange with you a time that I may visit you once with one of the Russian gentlemen.

I am enclosing for you a very interesting text about the modern Russian system of education, in which you will also find something about the "House of Scholars."

In requesting you give my kind compliments to your very esteemed wife, I am, with best wishes, devotedly yours,

Wilhelm Westphal.

231. "Emil Warburg as Researcher"

[*Einstein 1922I*]

SUBMITTED 13 June 1922
PUBLISHED 22 September 1922

IN: *Die Naturwissenschaften* 10 (1922): 823–828.

Emil Warburg as Researcher.[1]
By Albert Einstein, Berlin.

Last April *Emil Warburg* tendered his resignation as director of the Physika- [p. 823] lisch-Technische Reichsanstalt, a man who for fifty-five years has been successfully contributing with tenacious energy and multifarious talents toward the development of physics. Is it legitimate to extract the story of one individual from the organic structure and growth of science? Is his activity not so tightly woven into the work of predecessors and contemporaries that it would be regarded somewhat as chance whether one individual or another were the first to have taken a particular step? The content of a science can, no doubt, be understood and assessed without entering into the individual development of those who have created it. But with such a one-sidedly objective depiction, the individual steps sometimes appear to be guided by chance.[2] An understanding of how these steps were possible—indeed necessary—can only be attained from following the intellectual development of the individuals who defined the direction as they joined in the effort. Let us try to survey the work of this contemporary of ours from this point of view. We must, however, confine ourselves to what appears especially important to us today because the four weighty volumes of Warburg's original papers[3] lying here in front of me address the most disparate topics of physics and do not all let themselves be effortlessly subsumed under uniform aspects, which really is imperative for our survey here. For it, though, I refer to the list of publications, partly with brief indications of the content, at the end of this article, to facilitate for experts the use of E. Warburg's abundant scientific findings.

Warburg's first works (including the Latin dissertation 1868) concern themselves theoretically and experimentally with the mechanics of acoustic vibrations (Oscillations of rods, determination of the speed of sound in soft bodies by attaching them to virtually undamped vibrating systems. Reversible oscillatory change in the magnetization of iron rods through vibration deformations; heating from sound vibrations; muffling of tones from internal resistance in solid bodies).

In 1870 Warburg demonstrated with experiments on the discharge of mercury from glass capillary tubes that no slippage of observable amounts occurs along the glass as the mercury flows out. This work supplied the natural point of departure for an important investigation that Warburg, together with A. Kundt, had Helmholtz present to the Berl. Acad. of Sci. in 1875 (on friction and heat conduction in rarefied gases). Whereas for flowing liquids a noticeable slippage of the layer directly bordering the wall does not occur, for gases a perceptible slippage is required by the kinetic theory of gases in the case where the gas molecules' free length of path is not practically negligible against the dimensions of the vessel under consideration. According to the theory at the wall another flow rate prevails for the gas that would take place at a distance of 0.7λ (λ = free length of path in the gas) from the wall without the slippage phenomenon. Thus at the wall a discontinuous change in the rate of flow takes place that is larger, the greater the length of path, i.e., the smaller the gas density.

The explanation for this phenomenon is simple. The thermally agitated molecules that hit against the wall had last collided in a deeper layer, so they have a mean one-sided translational velocity (flow) parallel to the wall. After the collision with the wall, on average, they do not have any rate of flow. *On average*, therefore, the molecules directly next to the stationary wall have a flow rate other than zero (apparent slippage).

By an entirely similar consideration one finds that a jump in temperature must take place between the wall and the gas if a temperature gradient (heat flux) exists orthogonally to the wall. The temperature of the gas at the wall must be as prevails at a distance of 0.7 λ from the wall without a temperature jump.

The existence of both effects was experimentally proved beyond doubt by Kundt and Warburg: important arguments that the kinetic theory of gases corresponds to reality. It was the first time that a new phenomenon was predicted on the basis of the molecular theory of heat, in particular, a phenomenon that on the basis of the conception of continuous matter was as good as excluded. If energeticists at the end of the 19th century had appreciated these arguments sufficiently, they would have [p. 824] hardly been able to doubt seriously the deeper legitimacy of the molecular theory.

A year later the two authors found another important experimental proof arguing in favor of the kinetic theory of gas. They demonstrated that the heat capacity of mercury vapor was $\frac{3}{2}R$ per mole (R = constant of the gas equation). For if monatomic gas molecules have no rotational energy, therefore act like material points, then the entire thermal energy of a gas will just consist of the progressive motion of its molecule, which for its part uniquely determines the pressure at a given volume. This corresponds to the equation:

$$\text{thermal energy} = \frac{3}{2}pV = \frac{3}{2}RT.$$

The proof was given by measuring the velocity of sound according to Kundt's method.

The experimental work of the next few years 1872–79 is devoted to the study of external friction, particularly the study of elastic properties of solids deformed beyond their elasticity limit. These researches may have led Warburg by analogy to one of the finest fruits of his labors, namely to the proof that the cyclic magnetization of ferromagnetic substances is connected to a loss in mechanical or electromagnetic energy that manifests itself as hysteresis heat (1881). At that time he also found the quantitative relationship between this energy loss with the surface area of the hysteresis curve. Warburg calculated the potential energy of a permanent magnet with reference to a magnetized piece of iron as

$$\Phi = +\int dV\left(J_x\frac{\partial\varphi}{\partial x} + J_y\frac{\partial\varphi}{\partial y} + J_z\frac{\partial\varphi}{\partial z}\right) = -\int (Jh)dV,$$

where J denotes the magnetization, φ the potential of the permanent magnet, dV the volume element of the piece of iron. The mechanical work dA produced by an infinitely small motion of the permanent magnet therefore comes out as equal to the increase $d\varphi$ from φ for constant J:

$$dA = d\Phi_{(J)} = -\int (J, dh) dV ,$$

hence the mechanical work produced upon cyclic magnetic reversal per unit volume of the iron is equal to

$$A = -\int J dh ,$$

where vector J now should be regarded as a function of the vector h. Nowadays we tend to write:

$$A = +\int h dJ ,$$

which for a closed cycle of the magnetization comes down to the same thing, of course.

After such important advances made by the kinetic theory in the case of gases, the question of how far the theoretical conceptions held out for highly compressed gases was of high interest. One of the most notable conclusions of this theory supported by experiment, that the coefficient of friction was independent of density, was therefore tested by Warburg and Babo (1882) for carbon dioxide at high densities. The outcome was that although the coefficient of viscosity increased, it was only by about 9%, if the density rose to about 500 times the normal density (at atmospheric pressure and regular temperature). The conclusion from this is that the basic conceptions of the theory of gases apply up to high densities. We have no certain explanation for what this slight increase is based on.[4] Perhaps it is based on that in dense gases the molecule's apparent diameter is smaller than in less dense ones because the molecular forces by neighboring molecules on the one under consideration partly compensate each another.

From 1887 Warburg's research is concentrated on the study of electrical conduction in gaseous, fluid, and solid bodies, the inquiry into electromotive forces and chemical reactions produced by electrical processes in gases. These latter studies then led him to his pioneering articles in the area of photochemistry. Reading the papers on gaseous discharge, one is astonished by the amount of meticulous experimental research which initially was not yet guided by the hypothesis of ions. From among the abundance of these articles I pick out only those that strike me as of special importance.

In 1887 and 1888 Warburg and Tegelmeyer found that rock crystal heated to 200° conducts electrolytically, specifically, parallel but not orthogonally to the main axis. Initially using gold-leaf electrodes, they obtained from a high voltage a kind of polarization that had the effect of slowly diminishing the current when a voltage was applied. When using a sodium amalgam as the electrodes, this polarization was eliminated. In recent years these investigations, which are of importance in the study of the solid state of aggregation, have been successfully continued by Joffe. In 1890 a paper appeared by Warburg on galvanic polarization, the significance of which has perhaps not yet been fully recognized even today. Helmholtz is known to have proposed a theory for Lippman's capillary electrometer based on the following idea. At the boundary surface between mercury–dilute sulphuric acid, there is an electric double layer, one of which is occupied by the metal, the other by the electrolyte. A polarization current activated by an applied voltage alters the occupation density of this double layer in such a way that in this process the surface of the metal plays the role of an insulator. The observable surface tension of the mercury touching the electrolytes is composed of the actual (positive) surface tension T_0 of the surface layer and the negative voltage T of the electric double layer. The total tension $T_0 + T$ thus, according to him is at a maximum when T, thus also the electric double layer, disappears. One would consequently here have a means of making the difference in electric potential between mercury and electrolyte disappear and thus of conducting absolute measurements of the metal–electrolyte differences in potential.

[p. 825]

Warburg objects that a large part of the polarization current can very well be utilized to discharge hydrogen cathodically and that the change in the mercury's total surface tension $T_0 + T$ could very well be based on the change in the mercury's surface caused by the discharged hydrogen and hence on T_0. This interpretation also leads to a theory of the nature of polarization other than the purely physical one given by Helmholtz. Warburg underpinned his standpoint in detail in a number of papers and with this analysis seems to me to have taken a path-breaking step in the field of the electrochemistry of boundary layers, which is anything but concluded.

Two other important papers by Warburg are related to this problem, one from the year 1896 on the behavior of unpolarizable electrodes against alternating current and one (1901) on the polarization capacity of platinum. One "unpolarizable electrode" is, for inst., Cu upon dissolving into $CuSO_4$. Today we would characterize such electrodes as ones in which the difference in electric potential between metal and electrolyte at any moment is determined by the metal-ion concentration at the electrode. In this case, as Warburg showed, the entire polarization is attributable to the changes in concentration produced by electrolysis at the electrodes, limited by diffusion. The phase difference between the polarization e. m. f. and the current is in this case substantially (e.g., around 40°) smaller than $\frac{\pi}{2}$. Entirely different from polarizable electrodes, for ex., mercury-dilute sulfuric acid. In this case the phase delay of the polarization tension against the current at a high frequency of the alternating current is just a little smaller than $\frac{\pi}{2}$; the electrode thus behaves similarly to a high-capacity condenser. Warburg shows that this case can be interpreted as that products of electrolysis, e.g., hydrogen, are periodically electrolytically discharged from the (platinum) electrode and dissolved, where the difference in potential between electrode-electrolyte depends linearly to first approximation on the discharged amount. Without diffusion of this discharge (e.g., hydrogen) into the solution and into the interior of the electrode, the phase difference between current and tension would be $\frac{\pi}{2}$; but this diffusion reduces the phase difference. These processes are analyzed by Warburg in the second of the mentioned papers.

The numerous subtle analyses on the chemical effects of silent electric discharge will have to be acknowledged by others who can judge the mastery of experimental precision work better than I, likewise the precision analyses Warburg conducted together with physicists from the Physikalisch-Technische Reichsanstalt on Planck's radiation formula. Whoever wishes to gain insight into Warburg's generously inventive experimental skill, critical foresight, and indefatigable working energy will have to study the original contributions. But we must still give tribute to the photochemical papers of the last century, which—it is permissible to say without exaggeration—provided the initial basis for quantitative photochemistry. From gas reactions he showed in an entirely impeccable way—first in 1906 from gaseous hydrogen bromide—that the primary process constitutes the absorption by a molecule of the energy quantum $h\nu$ of the effective radiation. This primary process of absorption has on its own nothing to do with the subsequent chemical reactions, to which it merely supplies the energy. The molecule loaded with one absorbed quantum now has special reactive options. It can either disintegrate spontaneously (at sufficient magnitude of the quantum energy), whereupon the fission products continue to react with other molecules; or else the molecule, equipped with one absorbed quantum, can react chemically in a certain way with other molecules. Only in the case where these chemical reactions are unambiguously linked with quantum absorption will the number of converted moles per quantum be theoretically predictable, e.g., in the case of HBr, where one molecule H_2 and one molecule Br_2 is formed per quantum of absorbed radiation. The lateness of this important confirmation of quantum theory was caused, on one hand, by the great experimental difficulties (measurement of the absorbed ultraviolet radiation energy and the minuscule converted amounts; attaining the required gas purity), on the other hand, by the theoretical interpretation of the experimental finding.

These lines can only give a faint notion of the life work of such a multifaceted [p. 826] researcher. But perhaps it will inspire some fellow members of the profession to delve into one or the other of his original papers, the following itemization of which will perhaps be more welcome than the scant allusions given thus far about the content of a little part of it.

Scientific Articles by Emil Warburg[5]

[. . . See the documentary edition]

232. To Aurel Stodola

Berlin, 13 June 1922

Esteemed Professor Stodola,

It pleased me exceedingly that you have taken an interest in my paper of 1917,[1] as it relates to my special hobby that is currently finding little favor among physicists. They tend determinedly to uphold electrodynamics in a vacuum, which in my opinion leads to contradictions.[2] I do have to admit, though, that the more precise mechanism behind the elementary processes has remained completely obscure to me.

I imagine the production of heat by the absorption of visible or ultraviolet radiation in two ways:

1) An isolated electric elementary particle is unable to absorb because it has as its precondition an infraction of the law of conservation of momentum. Two colliding electric elementary particles certainly emit radiation at a loss of mechanical energy. So they also have to be able to absorb radiation at a gain of mechanical energy, i.e., the production of heat.

2) A molecule or atom can absorb in the Bohr manner (through electrons). If such a molecule collides with another, this energy can transform during the collision into kinetic energy (heat). This is related, e.g., to the fact that at rising temperatures fluorescence tends to diminish at the cost of heat production.

A quantitative execution of these things is possible but presently not attractive insofar as it requires quite a few detailed hypotheses.

Your kind words about my efforts in the area of international reconciliation pleased me very much. In the present state of excitement no one may retreat completely into their quiet little abode when occasion arises for assisting in the "reparations."

With cordial regards and in hopes of a happy reunion, yours.

233. To Thorstein G. Wereide[1]

Berlin, 13 June 1922

Esteemed Colleague,

Your inquiry puts me at somewhat of a loss.[2] Despite some deep qualms, I allowed Mr. Moszkowski, who is already over 70 years old and has lost his savings in the war, to elaborate on his personal relationship with me in a book.[3] But I could never bring myself to read the book. It is certainly not a small thing for a living person to be served up to the public as close to naked as possible. In any event I could never muster up the courage to collaborate personally on such an enterprise. You should be able to grasp this if you put yourself in my place just for a minute. One really ought to be content with presenting to the public what a person has published about objective ideas. If anything is said about the person himself, it should just be in order to link objective ideas to it. You will probably find the more superficial details about my career correct in Moszkowski's book.

Regards from your colleague.

234. Introductory Remarks to Hans Thirring, *L'Idée de la théorie de la relativité*

[*Einstein 1923a*]

Manuscript completed on or before 14 June 1922.

Published 1923

In: Hans Thirring. *L'Idée de la théorie de la relativité*. Maurice Solovine, trans. Paris: Gauthier Villars, 1923, p. [vii].

The book by Thirring[1] contains one of the best expositions of the theory of relativity addressed to the general public. The synoptic schema[2] added at the end, which illuminates the independence of the guiding ideas of the theory and the way in which it interconnects them, is particularly instructive.

(From a letter from M. Einstein to the translator.)[3]

Translation of the German draft:

The book by Thirring is one of the best executed popular descriptions of the theory. The appended diagram about the independent guiding ideas of the theory and the ways and means by which they are brought into relation with each other by the theory is particularly felicitous.

235. From Emile Borel

Paris, 14 June 1922

[Not selected for translation.]

236. From Max Born

Göttingen, 16 June 1922

Dear Einstein,

Already long while ago, a manuscript by Guillaume[1] was forwarded to me by Debye for a decision on its inclusion in the Physikalische Zeitschrift.[2] As reference is made therein to your Parisian lecture,[3] I did not want to print the paper right off, preferring first to get your approval. At the time, I wrote Debye about whether he agreed to this; however, as so frequently with him, I received no answer. Now Mr. Guillaume is complaining about the delay in the acceptance. I send you the paper now on my own authority with the request that you tell me whether I should print it, resp. whether you would like to add any comments that I could print along with it at the same time.[4] After such a long prior delay, please do settle this matter quickly.

Bohr and a thousand other physicists are here.[5] There is bustling activity and it is quite straining, especially since we Göttingers must continue to give our lectures and conduct our administrative business. That is why I make it brief and remain, with best regards between our households, yours,

Born.

237. To Max Born

[Berlin, on or after 16 June 1922][1]

Dear Born,

Guillaume is deranged and also got completely carried away in Paris.[2] I already put much effort into convincing him,[3] just as did Langevin. It's all futile. There's another such wretch, Mohorovičič. [4] I wouldn't have it printed. But if you do find it more convenient than rejecting, that's all right with me too.

Warm greetings, yours,

Einstein.

238. From Paul Ehrenfest

[Magdeburg,] 17 June 1922.

Dear Einstein,

In the end it was nice that you accompanied me up to the train.[1]—I otherwise easily feel so "abandoned." I just want to say one more thing to you: What makes such a meeting with you or Bohr (or such fellows as Busch)[2] so valuable and irreplaceable to me is not that I learn this or that thing (or see it through your eyes) but above all that such a calm force emanates from you. That *beneath* the hurry-scurry there is a solid ground whose technical units of time are higher, by a respectable number of powers of ten, than those in everyday life—others can find this in nature, I can do so better in a few special people.

Most affectionate greetings, yours,

P. E.

Jo[ffe] came 8 minutes before the train's departure—to my dismay not alone.[3] Regards to your wife, il[se]-mar-go-tse, and also to Frieda.[4]

239. To Hermann Anschütz-Kaempfe

[Berlin,] 18 June 1922

Dear Mr. Anschütz,

I am a little worried about the power supply.[1] I do not believe that the hard rubber can be enriched enough with carbon powder to conduct properly. It seems that it would be most natural to plate the aluminium sphere with a more precious metal [copper?] at the electrode points and *to saturate the solution with a salt of this same metal*. Have you already tried that? Or is there any means of applying a layer of graphite? Or is there any chemically impervious insulator (enamel?) that is so thinly coatable that the electrode can be fashioned into a condensor of sufficient capacity? (I estimate that the latter is probably not possible because the layer would have to be almost molecular in thickness.) In any case, this is a serious problem. Maybe platinized platinum[2] would be useful as an electrode; I do not know, though, whether the flaking off can be avoided.

I fear that a diminishing of the distance between the poles will be of little help because the characteristic in the equilibrium state would then flatten out.[3] But it is certainly worth a try. Thank you very much for repeating the heat-rotation experiment. My ponderings about the nature of the terrestrial field got stuck in improbabilities.[4]

I am glad that both of you are well. When I can get away for a while, I will happily come with my wife. This wish of hers does have to be fulfilled, one day.

Cordial regards to both of you, also from my wife, yours,

A. Einstein.

240. To Gustave Le Bon

Berlin, den 18. VI. 22

Sir —

It is right that my knowledge of the literature is relatively weak, but I always sought to do justice to all authors whose works I knew about.[1] The idea that mass and energy are the only true matter had already been proclaimed by many authors. But it is the theory of relativity alone that allows for true proof of this equivalence. If you would like to write to me about your manner of reasoning (for the formula $m = \dfrac{E}{c^2}$),[2] I would be very grateful to you. Finally, I assure you that crimes against intellectual property are personal affairs and not national.[3]

Very respectfully yours,

A. Einstein.

241. To Heinrich Zangger

18 June 1922

Dear friend Zangger,

You wrote me affectionately and at length, and some of it I was even—able to read. Don't take any heed, if [anyone] in Zurich is placing obstacles in your path. You always have pleasure in doing your thing well; it must give you independence from the whole buffoonery into which we have been born. I have largely attained this independence. Worshipped today, scorned or even crucified tomorrow, that is the fate of people whom—God knows why—the bored public has taken possession of. I have as much talent as an organizer as a cow does as a dancer! But if I had refused the election into that organization for intellectual cooperation, there would have been new tensions again because perhaps no one else from Germany could have encountered such trust at this moment.[1] I will resign the office again as soon as it can be decently passed on to somebody else. Also, I only went to Paris *after* hefty inner reservations, [in order to] serve true peace, which really did succeed. But I can't tell you how much I am yearning for solitude; and that is why I am going to Japan ([from] October), because that means 12 weeks of peace on the open sea. I am taking my wife along. My boys are coming here in two weeks' time. I have a cottage by the Havel [river] with a sailboat and we three plan to [camp] there like Indians.[2] I'm very excited about it.

Scientifically speaking, there isn't anything in particular going on. The gravitational field is still standing side-by-side with the electromagnetic one. What Weyl and Eddington did in this regard is certainly nice but not true.[3] Truth cannot be found by pure speculation; God Almighty goes His own ways. Why you think that we are on the track of the secretive quanta is beyond me.[4] To my mind at least, no more light has been cast on that, no matter how many individual findings it has been possible to interconnect within this field.

Cordial regards, yours,

A. Einstein.

242. From Eduard Einstein

18 June [1922]

Dear Papa,

We haven't written you for such a terribly long time now. But Mama was away so long and so was I, and now I have to catch up on school work, or rather, I should.[1] Albert,[2] too, is cramming terribly hard for the final exams!? The day before yesterday we went on a school excursion. We went to Habsburg castle and Vindonissa.[3] What's the situation with our summer vacation? Do you already have a sailboat?[4] I now have a new teacher, a Mr. Kolb.[5] He's much younger than my old teacher and everything's different. I'm now passionately reading Karl May books.[6] They're very exciting bandit stories. Could you maybe send me one? They're available at almost any bookstore. I've already read quite a few of them but not even a fifth of them all yet.

Do you know our new address yet? It is: 3 Büchner St., Oberstraß, Zurich 6. Was it nice in Paris? Write us sometime![7]

N.B. The photos arrived shortly before Mama's departure, so we forgot to write. I send you many greetings, yours,

Teddy

243. From Hans Albert Einstein

[Zurich,] 21 June [1922]

Dear Papa,

Your letter just arrived.[1] Now, we should first see how we can settle the bus[iness] part. Please do send us another one of those letters for the German consulate.[2] Now then: What's the story with your newest honorary office?[3] I think it would make no sense for us to come to Berlin if you have to go to Geneva just 8 days later. I think, if you could somehow find out when the meeting starts, do let us know and then something else should simply be planned for the vacation. In any case, I'll only go to the consulate when I've got your advice and the note in question. So, our vacation starts on Monday, 17 July, but we might be able to leave as soon as Saturday. Will [Anna] be able to make it again, incidentally?[4] I'll see to it that she comes! We're all wild about the vacation. Please write us so that we can finish up the preparations in time.

Many greetings from

Adn.

N. B. If you happen to have to go to Geneva precisely during the vacation, I think it makes no sense at all then for us to travel first to Berlin; it would probably then be best if you stayed here before or afterwards. For it seems that the costs of traveling there now have grown noticeably, whereas you have to take the trip to Geneva anyway. Moreover, it's vacation time here then and most "social obligations" fall flat in the water (or as we say, into H_2O).

Hoping for the castle!![5] Let's let the League of Nations be the League of Nations.

244. From Max(?) Kreutzer[1]

Berlin, 23 June 1922

[Not selected for translation.]

245. To Mathilde Rathenau[1]

[Berlin, after 24 June 1922][2]

Esteemed Mrs. Rathenau,

Without knowing you personally, I feel compelled to extend my hand to you in the face of terrible misfortune.[3] I became acquainted with this splendid person four years ago and the longer I knew him, the more his mind and personality grew in my eyes.[4] If there can be some consolation, it is that your son is one of the great personalities whose influence stems from their works and extends far beyond their corporeal existence. He will be recorded in the book of history not just as a great comprehender and mover but also as one of the great Jewish figures to offer up their lives to the ethical ideal of reconciliation among peoples.

I share from my heart in your deepest sorrow, as I myself feel the separation from him as an irreplaceable loss.

In expression of my warm sympathy, yours,

A. Einstein.

246. From Hans Albert Einstein

[Zurich, after 24 June 1922]

Dear Papa,

I've been longingly waiting the whole time for an answer from you. It appears you didn't get Mama's letter. It's about this:

I looked through "the program" from the poly[technics] in Munich and here and found as a requirement in both c. *6 months praxis* before or during the period of study.[2] Thereupon all reason for me not to travel with you to Japan was gone, because then I could go to Kiel in the summer to do my praxis.[3] Well, Mama already wrote about that. In short, under these conditions, if possible, please do take me along.

In any case, please inform me of your decision as quickly as possible because I otherwise have to register myself somewhere. The school leaving exam keeps me quite busy, essays are becoming due now, one after another. Incidentally, about those two accursed stars, I continued the calculations and found everything positioned as would be expected, i. e., they just remain on the plane of the picture, that is, observed from Zurich, until some cause comes along. So all's well. I hope you're well, too.

Many greetings

Adn.

247. From Eduard Einstein

[Zurich, after 24 June 1922][1]

Dear Papa,

I'm already stuck in the middle of school again now; but it's not all that bad, we have a substitute teacher who simply keeps us busy, but we aren't learning a thing. Yesterday, e.g., during arithmetic class I proved the Pythagorean theorem all by myself, namely, like this:

The 4 triangles are congruent (4th congruency th.), consequently:

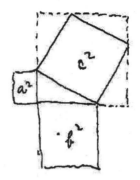

$$c^2 = (a + b)^2 - \langle + \rangle 2ab$$

$$(a + b)^2 = a^2 + 2ab + b^2$$

cancels out and the squares remain. Otherwise there isn't much of interest. The weather is much worse than in Berlin.

Many greetings from your

Teddy.

248. From Mileva Einstein-Marić

[Zurich, after 24 June 1922][1]

D[ear] A[lbert],

I hope my last letter made it to you! Albert would be very happy if he were allowed to travel with you and I would like to grant him this favor so much that I join his plea and again ask you to please take him along.[2] It would be as I suggested to you, that is, that he do his praxis in the summer, certainly not a wasted year for him; and he would surely also profit very much from being with you. So we ask you, please! Mainly also for a prompt reply.

A prof. from Belgrade asked me whether it was true that you plan to deliver lectures in B. for the benefit of starving Russians. Do you know anything about that? From another quarter I happened to hear that you are among those people whom certain elements—I don't know which—are plotting against, like against Rathenau.[3] I'm shocked! Must you really be living there when everything is so unstable and insecure? I found this news horrifying.

Please answer Albert as soon as possible.

K[in]d regards,

M[ileva].

249. From Mathilde Rathenau

[Berlin, after 24 June, 1922][1]

Esteemed Professor,

Among the thousands of letters, there are only a few that I answer personally—let this say more to you than words.[2] My son often told me about his conversations with you; I know what he thought of you and so please retain his memory in its old purity; he is second to none in the whole wide world. Only God knows what he meant to me!

Mathilde Rathenau

250. From Hermann Anschütz-Kaempfe

Kiel, 25 June 1922

Dear, esteemed Professor Einstein,

Many thanks for your letter of the 18th inst.; it is kind of you to occupy yourself so much in thinking about the floating sphere and all that is associated with it. Your suggestions regarding a metallic salt solution and the same metal as the electrode[1] had already seemed viable to us as well but had proved impossible for Cu as well as for Ag; the solution does not stay pure; electrode compounds evidently occur that do not reduce anymore.

Carbon in $H_2O + H_2SO_4$ has been the best by far, the latter c. 0.5%; the liquid stayed clear and unchanged with a current passing through it for months.

Such things are much more easily and productively discussed in person, of course; I do have another interesting question about the surveying-compass readings off the rotating mirror[2] that is mounted on the gyro's axis; this is busying me right now, as Schuler has to keep away from any kind of research for a longer period, for health reasons.[3]

It would be fine and nice if you came with your wife to see us; how about around July 6th? July 8th is anyway the 2nd date scheduled before the Regional Court of Appeal;[4] your presence then would surely have a clarifying and at the same time intimidating effect; for the other side in this matter, you simply are the bogeyman.

And by then I shall surely be able to demonstrate the rubber to you, which is made conductive on the surface with carbon; I think it works.[5]

Today we are all weighed down by Rathenau's murder;[6] I was reminded that you knew him well and prized him very much; how very far extremely exaggerated nationalistic sentiment leads people away from any sense of goodness!

At the moment we have some nice guests here from Austria; but at the beginning of July we shall be alone again; so it would be nice if you could visit us with your wife. After that, all the grandeur here will come to an end: in August the moving van is coming with its mouth agape to devour everything in order to spit it out again in Lautrach;[7] I wonder whether the whale in the Bible wasn't really a moving van?

For the present I and my wife are quietly looking forward to July 6th. Does it suit you and your wife? Cordial regards between households, yours,

Anschütz-Kaempfe.

251. From Paul Epstein

[Pasadena,] 26 June 1922

Highly esteemed Professor,

A few weeks ago your friendly letter, addressed to Millikan, Tolman, and me, arrived[1] and it was a great pleasure for me to receive this ⟨letter⟩ news from you along that route. Tolman is employed at our school; however, he is taking up his professorship only in the coming fall[2] and is currently staying in Washington.

As for me, I have influence on staffing issues only insofar as Millikan asks me for advice. Thus I could not do otherwise than forward your valued letter with my positive support to Millikan, who is currently sojourning in Belgium and France as an exchange professor.

I have meanwhile read that eleven members of the League of Nations Committee on Intellectual Cooperation have been elected.[3] I do not know, now, whether the principle was established only to elect scholars from nations that had joined the League of Nations. But I would regret it if that were the case. It is, of course, very fortunate that by your person the German academic machine has been incorporated into the world organization; for the whole matter would not have much purpose without Germany. But the absence of America does also seem to me to be very regrettable. America has the money and it is made generously available here for scientific purposes. The application of such large funds is not always entirely effective and I believe that it would be very useful if America became imbued with European ideas about the development of science here. There is no doubt that Europe could also gain very much from this collaboration. That is why I would very much welcome it if you would cede the still vacant twelfth position to an American, and specifically, as influential an individual as possible. Since Hale's health is extremely shaky,[4] I do not think that you could find a better man than Millikan, as concerns scientific reputation, practical judgment, and energy. Provided, of course, you need an exact scientist.

We are expecting Michelson here in a few days, who will be occupied with three projects: 1. Continuation of the interferometric determination of stellar diameters. 2. New measurement of the velocity of light according to Fizeau along a light path of c. 70 km.[5] 3. Sagnac's experiment with the Earth as a rotating disk.[6] This last is a product of the not-yet-overcome theory of condensation of the ether near Earth. M[ichelson] says: It is not at all astonishing that in Sagnac's experiment the ether is not dragged along by the air, but it is an entirely different matter when the whole Earth with its enormous mass is rotating, and then the ether could be dragged along. I think it is quite good that he is doing this experiment so that these speculations finally be eliminated for good.

I would have liked to come to Europe for the summer vacation but am prevented from doing so for private reasons. However, I do hope to be in a position next summer to feel the pulse of science.

Are you thinking of going to Holland this year? If so, do please give my most cordial regards to Lorentz and Ehrenfest.[7]

With most respectful greetings, I remain very devotedly yours,

Paul S. Epstein.

252. From Gustave Le Bon

Paris, 27 June 1922

[Not selected for translation.]

253. From Friedrich Sternthal[1]

Berlin, 28 June 1922

[Not selected for translation.]

254. From Emile Borel

Paris, 2[9] June 1922

[Not selected for translation.]

255. To Gustave Le Bon

Berlin, 30 June 1922

Sir,

Thank you for your letter of the 27th of this month.[1] It seems that your relation between mass and energy is irreconcilable with the corresponding conclusion in the theory of relativity because the latter proved that the coefficient of the equation of equivalence is equal to the square of the velocity of light. Concerning the rest of your letter, I must say it does not contain a proof or an argument supporting the relation you claim to be valid. Sincerely,

A. Einstein.

256. From Chaim Weizmann

[London,] W.C. 1, 77 Great Russell Street, 30 June 1922

Highly esteemed Professor,

Much to the dismay of all of us I am compelled to impose on you again in the matter of the funds collected by the New Century Club in Boston on the occasion of our joint visit there.[1] As the management of the Club was causing difficulties about the transfer of the collected sums—according to information I received, of the undersigned $20,000, about $4,000 has been deposited up to now—we suggested, as you will surely recall, that a special treasurer be appointed for the monies collected in America for the university and library, to whom the sums collected in Boston should be entrusted.[2] Unfortunately, as a result of local squabblings, this proposal did not lead to any result. As all of us right now urgently need whatever funds are available to the library for the purchase of some extraordinarily valuable Judaica libraries, above all the large collection of the deceased Baron Guenzburg in Petersburg, one of the rarest and richest Orientalist libraries existing today, which we can now take over at a bargain price, I would like to make another attempt to make the monies designated specifically to library purposes liquid for these purchases.[3] According to the advice of our just recently returned gentlemen from America, a result might perhaps be achieved if we jointly petitioned the management of the Boston Club, because they are unable to reach agreement on the business, to remit the funds directly to a special library account at the local Jewish Colonial Bank to which you and I together should have access. I wrote to this effect in the enclosed letter to the president of the Boston Club and ask you please, if you approve of this proposal, to sign the letter and return it to me for co-signing and forwarding to America.

With amicable regards to your est[eemed] wife, I remain, in hope of perhaps being able to see you again in the fall, yours very sincerely,

Ch. Weizmann

I hear that you are soon embarking on a voyage.[4] Please do write me a line before you leave and wish us luck for the next meeting of the League of Nations here on the 17th/in due time [*cum tempore*],[5] when the mandate for Palestine is supposed to be ratified.[6] I wish you much luck in your travels. Most cordially,

Ch. W[7]

257. To Hermann Anschütz-Kaempfe

[Berlin,] 1 July 1922

Dear Mr. Anschütz,

I am arriving in Kiel on Wednesday evening (5th July) with my wife and am very much looking forward to being able to spend a few days with you. My wife is also royally pleased that for once this wish of hers is being fulfilled. Then I can be present for the hearing on Thursday (as the bogeyman)[1] and also discuss the other problems with you. I misunderstood you about the conduction.[2] You don't want the rubber shell to be made conductive by impregnating the carbon through and through, of course, but only superficially. You will connect this surface layer with a conductor that permeates the rubber layer and then, in isolation, the aluminum sphere. Now I also think that a purely polarizing current in carbon is the kind of electrode that suffers the least modifications. The connection of the carbon layer to the conductor is the only thing that I don't yet clearly see before me, but I can imagine solutions to this problem do exist.

Rathenau's assassination deeply shocked me and generally caused a great stir. It is unfortunately doubtful[3] whether the Reich government will succeed in gaining mastery over all opposing elements. The army seems to be particularly unreliable. The old traditions of contempt for morality—fabricated for purposes of foreign policy—are now taking their toll inside the country. Yesterday I saw a play by Toller; this magnificent person is still sitting in a Bavarian jail.[4] Another sad sign of the times. He is forcibly prevented even from working intellectually. (Oh, nation of poets and thinkers, what has become of you!)[5]

Cordial greetings to you and your wife, yours,

A. Einstein.

258. To Walther Nernst[1]

[Berlin,] 1 July 1922

Dear Colleague,

You asked me about my opinion on conducting funeral services at the university on the occasion of Walther Rathenau's violent death.[2] My opinion is the following: Holding a funeral at the university upon the death of any minister is naturally intrinsically unjustified. Such a ceremony can only be perceived as a statement about political assassinations. Meddling by educational institutions in political affairs is reprehensible. However, ⟨it would not just⟩ here it is a matter of *strengthening moral attitudes in general*, that is, solely the preservation of values that stand above partisan debates. ⟨Silence by the university⟩.[3] The university should, in my opinion, take a clear stance in the sense that it definitely condemns politically motivated murder. (Students and professors should rise to speak.) It should clearly proclaim that assassination, even in the service of politics, is a reprehensible crime and that a humane society in which confidence in a respect for life is missing must necessarily fall apart. I am convinced that such a declaration, if it is resolute and unanimous, will exert an important ⟨and good⟩ influence towards helping public opinion back to health. Silence by the university would, in the current situation, *also* be taken as a statement.

With best regards, yours,

A. Einstein.

259. From Otto Gradenwitz[1]

Heidelberg, 1 July 1922

Highly esteemed Sir,

At Diels's funeral I saw you[2] and recognized you from photographs; and the impression I gained emboldened me to pose the following question:

I have known Philipp Lenard for 13 years; we have been on very frosty terms for almost 5 years. After what I have heard, I too am of the view that the source of his follies is that the incident about Röntgen's discovery of [X-]rays has destabilized him.[3]

The scandal of the last few days[4] had me thinking now: Would it not be fair and square to appease him by having science refer to Röntgen-Lenard rays?

Pardon such a well-meaning question from a complete stranger!

In utmost respect,

Otto Gradenwitz, Prof. of Law.

260. From Hermann Anschütz-Kaempfe

Kiel, 2 July 1922

Dear, esteemed Professor Einstein,

For formality's sake I add the information that the date was shifted once again by two days, therefore climbing to July 10th.[1] Kossel declined the opposing party[2] so now there isn't going to be any neutral expert present at all; the court now wants to rely exclusively on the privately engaged experts. If it should be impossible for you to come, then please do send me a telegram so that we can get a private expert in time; I can imagine that you have more important and better things to do.

The latest I have to report in the area of materials issues is that the tests with the carboniferous rubber are very promising;[3] I obtained a quite excellent product with astonishingly high conductivity—c. 1 Ω per cm^3.

I have also been thinking a lot about the surveying compass;[4] I am now trying to mount a mirror loosely, just by a carrier, onto the axis of the gyroscope, that is, onto one end of the axis, and I assume that during rotation it will have to set itself completely vertically to the gyro's axis; am I right about this, I wonder?

Thus the arrangement is:

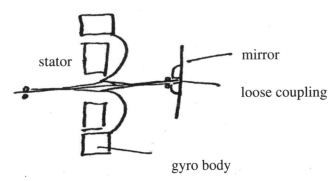

My wife and I are very much looking forward to being able to welcome you personally, and hopefully Mrs. Einstein, at last; we also want to order fine weather so that there is no danger of a chill.

You do know what pleasure you give us when you come. And a good new sailboat is also standing ready for you.

Most cordially from both of us to you and Mrs. Einstein, yours,

Anschütz-Kaempfe.

261. To Richard B. Haldane

[Berlin,] 3 July 1922

Highly esteemed Lord Haldane,

I thank you heartily for sending me your new book and want to tell you that I admire most highly your industrious energy and multifaceted range of thought.[1] What a pity that my inadequate knowledge of English does not let me read it with due precision.[2] Yesterday your young nephew, who is just about to travel to Göttingen, visited me; I had a very good time with that alert young man.[3]

Rathenau's murder is a great misfortune, not just because of the keen gap left behind by this exceptional man.[4] Dangerous domestic conflicts and critical damage to the weak trust abroad are the consequence. Where will this treacherous mental derangement lead us?

Cordial regards to you and your esteemed sister,[5] yours,

A. Einstein.

Eddington's extension of Weyl's theory really is based on a profound idea.[6] Yet I have tried in vain to extend its chain of reasoning so that it acquires the character of a closed theory.

262. To Marie Curie-Skłodowska

Berlin, 4 July 1922

Dear Mme. Curie,

You recently asked me about the committee of the League of Nations and I then informed you that I had accepted the nomination and asked you please to accept likewise.[1] My conception regarding the importance of such an endeavor is not the same as it was at that time. Unfortunately I now feel compelled to resign from the committee again and naturally feel obliged to inform you of this immediately.[2] I have the need to tell you about my reasons but ask you please to see that this does not become known to unauthorized persons. Not only on the occasion of Rathenau's[3] tragic death, but also on other occasions I perceived that very strong anti-Semitism prevails among those I to some extent have to represent at the League of Nations; and generally there is a mentality of a kind that makes me unsuited to be the representing and intermediary person.[4] I think you will surely understand this.

I remember with special pleasure the hours of harmony I spent with you, Langevin, and the other nice colleagues in Paris.[5] I am very particularly grateful to Langevin, whose touching solicitude I shall never forget. Weyl wrote similarly enthusiastically about him to me from Zurich.[6] Please convey to him my cordial greetings and accept my amicable regards to you personally from your.

263. To Eric Drummond

Berlin, 4 July 1922

Highly esteemed Sir,

In consideration of circumstances that became clear to me only since my acceptance of 30 May of this yr., I see myself unfortunately compelled[1] to decline the nomination to the "Committee on Intellectual Cooperation" retroactively, after all.[2] I would not like to fail to take this opportunity, however, also to express my warmest sympathy with your efforts toward nursing international relations back to health.

In great respect,

A. Einstein.

264. To Henry S. Hatfield

[Berlin,] 4 June [July] 1922[1]

Highly esteemed Mr. Hatfield,

I am very sorry for not having been able to answer you before now.[2] I am virtually convinced that your explanation is mistaken. The equipartition theorem of statistical mechanics demands very generally that the kinetic energy per degree of freedom always have the same value at a given ambient temperature. So there is no possibility for assigning to the surface a temperature different from the body's interior. Neither can the anomalies collected in quantum theory change anything here.

Best regards, yours.

265. From Sigmund Einstein[1]

[Baden-Baden,] 4 July 1922

Very esteemed Professor,

As fellow namesake and so humbly
my best regards I send to you;

I would not take this liberty
if I weren't called Einstein too.
At last, that far removed we're neither,
I cannot check the family tree;
our forefathers[2] certainly knew each other,
as all called themselves "Einstein," you see.
I also hark from the Swabians' fair land,
where I spent my youthful days;
I book to my "credit" as a businessman
that an "Einstein" already so much conveyed.
And because thus is named my "great namesake,"
who is ever on the heels of Science,
I scour my own brain in his wake
and greet him in clumsy verse in deference.
You won't be upset and hold it against me
otherwise I would be sincerely sorry;
Why this foolhardiness occurred to me
the dailies say now in a hurry.–
The clan on the right, of German-nationalist C[3]
to these anti-Semites, one and all,
there are too many high-minded Jews
so the greatest now would have to fall.
They don't ask about law and order,
always only have words of gall.
One sees, no means is beneath them,
not even murder gives them pause.
Not men of the world of politics alone
whose lives are targets of this band,
leading Jews are also stoned
who are loyal to Science in this land.

And as you also in the first row
belong among the greatest of the great,
as never before a genius bestowed
to teach the world something new to date,
who was gladly welcomed even in France
without spleen and without contempt,
as his "relative theory" grants
his contribution to international consent.
So, most esteemed Sir, please beware

as *Lenard*'s consorts[4] lie in wait,
guard yourself all-round with care
so we don't lament you as their bait.
If this works and you are safe
from the hateful pack of hounds,
then some use to the world I'll rate
to my joy beyond all bounds.

And if you aren't now on the brink
and do have a little time,
then this "cousin" would be tickled pink
with ever so brief a friendly line.

In respectful devotion,

Sigmund Einstein
Nuremberg, 43 Regensburger St.

266. To Max Planck

Kiel, 6 July 1922

Dear Colleague

This letter is not easy for me to write; but it really does have to be done. I must inform you that I cannot deliver the talk I promised for the Scientists' Convention, despite my earlier definite commitment.[1] For I have been warned by some thoroughly reliable persons (many of them, independently) against staying in Berlin[2] at present and ⟨generally⟩ particularly against making any kind of public appearances in Germany.[3] For, I am supposedly among the group of persons being targeted by nationalist assassins. I have no secure proof, of course; but the prevailing situation now makes it appear thoroughly credible.[4] If it had been an action of substantial professional importance, I would not have let myself be swayed by such motives, but a merely formal act is involved that someone (e.g., Laue)[5] could easily perform in my place.[6] The whole difficulty arises from the fact that newspapers mentioned my name too often and thereby mobilized the riffraff against me. So there is no helping it besides patience and—leaving town. I ask you one thing: Please take this little incident with humor, as I myself do.

   With amicable greetings, yours,

A. Einstein.

267. From Raymond de Rienzi[1]

Paris, 6 July 1922

[Not selected for translation.]

268. From Marie Curie-Skłodowska

Paris, 7 July 1922

Dear Mr. Einstein,

I received your letter, which caused me great disappointment.[1] The reason you offer for your abstention does not seem convincing to me. It is precisely because dangerous and harmful trends of public opinion exist that it is necessary to combat them; and in this regard you can exert an excellent influence *solely by virtue of your personal reputation*, without having to put up a fight for the cause of tolerance.

I believe that your friend Rathenau, whose sad lot I regretted, would have encouraged you at least to try to bring about peaceful international intellectual collaboration. Couldn't you still change your mind?

Your friends here have fond memories of you.

M. Curie.

269. From Gustave Le Bon

29. rue Vignon. Paris. 7 July 1922

[Not selected for translation.]

270. From George Jaffé

Leipzig, 26 Ferd. Rhode St. III, 8 July 1922

Highly esteemed Professor,

Today I take the liberty of informing you about a brief consideration on the "relativity of mass," because under no condition do I want to publish anything on this problem without first having it subjected to your critique.

I start from Schwarzschild's solution into the form:

$$ds^2 = C^2\left(1 - \frac{2m}{r}\right)dx^{0^2} - (dx^{1^2} + dx^{2^2} + dx^{3^2} + (h^2 - 1)dr^2)$$

(1)
$$h^2 = \frac{1}{1 - \dfrac{2m}{r}}, \qquad C \text{ still arbitrary.}$$

Through the substitution:

(2) $x^i = \bar{x}^i \cdot B\bar{r}^{-\varepsilon}, \qquad i = 1, 2, 3, \qquad \varepsilon, B$ constant,

(1) becomes:

(3) $ds^2 =$

$$= C^2\left(1 - \frac{2m}{B\bar{r}^{-\varepsilon}}\right)d\bar{x}^{0^2} - B^2\bar{r}^{-2\varepsilon}\left\{d\bar{x}^{1^2} + d\bar{x}^{2^2} + d\bar{x}^{3^2} + \left[\frac{1}{1 - \dfrac{2m}{B\bar{r}^{1-\varepsilon}}}(1 - \varepsilon)^2 - 1\right]d\bar{r}^2\right\},$$

If I now prescribe that for $\bar{r} = a$ the $g_{\mu\nu}$'s should have their pseudo-Euclidean normal values, then the constants define themselves for

(4) $C^2 = \dfrac{1}{1 - \dfrac{2m}{a}}, \qquad B = a^\varepsilon, \qquad \varepsilon = 1 - \sqrt{1 - \dfrac{2m}{a}}.$

Equation (3) with the values (4) thus represents a metric, in which at infinity the inertia of the masses (small test bodies) converges on 0 and the velocity of light approaches a finite value.[1]

As the simplest "picture of the world" one could address the solution that agrees with (3) and (4) for $r > a$ and with the normal values for $r < a$. Because of the discontinuity, in the differential quotient of g_{00} such a solution would be interpreted as an idealized mass shell. But one could just as well extend solution (3) for a constant $\dfrac{dg_{00}}{dr}$ into a material sphere with the given energy tensor. (I performed the calculation for an incompressible fluid.)

The essential difference from your considerations in your cosmological paper seems to me to lie less in that I leave the space for $r > a$ void of mass than that I give up the constraint $\sqrt{-g} = 1$.[2] Of your reservations, as far as I can see, only the "obliteration objection" [*Verödungseinwand*] remains standing, thus merely one objection supported by statistical considerations.[3] Nor do I place any particular value on the cosmological consequences, as your cylinder world is much more satisfying in this respect; on the contrary, it appears—to me, at least—interesting that with (3) and (4) one has the field of a mass that is the unique mass in the world without filling it up completely.

Now I would like to ask you earnestly to tell me briefly whether you could say you agree with my considerations or whether objections could be raised against them that have perhaps escaped me. If the latter is not the case, I allow myself to pose another question: whether you find the subject suitable for a brief announcement at the Scientists' Convention.[4] Otherwise I could also report about a "theory of the anisotropic radiation field" that I finished just recently.[5]

Requesting you please forgive this inconvenience and in most sincere admiration, I am very truly yours,

George Jaffé

271. From Max von Laue

Zehlendorf, 8 July 1922

Dear Einstein,

Planck sent me your letter from Kiel of the 6th of this mo.[1] and asks me to deliver the talk "relativity theory in physics" in your stead at Leipzig.[2] I can't shirk it, of course, and agree; but only with the idea that at a moment's notice you can always say: I'll do it after all. I'm still hoping that in our fast-moving times the entire situation will have changed enough in the 2½ months until the Scientists' Convention for your misgivings to evaporate.

I had otherwise already interpreted your avoidance of the colloquium and the proseminar[3] in the tenor of your letter, to be precise, from indications Westphal had made to me probably on the basis of official material.[4] I don't need to assure you that I regard the necessity of allowing such caution to govern as an exceedingly depressing sign of the political unruliness we are now seeing in so many quarters.

With warm greetings from household to household, yours,

M. Laue.

Please write me at your convenience about how you were thinking of arranging your talk. I might be able to use it.[5]

272. From Max Planck

Grunewald, Berlin, 8 July 1922.

Dear Colleague,

Your valued letter of the 6th of this mo.[1] struck me like a bolt out of the blue. So the rabble really has pushed it so far that you have to be worried about your personal safety! I'm only thinking in the second place of the loss that our centenary celebration suffers by the cancellation of your talk; although even this, as you can imagine, touches me very deeply. For I cannot regard your participation as a mere formal act. But I shall try to win over Laue as a substitute and thank you for yourself having pointed him out. That might perhaps contribute toward his deciding to take on the speech.

Above all, though, please accept confirmation of what actually goes without saying, that I entirely understand your retraction of your earlier commitment to me and have just one urgent wish, that you will soon be relieved of these utterly intolerable circumstances, which are perhaps less so for you than for your friends.

With warm greetings and in hope of seeing you again soon, yours,

Planck.

273. From Gilbert Murray

Zatscombe, Boar's Hill, Oxford, July 10th 1922.

[See documentary edition for the English letter.]

274. To Henri Barbusse

Berlin, 11 July 1922

Esteemed Mr. Barbusse,

You asked me in your letter of 8 May of this y[ea]r to tell you something about my Parisian stay.[1] The days in Paris are among my finest experiences that I shall always remember with pleasure and gratitude.[2] My fellow colleagues in Paris received me like an old friend, without the reserve I would have had to expect in the current domineering sentiment of political nationalism. Shared activities and shared interests immediately chased away the shades of the past. When we were pleasantly sitting together, we sometimes also spoke about political affairs. The fine thing about it was that I did not perceive any hatred or flush of victory, instead far more sorrow and trepidation. As regards assessing the causal dependence of the World War and the current political situation, in France (just as in Germany) a uniform view exists, characteristic, so to speak, of the country, that is taken with certainty and honesty as the only one possible. The discrepancy between the dominant views in the different countries relates less to a statement of the specific facts than to their evaluation. I expect little moral recovery in the two countries from this immersion into that sad past and the discussions about it. Collaboration between the two countries to restore the areas of destruction appears to me to be much more important.

Fruitful collaboration presumes trust and trust only flourishes from a fostering of personal relations. From this point of view my being invited by the faculty of the Collège de France signifies one first courageous step that will hopefully be followed by others on both sides.

I greatly regretted not having made your personal acquaintance. Your picture hangs near my desk beside that of my dear mother.[3]

With cordial regards, yours.

275. To Marie Curie-Skłodowska

Berlin, 11 July 1922

Dear Mme. Curie,

I can understand why you do not agree with my decision, indeed that you find it incomprehensible.[1] But you do not know the situation here sufficiently. There is indescribable anti-Semitism among intellectuals here that is particularly intensified in that, first, Jews generally play a disproportionately large role in public life compared to their numbers, and second, many of them (such as, e.g., I) are pursuing international goals. That is why, from a purely objective point of view, a Jew is unsuited to serve as a connecting link between the German and international intelligentsia. A man who has close and untarnished relations with the German intelligentsia should be chosen, who is regarded as a "real German." (I am thinking of men like Harnack or Planck,[2] naturally, without wanting to presume to make any suggestions in this regard.)

I draw from the above-described opinion *sine ira et studio*[3] the full consequences in that I decided to resign my position at the Academy as noiselessly as possible as well as the directorship of the Kaiser Wilhelm Institute of Physics and to settle down somewhere as a private individual. I cannot stay in Berlin in any event because there are indications of attempts on my life by the ultranationalists.[4] Whether it is really true is, of course, difficult to judge. In any case, I am using this situation to remove myself from clamorous Berlin, to which much trouble is connected for me in particular, in order to be able to work in peace again. The material prerequisites for this do exist for me.[5]

All this I tell you because I owe it to you. But please do not tell anyone besides Langevin[6] about it; otherwise unfavorable consequences would arise.

With cordial greetings, yours.

276. To Hermann Anschütz-Kaempfe

[Berlin,] 12 July 1922.

Dear Mr. Anschütz,

That was a fine week full of happy hopes in Kiel in your fairy-tale house.[1] The prospect of a downright normal, natural life in tranquility, connected with the welcome practical employment opportunity in the factory, enchants me.[2] Add to that the wonderful countryside, sailing—enviable. We unfortunately do have to abstain from purchasing the romantic Villa Esmarch, however. The Kiel citizens would perceive the purchase of such a historically so heavily encumbered building[3] by a Jew as a provocative act and take revenge on me somehow; if there's a will there is always another way. It is my firm conviction that purchasing the villa would lead to serious complications. In such unsettled times people are generally strange. A friend of ours was telling us what a huge fuss a socialist paper in Kiel made about the black-white-red flag on your little boat ("murderer's flag").[4] The climate in Kiel seems to make the people a little stormy, too . . . Among the masses one sometimes feels as if one is in the midst of a buffalo herd. Individually they are not mean-spirited, but one does have to take care not to get trampled.

Do not worry about our residence. In the worse case my wife will stay in Berlin until something suitable can be found. I suddenly changed my vacation plans this year and am going to see my boys in Zurich for a month.[5] But I will make the boys' mouths water about your heartfelt invitation to come to Kiel.

Cordial regards to you and your wife, yours,

A. Einstein.

277. To Debendra Nath Bannerjea

[Berlin,] 12 July 1922

Dear Sir,

⟨Unfortunately⟩ I cordially thank you for your kind letter.[1] Unfortunately I cannot receive you here in B[erli]n sometime in the coming weeks because I will be staying in Zurich during my children's school break (from about July 17th on). If you happen to be near that city and have some time available, I propose you send me a note to the address 3 Büchner St., Zurich. I unfortunately have to inform you that for reasons of domestic policy I saw myself induced to resign from the International Committee of Intellectual Cooperation, even though the goal pursued by it is close to my heart.[2] I became convinced that local intellectuals would not actually perceive me as their representative so it would be thoroughly in the interest of the cause if a personality from Germany more suitable in that role were elected to the c[ommittee].

With utmost respect, sincerely yours.

278. To Max von Laue

Berlin, 12 July 1922

Dear Laue,

I'm very grateful to you for having so readily taken on the talk in Leipzig, and therefore, I leave it definitively and confidently in your hands.[1] Now I have another request the fulfillment of which I hope will be amenable to you. I'm leaving in October, you know, for God knows how long and it is necessary that someone else in the meantime substitute for me as director of the K.-W.-Institute. My wish, now, is that you kindly take on this substitution from October 1st on for an indefinite period. Of course I would transfer to you the earnings connected with this post.

As I am, as it were, officially already away from Berlin, I'm not coming to the meeting tomorrow[2] and ask you, in the event of your approval, to inform the board of directors about the above-mentioned as a petition of mine. I also ask you please to declare that I approve of the grant of 40,000 marks to Messrs. Kallmann and Knipping for the purchase of the Hoffmann electrometer.[3]

With best wishes for your vacation, yours,

A. Einstein.

279. To Max Planck

Berlin, 12 July 1922

Dear Colleague,

Your letter, which I found upon my return, pleased me exceedingly, especially since I could justifiably have expected reproaches.[1] To my great joy Laue has already accepted, so no problems are arising from my belated retraction.[2] The English expedition of 1919 is ultimately to blame for this whole misery, by which the great crowd seized possession of me.[3] Ever since, I have become a kind of flag that various sorts of interests parade about. So it is no wonder if one is sometimes overcome by an illusory yearning for a "quiet nook"! I addressed a petition to the Academy to temporarily cease paying out my salary from October 1st on, because I am leaving on an extended trip.[4] I also asked Laue, conditional on the approval by the directorate of the K. -W. -Institute of Physics, to assume the director's post provisionally. I am not going to attend any more meetings because I am officially out of town and then really am going to travel at the end of this week to see my boys in Zurich, in order to spend the period of their summer vacation together with them.

Heartily wishing you a nice vacation, regards from your

P. S. I have now retroactively declined the representation on the League of Nations committee for the organization of intellectual endeavors after all for the reason[5] that under the prevailing conditions I am not the suitable person for this func[tion].[6]

280. To the Prussian Academy of Sciences

Berlin, den 12 July 1922

[Not selected for translation.]

281. To Pierre Comert[1]

[Berlin, between 12 and 19 July 1922][2]

Mr. Struck sent me a copy of the letter to you concerning my retraction.[3] In spirit, his letter agrees with our discussion.[4] But he did unconsciously exaggerate a few points of my report. My joining the committee did not give rise to any criticism that reached my ears; rather, I recognized the circumstances described in the letter along a more indirect route. The situation over here is such that a Jew is better served[5] by acting with restraint in all public things. I also have to admit that I do not feel like representing those who would certainly not elect me as their representative and with whom I do not concur on the issues concerned in the present case.[6] There can be no question of any direct animosity toward me by local intellectuals. Here my friend Struck's temperament ran away with him.

With great respect.

282. From Bernardo Attolico[1]

Geneva, 12 July 1922

[Not selected for translation.]

283. From Sanehiko Yamamoto

No. 1 1-chome Atogoshita-cho, Shibaku, Tokyo, Japan [between 12 July and 8 August 1922][1]

[Not selected for translation.]

284. To Otto Gradenwitz

Berlin, 13 July 1922

Esteemed Colleague,

The history of the discovery of X-rays is not known to me precisely enough for me to be able to pass reliable judgment on the subject.[1] I know that Mr. Lenard[2] raises some priority claims in this regard but do not know any physicist who finds them legitimate. In my view, such priority issues should generally be given little significance, especially considering that the factor of chance plays such an important role in the history of discoveries.

Thank you for sending me the pretty little card with the true Italian gesture.

Very respectfully.

285. To Gustave Le Bon

Berlin, 13 July 1922

Sir,–

Given my difficulty in procuring your original works, I am not able to form a clear idea for myself of your method.[1] To elucidate the question, please contact Prof. Langevin[2] (Collège de France) personally. I perfectly rely on his judgment.

Very respectfully,

A. Einstein.

286. To Gilbert Murray[1]

Berlin, 13 July 1922

Highly esteemed Sir,

My decision to withdraw again from the Committee on Intellectual Cooperation[2] certainly did not come about out of a lack of interest in the international goals. On the contrary, I know that my resignation can only serve these goals. I became firmly persuaded that I am not a suitable representative of German intellectuals because I was not regarded by them fully as *their* representative. My avowed internationalism, my Swiss citizenship, my Jewish nationality, taken together, produce the effect that, in political matters, would not procure the necessary measure of trust, which a representative of a nation must possess, in order to serve successfully as a liaison. I feel very definitely that the majority of German intellectuals would not have the feeling of being really represented on the League of Nations committee if I sat on the committee. When I accepted the nomination, the situation was not as clear to me as now. In expressing my conviction that upon adequate knowledge of the German circumstances you will understand and approve of my motives, with earnest wishes for fruitful work by the committee and generally all work in the service of international exchanges, I am very sincerely yours.

287. From Richard B. Haldane

Westminster, 28 Queen Anne's Gate, 14 July 1922

Highly esteemed and dear Professor,

It pleased me very much to receive your letter of the 3rd of this month.[1] And also that you were so kind to my nephew, who is now in Göttingen.[2]

Weyl writes me that he has a philosophical supplement to his book in hand. It will be interesting to see whether he has shed some light on the difficult problem of the significance of his conception.[3]

With best regards to your gracious wife, most devotedly yours,

Haldane

288. From Gerhard Kowalewski[1]

Dresden, 14 July 1922

[Not selected for translation.]

289. To George Jaffé

Berlin, 15 July 1922

Dear Colleague,

Through mere transformations, of course, nothing substantially new is gained, because only the quantity ds determines the behavior of the clocks and measuring rods. You can also see that the spatially infinite does not become a singularity, e.g., from the fact that the quantity g_{00} that defines the behavior of the clocks does not become singular but has the limit C.[1] This argument is perhaps more cogent than observations about the inertia of masses, because in order to assess them one has to put the equation of motion into Newtonian form, which necessarily adds a certain arbitrariness in the interpretation.[2] One would have to interpret $m\dfrac{dx_\nu}{ds}$ as the mo-

mentum, according to the equations of motion. Hence, for the case of rest

$$\frac{m}{C\sqrt{1-\dfrac{2}{B\tilde{r}^{1-\varepsilon}}}}\frac{dx_\nu}{dx_0}\,\nu = 1-3\,,\text{ therefore, in spacial infinity the quantity }\frac{m}{C}\text{ as the}$$

mass, which is a finite limit.

Amicable regards, yours.

290. From Hermann Anschütz-Kaempfe

Hamburg, 15 July 1922

Dear, esteemed Professor Einstein,

I am writing you from Hamburg where we are staying with my wife's[1] parents for Sunday.

Too bad that nothing will come of the sleepy house with the wild garden for you;[2] my opinion is that you are painting things too black; Kiel residents—the well-to-do ones, I mean—surely would not find anything objectionable about the purchase, and among the working class, which is quite heavily represented here, there would certainly not be anything to worry about from the elements you pointed out. My workers in the factory, for ex., are so attached to me that it really is not saying too much if I declare I am perfectly sure of them; they are always on my side, even on matters that do not agree with the union. The announcement about the so-c[alled] murderer flag came after the fact, it had not been there just a couple of days before; the article evidently originated from the left-wing radicals at the wharf, presumably from the Reichswerft; my workers keep well away from that sort of business.[3]

And once again, too bad that you cannot come and visit us with your boys. Please do tell them that we are expecting them in Lautrach[4] in the fall; they should inform us in time; our address is, as of September: town of Lautrach (near Memmingen), manor house.

Then there is also the happy news that the suit was won as brilliantly in the second instance as in the first; a good part of it is due to your contribution.[5]

I am now looking elsewhere for a nice dreamy house; I shall find something, all right; there still is time before next summer; maybe it isn't all that senseless as an interim arrangement, considering that your wife is finding it so hard to separate herself from Berlin, if we furnish you with very romantic temporary quarters over here.[6] And both of us hope that you will choose Lautrach as a meeting place with your boys.

Yesterday, a 3-phase motor, engaged at a very large ohmic resistance in the 3rd phase in the running transformer, stepped up from $0-20,000$; thus this problem has surely been solved to satisfaction; if one lets the transformer step up slowly, it works even better, of course;[7] now I am completely occupied with the mirror on the compass axis.[8] I look forward to the time when I, quite naturally, can come to you with my technical questions and worries; then I will enjoy my work twice as much.

Most cordial greetings from both of us to you and your wife, your forever true

Anschütz.

291. From Heinrich Zangger

[Zurich, between 15 and 25 July 1922][1]

[Not selected for translation.]

292. To Hermann Anschütz-Kaempfe

[Berlin,] 16 July 1922

Dear Mr. Anschütz,

Upon calm reflection I do think it is better to continue living here in Berlin.[1] My wife is quite ailing (recently) and she dreads the change to all her habits. She does not feel up to managing a house anymore.[2] Regarding me personally, I fear that after my return from Japan next spring[3] there will be little left for me to do at the factory. If the case should be otherwise after all, which would please me very much, 1 or 2 rooms in or near the factory would suffice for me; I would then be able to spend a substantial portion of the year there, e.g., coming to Kiel for a while at regular intervals. This would, of course, be possible without great complications. Furthermore, I would be glad if I could participate usefully in your work until the 1st of October. You may ridicule me for my fickleness, all right; I deserve it!

My boys will be coming here, after all,[4] but a little later on because Albert is indisposed. So nothing will probably come of a trip to Kiel with the boys, especially as your wife must also have some rest before the exertions of the move.[5] Mr. Licht informed me with jubilation about the suit victory;[6] the rascal[7] did not come through with his crafty tricks.

My Lisa (sailboat) is still lying, completely drunk, on the shore and is waiting for the untrustworthy shipbuilder, who made fun of us today, on top of it all, to take pity on her.[8] It's good that the boys don't encounter her in this disrespectful state.

Still full of memories from Kiel, cordial regards to you and your wife,

A. Einstein.

Dear Mr. Anschütz,

The good man just shifts the blame for the changed opinion regarding resettlement onto his wife. This is not entirely accurate. He reconsidered the matter in peace here and arrived at this conclusion. My husband has been strongly affected these past few weeks by Rathenau's murder (he loved him); the affair touched him very deeply and he just had the feeling: get away from here to work in tranquillity. I think he realizes that this thing with tranquillity is an illusion. He cannot duck out of sight better anywhere than here in Berlin; in a small town he is on a platter, he does not trouble himself about the people there, but they do about him. The main point of attraction in Kiel would have been you, Mr. Anschütz, and you are leaving! He is pleased about working in the factory, but he does believe that in about half a year the business will operate on its own and that his staying there would then not be absolutely necessary. He will come as often as you call him, and he comes so *willingly! Please don't ever* forget this. After the trip to Japan, he wants to give up his official positions over here; he should do so. Warm greetings to you and your sweet young wife, yours,

Elsa Einstein.

293. To Maurice Solovine

[Berlin,] 16 July 1922

Dear Solo!

Attached are the contracts.[1] I shall make the minor changes in the booklet, likewise shall write to Beck.[2] I am glad that you are now managing to go on your trip and will see your mother again, at last.[3] These are unsettled times here since the abhorrent murder of Rathenau. I, too, am constantly being warned, have abandoned my lecture and am officially absent, while being, in truth, here.[4] Anti-Semitism is very strong. The endless chicanery by the Entente will ultimately be blamed, yet again—on the Jews. There are complaints about major dirty tricks in industry; the destruction of manufacturing plants under the cloak of military usefulness.

Cordial regards and have a very good time, yours,

A. Einstein.

The Painlevé is interesting but what he says about relativity will scarcely be able to hold.[5]

294. To Chaim Weizmann

Berlin, 17 July 1922

Dear and esteemed Mr. Weizmann,

Enclosed I return to you the signed letter to Boston. Let's hope it will lead to the goal.[1] I hope likewise, with trepidation, that you will now finally succeed in bringing the mandate safely home.[2]

With regard to the article you want from me, goodwill is certainly not lacking. But I cannot figure out what the book's purpose would be and thus feel incapable of meeting your request if I do not receive any suitable information.[3] I am here in Berlin until October 1st.

With cordial greetings and my wishes for success in your difficult task, yours,

A. Einstein.

295. From George Jaffé

Leipzig, 26 Ferd. Rhode St. III, 17 July 1922

Highly esteemed Professor,

Today I received your nice letter.[1] I heartily thank you for having taken the trouble to reflect on my considerations. The more do I regret that I have to bother you again now; but I do not want to cast my musket aside before all my gunpowder has been spent.

I had contemplated precisely those two points that you argue as closely as my powers permit and therefore am so bold as to send you carbon copies of two papers in which I attempt to come to some clarity. The first is appearing in the *Physikalische Zeitschrift*,[2] the second is a draft of a presentation for the Scientists' Convention.

It was entirely clear to me that nothing substantially new can be gained by transformation; but it seems to me that through transformation one can turn a ds^2 that is constrained by special limiting conditions (or calculation procedures) back into a more general form, in which it is conformable to other limiting conditions. I chose the limiting conditions $g_{00} = C$, all the others $g_{\mu\nu} = 0$; I can just as well achieve that $g_{00} = \infty$ and all the others become $g_{\mu\nu} = 0$.[3] For this I only need to choose (in the notation of my preceding letter)[4] the second solution:

$$\ast) \qquad C^2 = \frac{1}{1 + \dfrac{2m}{a}}, \qquad B = -a^\varepsilon, \qquad \varepsilon = 1 + \sqrt{1 + \frac{2m}{a}}.$$

You demand singularity for the measurement of time.[5] It seemed to me that singularity in the measurement of space suffices, insofar as only the inertia of the masses vanishes; that is why I did not go into solution ($\ast$) at all, but I can still mend that.

Regarding the inertia of the masses, this point appeared to me important enough to make a statement about it in a preliminary note (paper I).[6] I would definitely like to believe that *only* the covariant components of the momentum are decisive;[7] I did not mention anything about this in my foregoing letter because I believed I was following along entirely in your footsteps.

Please forgive me for taking up so much of your time; but you will understand how important it is to me for my interpretation to appear acceptable to you. If a very short visit by me is more agreeable to you than reading these somewhat long-winded manuscripts, I would return to the kind invitation you made earlier and come and see you in person. I could arrange to come to Berlin for a few hours on the coming Saturday, Sunday, or Monday, but it goes without saying, only if such a visit and its purpose would not be inconvenient for you.

In assuring you of my most sincere admiration, I am yours very truly,

George Jaffé.

296. From Gilbert Murray

Yatscombe, Boar's Hill, Oxford. July 17th 1922.

Dear Professor Einstein,

Many thanks for your letter of the 13th. I quite understand your difficulties.[1] At the same time I would venture to make one observation which possibly may have weight with you. The Committee, as I understand it, is not intended to represent national points of view. It consists of individuals chosen for their own qualifications from various nations. I do not think it at all likely, for example, that the British universities would select me, and I am sure the Indians would not select Bannerjea.[2] Indeed, a committee like ours can only get to work satisfactorily if all the members are a good deal more international in their outlook than the average people of their country, intellectual or otherwise.

Pray forgive my pressing you again.[3]

Yours sincerely,

Gilbert Murray

Translator's note: Original written in English.

297. From Richard Eisenmann

Berlin N 24, 130 Friedrich St., 18 July 1922

Highly esteemed Professor,

Allow me to express my thanks to you again for the honor of your visit and examination of my apparatus.

No words need describe the gratification and satisfaction it gives me that they gained your approval.

You wanted to be so kind as to give me a written report and opinion. For this I permit myself to submit the facts to you.

The problem I posed for myself consisted in creating a device that makes feasible the option of having piano tones continue to resound and swell as long as you like, with or without the hammer mechanism, by which each tone sounds immediately and subsides rapidly. I wanted to make the forte-piano into a real piano-forte. The way I found the solution was to set the steel strings into constant resonant oscillations exactly synchronous with their eigenfrequencies by means of a pulsating current from electromagnets.

In the beginning I effected the synchronous current interruptions by means of tuning forks, tongues, or strings, but actually only theoretically. In 1889 I changed over to microphones, which had just been introduced at that time, which I redesigned as contact breakers for my purposes. Although I obtained a very simple apparatus that way, it was insufficient. Only isolated tones indicated what could and must be achievable. One exemplar built this way is set up in the Deutsches Museum in Munich.

For the rest, the apparatus was—and still is—composed of the contact-switch levers lying on top of the keys and a pedal that when pressed initially switches the current on and as it is pressed harder switches off resistances to intensify the current. The key-switches work likewise. It required very much thought and trials to design this to the satisfaction of the sensitive student pianist. Now he neither sees, nor hears, nor above all, feels it anymore.

A decade ago I decided to use a rotating breaker instead of the inadequate microphones. Before I took the step of carrying this out, I viewed various systems in which motors are employed. At the Telegraph Testing Station I saw a very finely crafted, neat engine system standing in the cellar while the experimenter had all the switch levers and meters right next to his seat on the second story. So I first designed the apparatus with this idea: Each of the 84 tones requires its own, dedicated, specifically designated breaker number. Besides that, tuning must be possible. Each of the 12 tones (and their octaves) therefore requires its special breaker. Its exact adjustment is provided by a motor-driven cone that is fed with minimal friction. The lowest A requires 27.5 oscillations per second = 1,650 per minute. The breaker has a diameter of 6.5 centimeters. At 1,650 revolutions, a cone of the equally small diameter of 6.5 cm therefore suffices. The cone I currently have has a diameter of 14 to 28 cm because the first breakers had a diameter of 13 cm.[1] I likewise superfluously have a 3 hp motor while one of ½ hp suffices, as it does not have to work much beyond the minor friction of the 12 breakers and setting the pace. Thus in its manufacture the entire apparatus is reduced to proportions slender enough to also be able to fit comfortably in a rear room of a middle-class apartment. Correctly adjusted spring action can prevent any shaking and loud noises. In order to achieve the goal of generating a fine, pure tone according to the method described here, I had to bridge a difficult obstacle. No engine runs absolutely constantly. And for my purposes not a single revolution more than planned is allowed. Otherwise the tones would immediately be out of tune. As there wasn't any suitable regulator, I invented the pendulum regulator, which forces the motor to rotate with the precision of a chronometer.[2] You had a look at this solution for yourself and it won your applause. Until this point, since the very first idea, I did everything without the least outside assistance, intellectual or material.

Now the research is far enough along for a man with capital and initiative to apply it commercially, whereby I place at his disposal my thirty-six years of experience. The regulator apparatus, owing to its simplicity and accuracy, will be easily introduced. And I am convinced that even the entire keyboard apparatus will very quickly develop into a necessity in musical life, once it has been demonstrated publicly a few times, as Arthur Schnabel once played it in my home.[3] The human ear quickly becomes accustomed to it. The new tone melds the fullness and volume of an organ-like but metallic orchestral sound to the coldness of the keyboard, just as Franz von Liszt himself had already wished.[4]

In utmost respect, I sign as yours truly,

Dr. Richard Eisenmann.

298. From Hermann Anschütz-Kaempfe

Kiel, 19 July 1922

Dear, esteemed Professor Einstein,

I find very sensible your suggestion of a small retreat near the factory, if possible in the romantic corner of the garden;[1] the more I think about it, the more it appeals to me; we saw in Hamburg, in the villa quarter where my wife's parents live,[2] a very quaint little cottage; it is nicknamed Diogenes's Tub and is full of flower boxes and has a low-slung roof that looks very homey.

It would, of course, be nice of you if you came here a little before our departure; there are still so many things I would like to show you in the laboratory. And if you happened to bring your boys along, you would be doing my wife a real favor as well. Nothing needs to be done inside the house itself until August 8th; and we will be shirking all the effort of the move off onto the packers;[3] in any case, my wife is energetically protesting the idea that she has to rest beforehand already: And the currants and gooseberries are just ripe for your boys and there will surely be fine weather at last for sailing as well.

And tipsy Lisa[4] can then be made properly ship-shape again during your stay here and will still be waiting when you return to the hut.

And please, dispel your wife's doubts about your being able to come yet again; it is, I believe, the first time that no more-or-less amusing job is waiting for you; the finer and freer the time here will be, just occasionally spiced with a couple of finicky technical problems that luckily entertain you.

I have to keep the last page for my wife, otherwise she'll scold.[5] So, for now, hoping to see you very soon, warmest regards to you and your wife, yours,

Anschütz-Kaempfe

299. From Gustave Le Bon

Paris, le 19 July 1922

[Not selected for translation.]

300. From Peter Pringsheim

Berlin W., 63 Lützow St., 19 July 1922

Dear Professor Einstein,

During a visit that Franck[1] paid us a couple of days ago in Berlin we discussed "our" problem more thoroughly and I think we arrived at some results.[2] At that time we tried to reach you by telephone; but as we learned that you are away[3] and I do not know when I shall have an opportunity to talk to you again, I would like to write you a little about it before I forget the matter and then first have to pick my way through again. The main point seems to me to be, if all indications are not deceiving us, that the experiment concerned again does not lead to a decision between instantaneous or longer-lasting emission, instead that not only the "position effect"[4] but also the "Lorentz effect" (broadening through [cutting off] of the wave train or through phase shift)[5] can likewise occur in instantaneous emission but then, of course, another explanation is needed. First of all, it is an experimental fact that some spectrum lines that exhibit virtually no Stark effect (and the position effect probably ought to be interpreted as such) are very strongly broadened by a rise in pressure.[6] Theoretically, however, two reasons could lead to a line broadening even for timeless electron transitions from an increase in collisions. The first is already treated in a publication by Franck, as yet unfamiliar to me, from the Festschrift issue of the Kaiser Wilhelm Society.[7] It is, of course, conceivable that a collision could make the natural value of an electron's mean lingering period on the excited orbit not be attained, but that the electron suddenly jumps back again and in so doing emits light. These, so to speak, forced acts of emission are, of course, always greatly disturbed; the more frequent they become, the more must a line broadening consequently manifest itself. A second possible consideration is the following: Collisions of excited molecules should always extinguish the luminosity in that the energy is always expended in a different way. Now, one can surely generally assume that, even excluding the Doppler effect, external disturbances, etc., each line has a finite breadth; it is classically calculated from the ⟨radiation⟩ damping; the greater the damping, therefore, the shorter the emissive process, the broader the line. Something similar may also exist for quanta and, I believe, was once even assumed by Bohr: The quantum orbits are not absolutely precise, they have a certain width that matches a corresponding spectrum-line width. The stability of the electron on the excited orbit and hence its lingering period is less, the greater the scattering is against the actual precise quantum orbit. So, again, the largest broadening corresponds to the fastest decay; whereas the transitions corresponding to the true center of the line, on average, occur the slowest. Thus, if every collision of an excited atom leads to an extinction of the luminosity, then the middle of the line is mainly weakened and an (apparent) line broadening is the consequence. I do not know if this second thought makes any real sense. The first one, though, does have some experimental backing that Franck also already mentions in a publication.[8] If one excites Na-vapor to fluorescence just with the D_1-line, then just the D_1-line occurs in the resonance radiation as well. If one raises the pressure by adding hydrogen, then the D_2-line also appears in the emission. Evidently, collisions between excited Na atoms and H_2 molecules shift electrons from the D_1-orbit to the D_2-orbit (the opposite for primary excitation with D_2 works just the same, incidentally). One could just as well imagine that as a result of such a collision the electron is thrown down directly onto the normal orbit (1.5 S-orbit)—the question now is whether it occurs while undergoing light emission; but if so, then surely under disturbed conditions.–

I think this is essentially what we agreed on. It would certainly be nice of you if you would let me know sometime what you think of it. That the luminosity can be completely extinguished by collisions is also proved, by the way, precisely for Hg resonance, which vanishes at Hg pressures that are too high. Surely the basic concept that must absolutely be adhered to is that whatever is all right for the emission process must also be valid for absorption. On the contrary, it is questionable how large the percentage of such destructive collisions is compared to the collisions evincing emission or by conveying the energy to another similarly excitable atom (Klein-Rosseland).[9] I hope I have not written too unclearly, either in the literal or figurative sense. With best regards,

Peter Pringsheim.

I just discovered in an older paper by Wood some information[10]—if completely correct—perhaps not without some interest: (1) By directing 1 atm. air to cold Hg-vapor, the absorption line is strongly weakened in the center but is broadened overall enough for the total absorption to remain constant. (2) 30 mm air already almost completely suppresses emission of the resonance line, i.e., most of the excited atoms experience the extinguishing collision before their lingering period has elapsed. (3) However, 30 mm air does not really reduce the width of the absorption line, that is, the absorption of a beam of resonance rays sent through the Hg-vapor cell is weakened just the same as if no air were there (hence with reemission) but with 30 mm air (without reemission). Does this mean: the timeless process of absorption is not noticeably [. . .ed] by the light molecules, which do not yet affect the configuration much; likewise, of course, for the individual process of emission itself (the emission line has not been measured but is surely just as little broadened), only that often mostly becomes impossible before it occurs, owing to the lifetime which is long compared to the emission process and not short anymore compared to the mean collision time? Perhaps all the earlier questionable points are consequently moot?

P. P.

301. From Hermann Struck

Königstein i. T., Falkensteinerweg 6, 19 July 1922

[Not selected for translation.]

302. To George Jaffé

Berlin, 22 July 1922

Dear Colleague,

To your letter of the 17th of this mo.[1] I would like to reply the following:

1) If one upholds that in the general theory of relativity

$$\frac{\partial}{\partial x_\nu}(T_\mu^\nu + t_\mu^\nu) = 0 \qquad (T_\mu^\nu + t_\mu^\nu) = A_\mu^\nu$$

should be the expression for the conservation laws[2] (which one definitely must), then

$$\int A_\mu^4 dx_1 dx_2 dx_3 = I_\mu$$

should be regarded as the expression for the energy momentum[3] of a finite, Galilean embedded system.[4]

Permitting only linear transformations, I_μ is a covariant vector, as

$$\frac{\partial}{\partial x_\nu}(T_\mu^\nu + t_\mu^\nu) \text{ is a tensor density (of first order)}$$

$$\frac{\partial}{\partial x_\nu}(T_\mu^\nu + t_\mu^\nu)dx_1 dx_2 dx_3 dx_4 \text{ is a first-order tensor (covariant vector)}$$

Likewise the integral. Through integration it follows that

$\left. |I_\mu| \right|_{t_1}^{t_2}$ is a covariant vector, hence also I_μ itself.

For the case of a mass point, therefore,

$$I_\mu = m_0 g_{\mu\alpha}\frac{dx_\alpha}{ds}$$

should be set, as you did. (Comp[are], however, what is said under 6.)

2) It agrees with this if in the static case you regard $m_{i\kappa} = \dfrac{m_0 \gamma_{i\kappa}}{\sqrt{g_{44} - \left(\dfrac{ds}{dx}\right)^2}}$

as the expression for the inertia of the test mass.[5] For your solution this expression disappears into infinity. The velocity of light measured in the coordinate scale also extends into infinity in all directions toward the value ∞.

3) Even so, I cannot view your consideration as a satisfactory solution to the cosmological problem at all, because the limiting condition $g_{44} = \infty$ has to be retained; otherwise one could not speak of a degeneration of the metric continuum at an infinite distance from the matter.[6] A clock at infinity would simply have to run infinitely rapidly, because the clock is slowed by neighboring matter and its running speed should be determined solely by the matter.[7] A spring with a mass attached is a clock that must operate faster the farther away it is from the bodies. For, what should a finite limiting velocity be determined by, if inertia is ultimately a kind of interaction? The incompatibility between this condition and the fact of small stellar velocities that demand a constant g_{44} at the limit constitutes one of my arguments for the necessity of spatial closure for the universe.[8]

4) You will also be able to persuade yourself of the inadmissibility of your way out in that you can move from the matter-free space $g_{\mu\nu} = \text{const.}$ $(ds^2 = dt^2 - dx_1^2 - dx_2^2 - dx_3^2)$ [9] by mere singularity-free transformations to an expression that behaves in infinity like yours.

For that, in your argument you need

$$x^i = \bar{x}^i B \bar{r}^{-\varepsilon} \quad [10]$$

only to be as continuously dependent on $[\bar{r}]$ so that it vanishes for small $\bar{r}$'s and continuously reaches the value ε for increasing $\bar{r}$'s. One thus does not need any masses to make the space behave as you prescribe. So you certainly cannot view this behavior as an expression of the relativity of inertia.

5) The relativity of inertia, or a linkage between the (quasi-static) metric field and the existence of matter can only be achieved for a spatially finite, closed world.[11]

6) I don't want to uphold the remark I made in my earlier letter about the inertial mass for the reasons mentioned.[12] One could do so, however, in that the equation for the geodetic line be interpreted in the Newtonian manner on the justification that the concept of inertial mass actually only makes clear sense from the standpoint of the Newtonian approach and that *for this approach* there would therefore be little justification for defining inertial mass from the momentum concept of the general theory of relativity.[13]

I do not think it necessary for you to come here personally, especially since I myself [am] not in Berlin and my two sons are visiting.[14] I think we can communicate very well by letter.

Kind regards, yours.

303. To Erich Marx-Weinbaum[1]

Berlin, 22 July 1922

Dear Mr. Marx,

I remember you very well. May I say a candid word to you? I observe with great displeasure imprudent statements about my endangerment figuring in our Jewish press.[2] I went to the bottom of it and saw that people in fact do not know anything. A certain danger may well exist, as for all Jews of whom public awareness has taken possession. But it is clear that such notices only magnify the animosity against me and with it the danger. For the time being I make do with keeping away from all things that occupy the German public.[3] A change of residence is not necessarily a protection because a half-wit and a revolver can be found anywhere.

With kind regards to you and my cousin, to whom I send my best congratulations for her passed exams,[4] I am yours.

304. To Wolfgang Ostwald[1]

Berlin, 22 July 1922

Esteemed Colleague,

It goes without saying that the principal issues of colloid chemistry interest me most keenly; and it vexes me to have to assume a negative stance before you.[2] The deeper reason why I do not want to bite the bait is that I see serious dangers in such a fine specialization of scientific activity and that I know very well, besides, that I do not have the time to keep abreast of the meetings and publications of a society seeking such specialized goals. The fragmentation of my work is already far too advanced as it is without my being able to stave it off successfully.

Hoping that you appreciate and approve of this standpoint dictated by a kind of inner plight, I am very sincerely yours.

305. From Chenzu Wei

Berlin, 22 July 1922

Esteemed Professor,

With reference to your very valued letter of May 3rd of this yr., I have the honor of informing you herewith that I just received a telegram from the president of the Imperial University in Peking,[1] whom I had contacted regarding the conditions you had proposed,[2] the content of which is that the university gladly accepts your conditions.[3]

The latter would accordingly be the following: an honorarium of 1,000 American dollars, payment of the travel costs Tokyo–Peking and Peking–Hong Kong for you and your wife, as well as hotel costs in Peking. At the same time the university authorities express their joy at being permitted to welcome you in Peking.

In utmost respect, I am very sincerely yours,

Wei Chenzu
Chinese envoy.

306. To Hermann Anschütz-Kaempfe

[Berlin,] 25 July 1922

Dear Mr. Anschütz,

I find the Diogenes Tub plan wonderful.[1] But it really should be small, too, to suit its name. I'm as pleased as a child about it. The boys are here and are lodging in my Spandau Castle.[2] I am commuting back and forth between the city apartment and the castle which, contrary to my yacht, is proving to be watertight. The latter will soon be seaworthy, now;[3] I don't know whether we should baptize it "dupe" [*Reinfall*] or "big man" [*Grossmann*].[4] We unfortunately cannot accept your kind invitation because my wife is quite seriously ill. Albert tells me that there's no time between the end of the school-leaving exams and the beginning of the semester at the Polytechnic. But I think he can skip one or two weeks to be able to visit you in Lautrach without placing his technical salvation in jeopardy.[5]

From the second to last letter of yours I see that you decided on a third ⟨contact⟩ electrode on the sphere now, after all, in order to ensure the start-up;[6] or did I misunderstand your comment? Perhaps in order to ensure the start-up of the ⟨anchor⟩

rotor at full frequency it would also suffice if it were given a little more resistance. Maybe this could be done without increasing the energy consumption beyond the permitted level.

Many regards to you and your wife,[7] also from my wife and from my boys. Yours,

A. Einstein.

307. To Sigmund Einstein[1]

Berlin, 25 July 1922

Esteemed Mr. Namesake

Outrageous injustice it would be
this doggerel answerless to leave![2]
Especially to repair the error—
not everyone's called Einstein, either.
You probably originate from Buchau,
just like my old man,[3] rest his soul.
So many friends explained to me
my corpus in great danger be,
heroes prowling about with a plan[4]
to turn the tap off of my life lamp.
It may well be. But unfortunately
the likes of us, shameless free gypsies,
cannot stand lofty protection.
Rather pitch my tent in hid location,
perhaps even in a foreign country;
of which, praise be God, there are aplenty.
Now, my namesake, thanks for the favor;
moreover, a paper lifesaver.

(sig.) A. Einstein.

308. To Gerhard Kowalewski

Berlin, 25 July 1922

Dear Colleague,

I was not yet able to read the booklet by your brother but hope to be able to catch up on it soon.[1]

Mr. Dember is not familiar to me from his papers;[2] if you have them sent to me, I shall be happy to evaluate them. Mr. Gans in La Plata is, in my opinion, one of the most prominent and versatile German physicists.[3] He conducted theoretical and experimental research under difficult conditions in entirely unscientific surroundings mainly in the field of magnetism. Other excellent German experimental physicists I would also like to name are Wagner in Munich[4] (precision measurements on x-rays) and Füchtbauer in Tübingen,[5] who to my knowledge still have no professorships.

Pardon the brevity owing to my oppressive load of correspondence and kind regards to you from your.

309. To Gilbert Murray

Berlin, 25 July 1922

Highly esteemed Mr. Murray,

I thoroughly appreciate the reasons you indicated. However, even if elected members from the various countries were not viewed as direct representatives of their countries, they still must act as psychological links between the Committee and the individual countries.[1] Now, my situation is, as I already wrote in the last letter,[2] that through my Swiss citizenship, my activities in Jewish affairs and my Jewish nationality in general, and because of earlier political statements, I am being regarded by the guild of local intellectuals as so alien that one absolutely could not feel that Germany was being represented de facto on the Committee. This alienation goes so far that it even made me feel obliged to cancel a talk already arranged for the centennial jubilee of the German scientists' association in Leipzig.[3] Under such conditions, apart from the already mentioned objective reasons, it is understandable from my personal perspective that I do not feel like accepting the role envisioned for me, which is, after all, naturally perceived as a kind of representation of German intellectuals.

Hoping that you comprehend and grant me my point of view, I am, in utmost respect, very sincerely yours.

310. From George Jaffé

Leipzig, 26 Ferd. Rhode Str. III, 26 July 1922

Highly esteemed Professor,

First of all I would most heartily like to thank you for having taken the trouble to study both my manuscripts and to refute my consideration so thoroughly. Now, afterwards,[1] I most acutely regret having imposed upon your time so inordinately.

I cannot evade the conclusiveness of your arguments and now see that the condition $g_{44} = \infty$ has to be set.[2] I must admit, however, that something unsatisfying remains, because although the formal condition $m_{ik} = 0$ makes the inertia of the masses vanish, it does not account for the relativity of the masses. One would probably have to "prohibit" the use of coordinate systems in which the determinant (g) vanishes—even if only at infinity.

As you were so kind as to delve into my considerations, I would like to submit to you another argument that—if another error has not crept in—would at least show that one can make the relativity of masses agree formally with the gravitation equations of the first kind.

I proceed from the spherically symmetric solution in the form:

$$1) \qquad ds^2 = -\left(1 + \frac{m}{2r}\right)^4 [dx^{1^2} + dx^{2^2} + dx^{3^2}] + \frac{\left(1 - \frac{m}{2r}\right)^2}{\left(1 + \frac{m}{2r}\right)^2} dx^{4^2}$$

(H. Weyl, *Ann. d. Phys.*, 54, p. 132;[3] in the Pauli Enc., page 730, formula 421^b, there is a misprint[4]) and transform it by the substitution:

$$x^i = \frac{\bar{x}^i}{4}, \qquad i = 1, 2, 3, \qquad t = \frac{2}{\varepsilon}\bar{t}, \qquad \varepsilon = 1 - \frac{2m}{a} > 0.$$

Thus we get

$$2) \qquad ds^2 = -\left(\frac{1 + \frac{2m}{r}}{2}\right)^2 [dx^{1^2} + dx^{2^2} + dx^{3^2}] + \left(\frac{2}{\varepsilon}\right)^2 \frac{\left(1 - \frac{2m}{r}\right)^2}{\left(1 + \frac{2m}{r}\right)^2} dx^{4^2}.$$

Now I consider the solution that results for $r > a$ with (2) and for $r < a$ agrees with the continuously attained result

$$3) \qquad ds^2 = -\left(\frac{2 - \varepsilon}{2}\right)^2 [dx^{1^2} + dx^{2^2} + dx^{3^2}] + \left(\frac{2}{2 - \varepsilon}\right)^2 dt^{4^2}.$$

It describes the field of a hollow massive sphere at resting mass m and radius a, where the velocity of light (measured in natural units of length and cosmic time) has the large value $\frac{2}{\varepsilon}$ at infinity if ε is small.

If one now lets ε go to the limit 0, one obtains in the limiting case pseudo-Euclidean conditions within the interior and in the exterior space Euclidean metrics, infinite velocity of light, and vanishing inertia of the masses! It seems particularly remarkable to me that at the limiting case $\varepsilon = 0$ the velocity of light rises to ∞ immediately outside the hollow sphere and that would remain so if one exchanged the hollow sphere for a massive sphere with a finite spatial density. The critical value $\frac{2m}{a} = 1$, according to this consideration, would be the one that makes the velocity of light and the inertia in the sphere's interior finite *without the influence of other masses.*

I originally started out with considerations of this kind, because I wanted to show that the Lorentz contraction (just like oblation in rotations) can be regarded as the effect of distant masses. I later moved to the attempt you know, because I wanted to smooth the overly abrupt transition at $r = a$. Now it almost seems as if the earlier consideration had greater legitimacy!

May I also ask you to send the two carbon copies back to me, at your convenience (there is absolutely no hurry!)? The second must make its way into the wastepaper basket.

In asking you again to forgive me for the disturbance and in utmost admiration and gratitude, I am yours very truly,

George Jaffé.

311. To Richard Eisenmann

Berlin, 27 July 1922

Highly esteemed Doctor Eisenmann,

After looking at your device for sustained stimulation of piano sounds and having admired the work you achieved, I must write you briefly to give you my view on it.[1]

1) The tones generated by your device have an extraordinary artistic attraction. Your method ought to lead to a valuable enrichment of the means of musical expression.

2) With your device you overcame all the principal problems standing in the way of electromagnetic stimulation of strings in such a way that practical construction of this device should not pose serious difficulty in the hands of a skilled design engineer.

3) Your main accomplishment, which seems to me to be of considerable technical importance, even disregarding the problem of sound production, is the solution to the problem of regulating the rotational speed of a rotating motor by a contact pendulum in such a way that the regularity of its motion absolutely corresponds to that of the pendulum.[2] You thus manage to have the strings stimulated at exactly

the right frequency without need for any feedback by the string motions to the current-producing mechanism. I believe that your regulating method could be applied with success in disparate fields of precision measurement.

With utmost respect,

A. Einstein.

312. To Hendrik K. de Haas[1]

Berlin, 27 July 1922

Highly esteemed Colleague,

I fully appreciate the great honor and favor that lies behind the invitation you relayed to me by the Battaafsch Genootschap[2] and the Rotterdam Naturkundig Genootschap.[3] To my great regret I cannot accept this invitation for the following reason:

The Society of [German] Scientists and Physicians has its annual convention in September in Leipzig, which is simultaneously the memorial celebration of this society's 100th anniversary. It had already invited me a year ago to deliver the keynote speech. I had to cancel this speech, for reasons of personal security, as a consequence of the agitated local conditions that everyone became aware of through Rathenau's assassination.[4] It would now, without a doubt, be perceived among the ranks of German scientists as an injury, indeed, a virtual affront if I accepted the keynote speech for an analogous event abroad scheduled for roughly the same time. In asking you to please give this reason full acknowledgment and to convey to the directorate my cordial thanks for the received invitation, I am very sincerely yours.

313. From Chaim Weizmann

27 July 1922

Dear Professor Einstein,

Many thanks for your letter of the 17th inst.[1] with reference to your article or introductory note for the *New Palestine*. I explained that this is a volume dealing with the problems of the new Palestine, and I thought it would be appropriate if you would be good enough to contribute a few words about the university. Its effect on the world at large; what you said in America, for example.

I should be very glad if you would be good enough to write something on these lines.

Kindest regards, yours very sincerely,

Ch. W.

Translator's note: Original written in English.

314. To Eric Drummond

Berlin, 29 July 1922

Highly esteemed Sir,

The necessity of settling a number of urgent matters before my imminent trip to Japan makes it, to my great regret, impossible for me to appear for the first conference of the Committee on Intellectual Cooperation in Geneva.[1] In this situation I am nonetheless consoled by the circumstance that this inability to participate at the meeting will make any loss seem that much less acute, owing to my being prevented by the above-mentioned trip from continuous collaboration for the coming half year. However, I do hope to repair this omission all the more diligently after this period has come to an end. In expressing the hope that the Committee will be able to accomplish much during this first half year, I remain, in assuring you of my great esteem,

A. Einstein.

315. "Quantum Theoretical Comments on the Experiment of Stern and Gerlach"

[*Einstein and Ehrenfest 1922*]

Manuscript completed before 30 July 1922.[1]

DATED May–June 1922
RECEIVED 21 August 1922
PUBLISHED 16 September 1922

IN: *Zeitschrift für Physik* 11 (1922): 31–34.

§ 1. O. Stern and W. Gerlach[1] allowed a vaporized beam of silver atoms to fly [p. 31] through a magnetic field in order to determine whether the atoms have a magnetic moment and—if so—what orientation they manifest while traversing the magnetic field. Their experiment gives us a very important result: while traversing the field the magnetic moments of all atoms coincide with the direction of the lines of force,

[1] *Z[eit]s[chrift] für Phys.* 9, 349, 1922.[2]

and specifically, for about half of the atoms, in line with the field; for the other half, opposing it.[3] The question of how the atoms arrive at this orientation naturally poses itself.

§ 2. It is, above all, observable that atoms do not experience any collisions as they enter the deflective magnetic field—the last collisions they experience are in the vaporizing chamber of the small melting furnace.[4]

We initially ask ourselves how magnetic atoms change their orientation at all under the influence of a magnetic field. As long as one ignores the emission and absorption of radiation, the collisions, or other similar influences, the atoms execute a precession (Larmor rotation) in the magnetic field around the direction of the field.[5] If the direction of the field changes slowly compared to the rapidity of the precessional motion, then the angle of the precessional motion remains unchanged.[6] Accordingly, an adjustment to the tilts required by quantum theory (O and π for the silver atom from the experiment by Stern and Gerlach)[7] cannot take place without external influences of the type of radiation or collisions.

§ 3. The most obvious explanation for the experimental finding seems at first instance to be that the atoms' adjustments take place upon entry in the electromagnet's field, namely through an exchange of radiation. Then not only a release of energy is necessary but also an intake of energy from the radiation field, namely the latter for the atoms that set themselves antiparallel to the force lines. How quickly [p. 32] does the repositioning of the atoms' moments now happen under the influence of the radiation (from room temperature)?—The required time is relatively reliably estimable for the case of transitions from quantum state to quantum state. For we know that in cases of this kind this time for the transition of an assemblage of atoms agrees—at least in order of magnitude—with that of a corresponding classical model. In our case of a precessing atom with a magnetic moment this would be a radiating magnetic dipole with its conic rotation. The adjustment time (for a field strength of 10,000 gauss) would be of the order of 10^{11} sec, provided the *emission* of the precessive motion alone was effective. But if one takes into account the influence of the surrounding room-temperature radiation ["*positive and negative radiation influx*"[(1)]], it shortens to about 10^9 sec.[8]

[p. 32]   (1) Comp. A. Einstein, Zur Quantentheorie der Strahlung. *Phys. Z[eit]s[chrift]* **18**, 121, 1917, § 2.[10]

Thus these are times of an order of magnitude that at least do not come into consideration at all for the experiment because there the adjustment must take place in a time of less than 10^{-4} sec.[9]

§ 4. If one tries to find a way out of this dilemma, two alternative assumptions initially present themselves:

A. The real mechanism is such that atoms never can get into a state in which they are not *fully* quantized.

B. For rapid influx, states result that violate the quantum rule regarding orientation; the adjustments demanded by the quantum rule are produced by emission and absorption with a reaction speed that is extraordinarily much greater than for transitions from quantum state to quantum state.

A decision a priori between these two alternatives does not currently appear to be possible; yet it is appropriate to take a clear look at their principal difference and the characteristic problems that each of them leads to.

§ 5. *Discussion of Alternative A.* 1. What it requires is particularly well exemplified by the Stern-Gerlach experiment: In the vaporizing chamber of the small melting furnace, each silver atom is fully quantized after each collision, so its magnetic axis is oriented according to the magnetic field prevailing there, weak though it may be. After its last collision during its flight through the various parts of the field, [p. 33] its orientation constantly conforms to the field direction of the location concerned.[(1)]

2. In the process, one portion of the (one-quantum) moments will be set parallel, another antiparallel to the field, where the statistical distribution will be dominated by the temperature and the field strength in the small furnace's vaporizing chamber and certainly not by the (radiation) temperature and field strength in the space that continues to be traversed!

3. One would accordingly have to decide to assume the following: Even weak fields must be decisive to the orientation immediately after the collision (i.e., the effect of very strong fields). In changes, e.g., of the direction of the magnetic field, which are arbitrarily quick compared to the Larmor rotation, the atom's magnetic axis should therefore follow the direction of the field as fully as for arbitrarily slow changes. More generally: For an arbitrarily rapid change in the external conditions of a mechanical system, this system would have to adjust itself to the same final state as for an infinitely slow (adiabatic) execution of the change in the external

---

[(1)] Dr. G. Breit already suggested such an assumption on the occasion of a discussion at the [p. 33] physics colloquium at Leyden.[11]

conditions. That this calls for a violation of the mechanical equations can be easily demonstrated with examples.[1]

§ 6. *Discussion of Alternative B*. 1. For the Stern-Gerlach experiment the following picture would result: In the small furnace's vaporizing chamber an atom's magnetic axis is oriented arbitrarily with respect to the weak field existing there immediately after each collision. The orientation takes place through infrared radiation, that is, through emission and positive and negative induced radiation and a parallel and antiparallel adjustment to the field. Thereby the precondition is essential that such transitions from *nonquantum* states to quantum states correspond to [p. 34] probabilities of a much higher order of magnitude[2] than for transitions from *quantum* states to quantum states. After the last collision, the orientation of the axis aligns itself quasi-adiabatically to the changing field directions as it flies through the various parts of the field, whereby each of the very small angular defections occurring are compensated by an extremely weak exchange of radiation of a highly infrared frequency (very much more infrared than the precession frequency).

2. The static distribution between parallel and antiparallel orientations to the field would also in this case be essentially determined by the temperature and field strength in the small melting furnace!

3. According to alternative B, a monatomic vapor whose atoms carry a magnetic moment would emit and absorb within the magnetic field on the long-wave side of the frequency of the precessive motion; thus in a suitable magnetic field within the range of electric waves.

4. It is characteristic of alternative B to make conformance to the quantum states dependent on the possibility of radiant absorption and emission. So a principal distinction is made between purely mechanical systems and ones capable of radiation. For ex., the axis of rotation of a symmetrical gravitational gyroscope could only attain a quantum adjustment with respect to the gravitational field if the gyroscope is suitably electrically charged. If one wanted to extend hypothesis B about the

[p. 33]   [1] One somewhat fictitious example: An adiabatic shortening of the string length of a gravitational pendulum is known to change the frequency ν and the energy ε conformly in such a way that the quantum rule stays satisfied. On the contrary, if one shortens the string length quickly, e.g., at the vertical position, the ν becomes larger while, according to mechanics, energy is not added. Alternative A therefore requires a supply of work that is mechanically inconceivable.— Second example: A magnetic atom in a weak magnetic field. During an infinitely slow rotation of the field (infinitely slow compared to the precession velocity), pursuant to the laws of mechanics the atom's magnetic axis follows the direction of the field. If this should happen likewise upon rapid alteration of the field's direction, then there is a mechanically inconceivable change in the angular momentum.

[p. 34]   [2] Corresponding to an adjustment time of 10^{-4} instead of 10^9 sec.

adjustment respecting orientation fully to the adjustment into quantum states in general, i.e., therefore, e.g., also only allow in the case of suitable electric charges a spontaneous adjustment to the quantum orbits for the oscillations of a crystal lattice and the rotations of a molecule, then one would come into evident conflict with experience regarding the specific heats, e.g., of diamond and gaseous H_2.

§ 7. The enumerated problems show how unsatisfactory both interpretation attempts discussed here are for the results found by Stern and Gerlach. Bohr's idea—that absolutely no definite quantization takes place in more complicated fields—has been left out of the discussion.

Leyden-Berlin, May–June 1922.

316. From Paul Ehrenfest

[Leyden,] 30 July 1922

Dear, dear Einstein,

Since seeing you, I was in Göttingen (Bohr);[1] then Joffe was here with me (under my whip, finished writing his paper, thank God, that had been lying around for 15 years).[2]– Then my American brother was visiting for 12 days with (very nice!) wife and eldest daughter (very nice, *clever* girl, but floating completely on superficiality).[3] Just today they are in Berlin in transit, Hamburg–Berlin–Vienna.– My brother is a talented, competent fellow—looks *immensely* similar to me (looks just as old as me although 10 years older) but he's an "instant-future" person, through and through—very very uncontemplative. Very hands-on, kindhearted, helpful.—Well, that was a brisk whirlwind blowing through our house.– As I'm supposed to see [him] again in Vienna (with my 3 other brothers[4]–), I had to rest a little and therefore did not accompany him to Hamburg and Berlin.

When I picked him up in Antwerp, I saw the old Flemish paintings there—particularly Rogier van der Weyden and Quentin Matheys left an enormous impression on me!!– A (self?)-portrait of Rogier van der Weyden should be hanging among the Kaufmann collection in Berlin.—I saw it in reproduction.—That's *magnificent*. Do go and look at it![5]

I plan to travel the day after tomorrow (1 August) via Frankfurt to Vienna.– From there I'll send the manuscript on "Stern–Gerlach"—which is ready for *press*, *to you* for publication in *Zschr. d. Physik*.– I am sending it to you because I added 5 lines right at the end that would take a weight off my chest, but that you perhaps want to discard. Please let it stand if you ever can bring yourself to do so. The rest of the entire manuscript is *exactly* as we had discussed.[6]

Our [Au]-ion sample is more concentrated still—circa 2.5×10^{-3} because first Methuen sent circa 1.1×10^{-3} grams and we could obtain 1.0×10^{-3} more here as well. We won't be able to produce any more here for half a year, however, because we first have to obtain the raw material from Japan, as you know.[7]

If you happen to have photos of little Albert and Eduard, I would like to get them.– Give my best regards to both the boys and tell them about us in such a way that when we see them later they will meet us with fondness!

At the energetic advice of doctors and especially also of my brother, my wife finally decided—provided it is possible—to place Wassik in an institution. Everyone is unfortunately unanimous that he is a member of a very well-known type of idiot ("mongoloid") that between the ages of 5 and 15 still possesses some capacity for development if appropriately treated. In favorable cases the final state then corresponds to the intelligence of a normal 5-year old child.– Whether it's possible to place him (we are corresponding at the moment with a developmental school near Jena) and whether it's not yet too late to save my wife's health, I do not know.[8]

Warm regards to your wife—Ilse—Margotkins [*Margoterl*] from me!– I'm afraid I'm not going to be seeing you again for a very long time now.[9] *You* don't need anybody—but I need you very much! Yours,

P. Ehrenfest.

317. "In Memoriam Walther Rathenau"

[*Einstein 1922i*]

PUBLISHED August 1922

IN: *Neue Rundschau* 33 (1922): 815–816.

[p. 815]   My feelings about Rathenau were—and are—those of delighted admiration and gratitude for his giving me hope and consolation in the current dismal situation of

Europe and for his sharing with me unforgettable hours as a prescient and warm person.[1] His purview of the grand economic correlations, his psychological appreciation of the peculiarities of nations, of all classes of people, his knowledge of individual persons was admirable. And he loved everyone, even though he knew them, as one who has the strength to say yes to this life. A priceless mixture of earnestness and genuine Berlin humor made his speech a unique pleasure when he chatted at table among friends. It is easy enough to be an idealist when one is living in Cloud-Cuckoo-Land; but he was an idealist even though he was living on Earth and knew its odor as others rarely do.

I regretted that he became a minister. Given the attitude of a large majority of the educated class in Germany toward Jews, it is my conviction that proud reserve by Jews in public life would be the natural thing.[2] Still, I would not have thought that [p. 816] hate, delusion, and ingratitude would go so far. To those, however, who have been guiding the ethical upbringing of the German people in the last fifty years,[3] I would like to call out: By their fruits ye shall know them.

Albert Einstein.

318. "On the Present Crisis of Theoretical Physics"[1]

[*Einstein 1922o*]

DATED August 1922

IN: *Kaizo* 4, no. 12 (December 1922): 1–8.

The goal of theoretical physics is to create a logical system of concepts based on [p. 1] the fewest possible mutually independent hypotheses, allowing a causal understanding of the entire complex of physical processes.[2] In response to the question of how this scientific system developed and grew, in the time before Maxwell one would have been able to give the following answer.

One first grounds the immovable basis of all exact science on uncontested facts of human observation or of human thought, namely on geometry and analysis. The Greeks already did this, so those coming afterwards had little left to do of principal novelty besides developing infinitesimal calculus. Then came the establishment of the actual fundamental laws of physics, i.e., the fundamental laws of mechanics by Galileo, Newton, and their contemporaries. Until around the end of the 19th century, physicists were convinced that these fundamental laws of mechanics would have to be the basis of absolutely all theoretical physics, i.e., that every physical theory would ultimately have to be based on mechanics.

So the basis of physics seemed to have been finally erected and a theoretical physicist's work seemed to be to conform the theory through specialization and differentiation to the ever-growing abundance of investigated phenomena. No one thought about the possibility that the foundations of physics would have to be changed. Then, as a consequence of Faraday's and Maxwell's researches, it gradually became obvious that the foundations of mechanics failed for electromagnetic phenomena. This change took place in a number of stages. Initially both the above-named pioneers realized that electromagnetic processes cannot be represented with the theory of unmediated forces by action-at-a-distance.

[p. 2]

According to Newton, all forces that can produce accelerations of material points are attributable to an instantaneous influence that the rest of the material points individually exert on the observed material point. As an alternative to this theory of unmediated action-at-a-distance, Maxwell and Faraday posited the theory of electromagnetic fields. According to this theory, all dissemination of electric force is based not on instantaneous action-at-a-distance but on a propagation-conducive state of space or the ether: the electromagnetic field.[3] The energy-endowed field, describable by continuous space functions, assumed physical reality alongside the movable mass point that according to Newton's theory had figured as the sole carrier of energy. It is known that Hertz helped bring about general recognition of this conception through his experiments on the propagation of electric force.[4]

At first, the extent of the revolutionary character of field theory was not fully realized.[5] Maxwell himself was still convinced that electrodynamic processes were conceivable as processes of motion by the ether; he even used mechanics to forge ahead to the field equations. In the interim, however, it became more and more clear that it was impossible to derive the electromagnetic equations from mechanical ones. In search of a unified basis for physics one was eventually compelled to try the opposite and derive the mechanical equations from electromagnetic ones.[6] This attempt suggested itself even more after J. J. Thomson realized that an electromagnetic inertia existed for electrically charged bodies and after M. Abraham showed that the inertia of electrons was interpretable purely

[p. 3] electromagnetically.[7] With the attribution of inertia to electromagnetic processes, a complete revolution in the foundations of physics had come to pass, at least in principle. Instead of the mass point as the ultimate of reality, the electromagnetic field entered as a fundamental elementary building block of theoretical physics. A theoretical construction of matter on a purely electromagnetic basis is generally known to have succeeded to a certain degree. In particular, we know today that the cohesive forces are of a purely electromagnetic nature.

These are not all the fruits of the Faraday-Maxwellian field theory. Recognition of the covariance of Maxwell's electromagnetic equations with Lorentz transformations led to the special theory of relativity and hence to an awareness of the equivalence between inertia and energy;[8] the extension of field theory to gravitation, taking into account the identity of inertial and ponderous mass, led to the general theory of relativity. With the general theory of relativity one pillar of Newton's theory sank away that had hitherto always been believed to be one of the necessary foundations of all science, namely, Euclidean geometry.[9] It had emerged in early ages from primitive experiments on solids and had been silently assumed by physicists to be an exactly pertinent law for the orientation of solid bodies of even temperature not subjected to external influences; but now, by reason of important considerations based indirectly on experiment, a doctrine already proposed by Gauss and Riemann had to replace it. With the general theory of relativity the developmental phase of theoretical physics founded by Faraday and Maxwell seems to have come to a close.[10]

In the last two decades it has been recognized that even a basis for physics characterized by the Faraday-Maxwellian field theory cannot hold up to experience any   [p. 4] better than the mechanics grounded on it. It is rather to be expected that scientific progress will demand a fundamental change no less profound than the one we have summarized under the name "field theory." As we are still far away from a logically clear foundation, however, we must content ourselves here with showing how the foundation has thus far proven inadequate and how well account has been taken of important groups of physical phenomena through successful yet still groping attempts subsumed under the term "quantum theory."

Quantum theory found its origins in the theory of thermal radiation, for which a unification of mechanics and electromagnetic field theory delivers a law irreconcilable with experience and even intrinsically irrational. The fundamental problem of heat radiation can be formulated as follows. Thermodynamics teaches that in the interior of a cavity surrounded by opaque bodies at temperature T, there is a radiation whose composition is entirely independent of the nature of the bodies forming the cavity's walls. If ρ is the monochromatic radiation density, i.e., $\rho\,dv$ is the radiation energy in the cavity per unit volume whose frequency lies between v and $v + dv$, then ρ is a very specific function of v and T. It is not determinable by purely thermodynamic considerations; its derivation rather presupposes an insight into the nature of the process of generation and absorption of the radiation:

Classical mechanics connected with Maxwellian electrodynamics yields for ρ an expression of the form:

$$\rho = \alpha v^2 T, \dots \tag{1}$$

[p. 5]   which cannot be right because it produces an infinitely large value for the total den-

sity of the cavity radiation $\int_{v=0}^{v=\infty} \rho \, dv$. Planck found the expression corroborating all

foregoing observations, in conformance with experience:[11]

$$\rho = \frac{8\pi h v^3}{c^3} \cdot \frac{1}{e^{\frac{hv}{kT}} - 1}, \dots \tag{2}$$

where k means a constant connected to the absolute size of atoms and h a natural constant hitherto unknown in physics, which one may well describe as the fundamental constant of quantum theory. In 1900 Planck offered a theory for this formula that implicitly contains a hypothesis irreconcilable with physics up to that time, which we can retrospectively interpret, supported by the experimental and theoretical research of the past two decades. Wherever a sine-type oscillating process of frequency v exists in nature, its energy always carries an integral multiple of hv; intermediate energy values do not occur in nature for sine-type oscillating processes.

On the basis of this hypothesis it was possible to derive correctly not only Planck's formula (2) of heat radiation but also the law of the specific heats of crystalline solids.[12] But these derivations are all intrinsically contradictory: While making use of this new hypothesis they always rely on the foundations of classical physics, which is not compatible with it.

Considering all the major advances that Maxwell's electrodynamics and Newton's mechanics have made in physics and their present indispensability, one is obliged to doubt the basic hypothesis of quantum theory as much as possible. But phenomena do exist that directly confirm quantum theory even though its incompatibility with the foundations of classical physics is immediately clear.

[p. 6]   The energy density of radiation emitted from a radiating source diminishes, according to Maxwell's theory, with the reciprocal square of the distance. The energy available at one place for processes of absorption per unit time would thus have to diminish infinitely with distance. As, for ex., the chemical decomposition of a molecule or the release of an electron from an atom requires a definite amount of energy, similarly, radiation that has been sufficiently weakened by propagation away from the light source should no longer be capable of generating such a chemical process. Experience shows, on the contrary,[13] that the chemical and photo-electric effectiveness of the radiation is completely independent of its density; the total chemical action of radiation that is permeated by matter is only dependent on its total energy but not at all on its spatial energy density. Experiments by E. Warburg have furthermore demonstrated that the energy absorbed per chemical elementary process is always equal to $h\nu$, independent of the spatial energy of the radiation.[14] This result also follows from experiments on the photoelectric effect and on the generation of cathode rays from X-rays.

We know today that this energy really does originate from the radiation and that it is not, for example, gradually accumulated. The absorption of light constitutes indivisible elementary processes, during each one of which the energy $h\nu$ is completely transformed. We know nothing about the details of such an elementary process. If just the energetic properties of radiation were known, we would then see ourselves obliged to postulate a kind of molecular theory of radiation of the kind of Newton's emission theory of light.[15] But finding explanations for diffraction and interference processes poses insurmountable obstacles. Moreover, it should probably be kept in mind that the field theory of radiation is no more false than the [p. 7] theory of elastic waves in solids, which establishes their thermal content; for, both theories collide to the same degree with the quantum relation and must be combined with the latter in the same way in order to arrive at an appropriate interpretation of the results of experience.

The grand development of our knowledge of the composition of atoms, which we essentially owe to the great masters Rutherford and Bohr, has led to a highly significant generalization of the quantum rule, which we now want to consider. Even prior to the Rutherford-Bohr theory,[16] the assumption was that absorption or emission of a spectral line would have to correspond to a transition of an atom or molecule from *one* preferred state into another. As the states of the elementary forms were certainly not interpretable as sine-type oscillations, the problem arose of how to extend quantum theory to mechanical systems of more general character, which Bohr's, Sommerfeld's, Epstein's, and Schwarzschild's investigations have succeeded in doing, step by step.[17] The results obtained by these researchers are being elevated to secure findings through precise confirmation in the field of spectroscopy.

If a mechanical system is describable by coordinates q_ν, which experience cyclic changes over time, and if for each degree of freedom ν the momentum p_ν $\left(= \dfrac{\partial L(q \cdot p)}{\partial q_\nu}\right)$ belonging to q_ν is describable as a function of exclusively the one coordinate q_ν, then for any ν the integral over one cycle $\int p_\nu dq_\nu$ is an integral multiple of Planck's constant h. Thus the "permissible" states according to quantum theory are defined for so-called "quasi-periodic" mechanical systems. This general rule has proven its value in so very subtle and varied special cases (e.g., Epstein's theory of the Stark effect)[18] that its general correctness has become quite probable.

[p. 8]   From the general theoretical point of view, the following is especially puzzling. On one hand—as already pointed out—mechanics does not appear to be generally valid, as statistical mechanics, which is based on it, leads to results that contradict experience (e.g., specific heats of solids). On the other hand, the mechanical laws most astonishingly stand the test within the scope of validity of the above rule. Are there only supposed to be quasi-periodic elementary processes in nature or more generally only such mechanical systems as have as many complete integrals as degrees of freedom? Such a thought seems absurd in view of the kinetic theory of gases. The problem of how the validity of classical mechanics (and electrodynamics) is restricted by the demands of quantization is shrouded in as deep darkness today as it was 15 years ago.

It has often been noted that in the present state of our knowledge, the possibility of representing natural laws by differential equations appears doubtful.[19] Indeed, according to the quantum rule just indicated, a complete cycle of the system must be considered in order to be able to judge whether a particular state of the system is quantum-theoretically permissible or not. In order to really do justice to the quantum relations, a new mathematical language seems to be necessary; in any case it seems preposterous to express the laws through a combination of differential laws and integral conditions as we do today. Once again the foundations of theoretical physics are shaken and experience is calling for the expression of a higher level of lawfulness. When will the saving idea be granted us? Happy are those who may live to see it.

319. From Hantaro Nagaoka

Science College, Imperial University, Tokyo, Japan, 2 August 1922

Highly esteemed Colleague,

Your valued letter of the 22nd of May pleased me greatly,[1] because your arrival in Tokyo is already beyond doubt and your hope of engaging in closer relations with scholars of the Far East has become clear. Sadly, I fear that you will ultimately say: "There are no scholars in the Far East!"

News about your oriental voyage has much roused the interest of the Japanese public; I have often been asked to give popular talks about the principle of relativity.[2] Apparently people have a great curiosity about the concept of space and time, but misinterpretations occur very frequently. The publication of one book about the principle of relativity by a mechanic from a battleship wharf is remarkable; he has mastered the principle quite well and it seems as if the book is addressed to specialists.[3] Against this brighter side there is also a more shady side. One graduate of the university misunderstood the principle and wrote a polemical text that is printed in the proceedings of a scholarly society.[4] The confused state of the younger scientific world is not so easily improved; it is useless to go on about it.

The members of the Academy are unanimous in extending their welcome and wholehearted hospitality.

If you come here in November, the chrysanthemum blossoms will be gone but the fall foliage, especially the maples, will be brightly colored in a manner unique to this island kingdom; the contrast of variegated painted fields against the blue sky is delightful in that season; hopefully you will have a good opportunity to enjoy the beautiful countryside.

In expectation of your arrival and in expression of my utmost respect, I remain your most devoted

H. Nagaoka

320. From Max Born

Göttingen, 6 August 1922

Dear Einstein,

A physicist who now lives in Holland was here recently; she told us that Michelson's experiment had been repeated in America, that is, with positive outcome.[1] H. A. Lorentz supposedly brought that back with him. Is that right? Michelson's experiment is among those "practically" [a]-priori things; I don't believe a word of that rumor. If you have time for a postcard, though, all of us here would be very grateful.

Franck and Courant told me about you.[2] We had many worries with the appointment matters. Pohl decided to stay at Göttingen.[3] This relieves us of the agonizing choice. But now I am trembling about Franck going to Berlin. I do heartily wish he gets the call; but he would be foolish to accept it. Courant says that you think so too.

Scientifically speaking, I don't have anything substantial to report. I am laboring with my assistant Hückel[4] over the quantization of polyatomic molecules to calculate infrared bands (e.g., H_2O). We have the right approximation method but the computation is very complicated. My encyclopedia article will probably be finished this month; I'm fed up to the teeth with it.[5] I have been mulling over all sorts of things about the quantum theory of molecule formation; a brief notice in the *Naturwissenschaften* on the H_2 molecule contains a few results for experts.[6] But the clearer they become, the crazier the whole system seems to appear. As regards the matters of principle, I haven't come a jot further along.

My wife and the children are well. The girls[7] are in the countryside with a former maid of ours; but they are coming back soon. We are staying here until mid-September, then we're going to Leipzig and from there to Italy. We got 22 English pounds for the translation of my book, you see, and converted them into lire.[8] It won't go very far, of course, but we are very much looking forward to this little trip down south.

With warm regards also from my wife to all of you, yours,

M. Born.[9]

321. From Heinrich Zangger

[Zurich–Basel, between 6 and 28 August 1922][1]

[Not selected for translation.]

322. From Michele Besso

Bern, *3a Fichten Way*, 8 August 1922

Dear Albert,

Your letter to the Committee on Intellectual Cooperation[1] communicated by the press responds at the same time to my letter from about 8 weeks ago; whether it reached you I do not know.[2]

So then, may the waves carry you safely to the land of the rising sun and back again! I would be no less surprised if the voyage there brought you the grandest that life abroad can offer a teacher of humanity, than if, on the contrary, icy solitude gripped you over there more than anywhere else. In any case, many of the elite in G[ermany] will envy you for being able to turn your back on this dismal Europe for a while (whose troubles, ⟨by⟩ also as a consequence of the contrast, may nowhere be more perceptible than where you are).

The question that leaves me no peace, time and again, is whether one should strive after resigned or contemplative serenity ατοραξια[3], in which even the seemingly greatest enormity appears merely as a wave, somewhat larger than its sisters; or whether one should, as far as one is able and knows how, preserve the will's vigor . . .

Four weeks ago my family doctor "diagnosed" a noticeable cardiac insufficiency and prognosticated for me a maximum of 10 years of life; two weeks later another doctor (in Bühlerhöhe near Baden-Baden, where I visited my brother Vittorio)[4] could find nothing. I would, however, gladly sacrifice the 10, or even the 20 or 25 years, to toil relentlessly as long as the day lasts—if I did not . . . almost certainly know that I do not know in which direction.

From Bühlerhöhe I sent out a book to you by commission of my sister-in-law Paula Winteler.[5] It is the work of her younger brother[6] who suddenly rejected life in 1910. A remarkably free and rich personality; he was so at ease with the Greek sages that he more quickly and confidently found the Greek word than the Italian; he had his body under extraordinary control as a swimmer and climber, and developed a definite talent in sketching. The crucial questions of life bore very heavily down on him even so, or as a consequence. See, for example, pages 229–230 of the book, which lies open before me now.[7]

Vero with wife and child are (since Saturday) in Zurich with the parents-in-law. He had productive practical training this past half-year, but is now looking for a more disciplined boss than he has had up to now.[8] As I wrote you in that letter of a few weeks ago, it would please me very much if he could gain a foothold precisely within the League of Nations. Every person has to have his connections. A scientist belongs to the powerful circle that created the mathematical and observational- experimental method. The normal person belongs to his family, his nation. He who, like us, does not have any clear bond of this type, must—and perhaps even can—connect himself as highly as he is capable of grasping.

Did I tell you that Chavan gave me a wonderful enlargement of a photograph of you taken in his laboratory in 1906 or 1908?[9] Also that he told me how happy he would be to be able to extend to you his hospitality in Bern? We can do so now, too, by the way.

So *auf Wiedersehen*, before or after the world tour!

In all our names, yours,

Michele.

323. From Henry N. Brailsford[1]

London, S.W. 1, 67 St. George's Square, 10 August 1922

Highly esteemed Professor,

I permit myself to approach you with a request, urgently asking you please not to turn me away without serious reflection. *The Labour Leader*, which you might know as the name of a Socialist weekly paper, will be appearing on 5 October in a new, enlarged, and as I hope, much better form under my direction.[2] This paper has always been instilled with bold pacifism, even during the blackest period of war, and I hardly need to say that the development of friendly relations between our two countries will be one of its main goals.

Now I have the ambition of being able to publish something from you in the first issue or one of the earliest ones. You are, of course, the only German whose every word is repeated in every paper in both hemispheres. If you would be willing to say something very specific, yet tactful, about the economic misery of the German people and the lack of understanding on the part of the Powers about the effect of their policy, which destroys the prospects for the Republic and the hopes for peace, then I believe that your words would be more effective than those of any other person.[3]

If you would be so kind as to write an article or grant an interview, I would take it upon myself not to use it in any egoistic or narrow-minded way. This means, if you were willing to send it to me so that I could publish it in my new paper, I would release it at the same time to any English daily paper of importance for quotation. It is also supposed to be sent to America. Unfortunately, I cannot offer you an honorarium suited to your high rank. As a rule, we pay five or four pounds, depending on the length. But it would undoubtedly be possible to obtain a much larger honorarium for you from America. I would try to secure publication in France as well.

Should you not feel called to make such a statement, I would seek from you something much lighter—perhaps the simplest of articles about current prospects of scientific discoveries, or perhaps about the effect of scientific discoveries on the evolving structure of society—any such topic of a general character.

You might be interested to hear that our paper will be publishing a scientific article every week and, contrary to general custom, we will be applying to scientists instead of journalists for them. Altogether, the purpose of the paper will be to prod intellectual ambitions and to raise the intellectual norms of the English Socialist and labor movement.

In allowing myself the hope that you will receive my petition with kind favor, I remain, in great respect, yours truly,

H. N. Brailsford.

324. From Moritz Schlick

Kiel, 13 August 1922

Highly esteemed, dear Professor,

Shipping the enclosed copy of the new edition of my short work *Space and Time* [*Raum und Zeit*] gives me welcome opportunity to write you a few lines.[1] This booklet has been altered somewhat in a few places and expanded; if you would like to see any improvements made to these corrections, I would be grateful from my heart for a brief message; it would be specially valuable for me to know whether you agree with the comments about the rotation problem on p. 77.

I would so much have liked to hand you this copy in person during your last stay in Kiel, and it was very painful for me not to have the privilege of seeing you during that short visit.[2] Yet you were surely also seeking relaxation from the strenuous Berlin duties during those few days and so I did not dare to look you up when I learned of your presence here. Now the thought that I am not going to be able to meet you in September in Leipzig, either, hurts particularly much. I was very dismayed when I heard from Planck and Laue about your cancellation[3] and you can imagine how overwhelmed with disgust I was that (as Planck put it) a band of murderers was disrupting the program of the Leipzig convention. What a sad chapter this is. It reminds me of another affair that I would like to report to you about today. It concerns my successor here at Kiel (for I have now finally decided to go to Vienna as of the coming semester). At that time you were so kind as to draw my attention to Wertheimer at W. Koehler's suggestions.[4] I first thought that W. would probably not be able to come into consideration because the University of Kiel could not offer him any institute for experimental psychology, but Koehler then reassured me about this point and persuaded me that Wertheimer was the right man to appoint even to a purely philosophical chair and he was not absolutely dependent on having an experimental institute. I then warmly supported him and it was possible to place him on the committee's list of candidates. The faculty as a whole then struck him again (I remain silent about the suspected reasons), generally dismissing the committee's nominations, and in the end a single name remained on the list for submission to the Ministry: that of the Giessen full professor, E. von Aster.[5] I consider this choice entirely fortunate because Aster is very capable and diligent, but I do not think nominating him as the sole candidate is justified and so I enclosed a separate vote with the faculty's letter to the minister in which I drew special attention to Wertheimer. I hope from my heart that it will help him somehow. Besides Wertheimer, I also supported H. Reichenbach very much; but he found little favor with the faculty even though my fellow colleague Scholz was on my side.[6] The faculty seems to be striving less for a philosopher than for a certified philosophy professor. In that case I dispensed with submitting a separate vote ⟨at least⟩ because of the futility of the affair. All these reports are completely confidential, of course.

It will be quite hard for me to go to Vienna, not only because the future in Austria looks so dark, but also because I ultimately felt extremely comfortable among my colleagues and students here. However, the Viennese climate is better and the duties for a teacher of philosophy are greater.

I wholeheartedly wish you a successful journey to Japan and hope to see you very soon again in Vienna and remain, with best wishes to you and for your well-being, in respect and gratitude, yours,

M. Schlick.

Address: 23 Orléans St., Rostock.

325. To Jacques Loeb[1]

[Berlin,] 14 August 1922

Esteemed Prof. Loeb,

I read much of your book [attentively] and admire the stringency of your chains of reasoning and the beauty and multifariousness of the connections.[2] The only thing that has not yet become clear to me is how it happens that the colloid molecule just acts *either* as an acid *or* as a base. But I hope to have penetrated that completely by the time you come here. I hope to see you here in Berlin in September.[3] I canceled my talk in Leipzig[4] because I am somewhat in jeopardy under the currently prevailing political unrest in Germany, so I have to stay nicely [in] hiding, which, however, is entirely pleasant for me. At the beginning of October I shall be going with my wife to Japan and think that everything will have calmed down again when I return half a year later. I had a fine month's vacation together with my boys in a small garden hut near Berlin on the river bank, a kind of Indian adventure.[5] The political and economic conditions in Europe are becoming increasingly convoluted; you will see many interesting things but fewer positive ones during your travels. Mr. Flexner did not come and see me.[6] I hope that you did not consign your money to organizations bound by a thousand considerations, rather directly to those who know properly what to do with it. Otherwise it goes in ineffective little portions toward supporting superfluous mediocrities.[7]

Happy to soon see you again, I am, with best regards from both of us, your admiring

A. Einstein.

326. From Helene Stöcker[1]

Nikolassee-Berlin, 14 August 1922

[Not selected for translation.]

327. From David Hilbert

Göttingen, 15 August 1922

Dear Colleague,

The daughter of my old patron and friend who recently died at an advanced age, Leo Königsberger (Heidelberg),[1] is married to a physician in China, Dr. Pfister.[2] She asks me to convey to you the enclosed invitation from her husband.[3]

As I hear, you are not going to Leipzig;[4] I regret it the more so, since I have not spoken to you for so long[5] and do not see any occasion to visit you myself in Berlin. Thus I take this written route to wish you a successful and productive journey replete [with] impressions and pleasures!

Most cordial regards, yours,

D. Hilbert.

328. From Hermann Anschütz-Kaempfe

Lautrach manor, 20 August 1922

Dear, esteemed Professor Einstein,

Many thanks for your letter of the 14th and your postcard of the 17th of this mo.

Single-phase motors work best with inductive *and* ohmic resistance, so it would perhaps be advisable not to dispense with the ohmic one, after all. The energy loss is not such that it would tip the balance much; it is merely a question of heat gain, which the carrying off seems to have solved. I had the next device sketched so that I install the blow magnets inside the sphere and the two half-shells (in which the electro-inductive resistance sits) perfectly symmetrically close around the sphere; this way I obtain a somewhat better efficiency, because I can make the spherical shell out of copper instead of aluminum; additionally, the danger that the force lines penetrate through the sphere and cause disturbances in the gyro's iron parts is avoided. One disadvantage is the larger flow of current between the electrodes, but the good suitability of the graphite-rubber coating now makes it worth risking. Perhaps it is also possible to use the two blow magnets as inductive resistors for the two gyros at the same time; it is possible that with this arrangement one can even make do with a sole blow magnet, if the sphere with the water at +4° has a minimum weight of 50 gr., which then can rise to c. 120 gr. at c. 50°. The blow magnet in the sphere can be much smaller because the centering is considerably simplified; it is similar to an inner tooth wheel against an outer tooth wheel; I got a quite good centering already from a coil magnet of c. 3° iron thickness; in the application I chose a ring coil drawn out to 20°.

I chose 2 gyros but set the axes of the gyro caps vertically, after all, because the ties between the sphere and the gyros seem less doubtful in this arrangement; I fear a rolling error could occur if I lay the bearings of the gyro-cap axes in the equator, because then the sphere is only attached by the springs to the gyros and the lower-lying weight of the two gyros will enable it to perform oscillations that are too close to the rhythm of the ship's accelerations; with the vertical arrangement there are no forces on a ship that could act rotationally on the sphere.

Your suggestion about using two counter-rotating gyros as an artificial horizon is already familiar to me; first, for the artificial horizon of airplanes, *one* gyro probably suffices that has an appropriately long oscillation period, e.g., 15 minutes. The accuracy required is not great, approx. one degree. The situation is different for an artif. horizon for ships to substitute a sextant; other ways will probably have to be chosen; for this perhaps you suggestion would be best. I look forward to being able to consult with you about this next summer.

When is your Albert arriving in Munich to matriculate?[1] Hopefully we will already be there then; we would like to accommodate him in our house upstairs; I hope the housing office won't scratch out this plan of ours. In any event, he will initially be living with us, a guestroom is here for him in any case, even if we should happen not to be here yet. We would like to be in Munich, depending on the weather, roughly around October 20th; but he can come earlier as well; we look forward to seeing him.

Then in November I shall go to Kiel to have a look at how the new sphere is doing,[2] and your Diogenes Tub,[3] too, the rough shell of which should be ready by then.

Here in Lautrach it still looks bad:[4] painters, decorators, and masons dominate; with some trouble we managed to wrest a very unlivable room from the workmen as a temporary refuge; but now things are getting better by the day. It is so peaceful and quiet here; you will enjoy it if you are yearning to escape from the Berlin hubbub and lodge here in peaceful tranquillity.

My wife and I send our best regards between households, yours,

Anschütz-Kaempfe.

329. To Paul Ehrenfest

[Berlin, on or after 21 August 1922][1]

Dear Ehrenfest,

The bad prognosis about little Wassi made me feel very sorry.[2] If it rests on a sound basis, I thoroughly approve of your plan to hand the child over to impersonal care. Valuable people should not be sacrificed for causes without any prospects, not even in this case. In that event it is good if you act quickly. I am glad that you are enjoying your brother[3] and his family so much and that this energetic man is a support for you in your dire trouble.

I sent our note out to Mr. Scheel.[4] I left your last sentence standing unchanged even though I did not understand the suggestion at all. I can't understand why the adjustment to the magnetic field should be more comprehensible by assuming in principle that the quantization was imprecisely realized. It would be good if you could elaborate on this to appease my literary conscience.[5]

Dear Ehrenfest, I need your friendship as much, perhaps even more, than you do mine; for my personal relationships are much feebler and sparser than yours and I have difficulty finding human contacts that make me feel good.[6] I spent a happy month with my boys in the hut;[7] it was a true camping experience [*Indianerleben*]; I told them a lot about you. Albert is a strong, independent fellow, but has little sense for the subtleties of human ways and little ability to empathize. The little one, by contrast, is softer, comical, fine, flexibly-minded, and sensitive; he would happily fit in well among you. I have to show him to all of you sometime when it's convenient.

There is a very intelligent paper by Tetrode in the *Zeitschrift für Physik* about the quantum problem.[8] Maybe he's right; either way, by this paper he demonstrates that he is a first-ranking mind. Nothing has been as fundamentally gripping for me in a long time. We, incidentally, also once discussed the possible importance of the relativistic distance 0 (?) in solving the quantum problems. The weaknesses in his approach seem to me to be:

1) The law of electrodynamics is blamed for everything whereas there surely must be a formulation of the quantum laws in this subfield of mechanics.

2) In principle working with actions-at-a-distance is questionable, because it certainly cannot be applied to the general theory of rel[ativity].

Heartfelt greetings to all of you from your

Einstein.

My tour starts at the beginning of October.[9] My wife was quite ill but is recovered enough to probably be able to accompany me.[10] She sends her best regards.

330. From Maja Winteler-Einstein

Florence, Colonnata Quinto, 5 Strozzi Street, 25 August 1922

My dear Albert,

I can imagine what a glorious life you led with your boys in solitude.[1] Tete's greeting pleased me very much. I would so much like to get to know him as well. I only saw him once when he was 4 years old. Is it really not doable to have the boys come and see us once as well? So, Albert is already going to college. He'll probably be going to Munich in October. What's he studying, anyway? I'd think engineering?[2]

I just hope that the business with your being in danger isn't really all that serious.[3] Even so, I'm not at ease as long as you're in Berlin. The mood is very excited here as well but the battle is just being played out quite openly between fascists and communists, so other mortals are protected from revolver shootings and fist fights. We are only roughly aware of what's going on from the papers.

I am very sorry that Elsa had to go through so much. Hopefully she's completely healed now.[4] I envy her from the bottom of my heart for the fine journey. The long voyage by sea will be good for her health. Do come and see us on your return trip. That way you'd be able to get the best idea of what our life is like. We never had it so nice and enjoy our luck each day anew. Pauli[5] is trying to gain a foothold in the Swiss colony, in order to be able gradually to find work as a lawyer. This fall he will be giving a talk there, which he is diligently working on. I take care of the household, cook, mend, and do whatever else when the occasion arises.

I would now like to beg you please to instruct us very exactly *before your departure* about what we should do with the things that will be coming here in late fall.[6] Where should they go? And should the thing that you are getting directly from us be brought to the same place?[7] I would also be very grateful if you could provide us with Uncle Jakob's address so that we can also immediately send his blessing directly to him.[8]

So, please, answer all these questions before your departure. Do let us also hear from you during your trip. You can't imagine how interested I am in the impressions you'll be having.

We send all of you our heartfelt greetings. I give you another sisterly kiss,

Maja.

331. To Maximilian Pfister

Berlin, 28 August 1922

Esteemed Doctor,

It is possible that I shall be able to deliver a few lectures in China.[1] Up to now I have been invited by the University of Peking.[2] With major internal difficulties prevailing in China,[3] however, I do not know yet whether I can really accept this invitation. I have only two to three weeks time for a possible stay in China, so (besides perhaps Peking) only locations lying along the coast could come into consideration. At the moment, though, I still cannot decide, because the matter with the University of Peking has not yet been settled. I would just like to make the following comments:

1) I cannot lecture in the English language, but I know an intelligent colleague (Mr. Rusch) who has been teaching theoretical physics for many long years in Tianjin and could serve very well as an interpreter.[4]

2) Only lectures for an audience of some scientific background come into consideration (medical doctors, engineers, teachers, etc.), because I know from experience that entirely lay listeners cannot understand anything.

If you would like to return to your plan, please send proposals to me in Tokyo (university), if possible in consultation with Peking, that are sufficiently detailed for me to be able to decide on that basis and draw up my agenda. Please give Mr. Robertson my thanks for his letter and inform him about the content of my letter.[5]

I thank you now most sincerely for your kind invitation.

Very respectfully,

A. Einstein

332. From Paul Dienes[1]

Kremsmünster (Austria), 28 August 1922

[Not selected for translation.]

333. From Paul Ehrenfest

[Leyden,] 29 August 1922

Dear Einstein,

I just came back from Vienna a couple of days ago, where I was reunited with my 4 brothers after separations of 23 years and 10 years, resp.[1] It was very interesting and in some respects a *very* profound experience for me.—It stirred up the deep question: What does the division by Nature into "individuals" mean?– Before your departure I'll write you another longer letter—for the moment, just this— mainly intended for Ilse-Margot.

My wife is traveling on the 4/5 Sept. with Tanitchka to Jena to deliver Wassik to the "Trüpers Educational Establishment" there. (The *people* and facilities there are quite excellent, as I was just able to see from a personal visit.)[2]—From Jena, my wife is coming with Tanitchka *circa* Sept. 7 or 8 to Berlin to see her Russian friends there.—In case Ilse would like to write to her in Jena, then to the address: Trüper's Educational Establishment, Jena.—My wife will contact you by telephone upon arrival in Berlin. The two of them want to go somewhere by the Baltic Sea after a couple of days in Berlin.—Tanitchka is supposed to be back here by 30 Nov. (beginning of the university lect.—she passed her exam well!!!),[3] my wife is staying longer in Germ[any].

Very warm regards to you all.

334. To Richard B. Haldane

Berlin, 30 August 1922

Esteemed Lord Haldane,

This short letter will draw a smile from you, but a kindly one, I am sure. For I would like to take the liberty of directing your kind attention to the enclosed article by Rechberg from the *Berliner Tageblatt*. It seems to me to contain a favorable solution to the reparations problem, perhaps the only one making harmonious developments for the future possible.[1] From French friends I hear that Poincaré is certainly not averse to such a plan but that there is considerable resistance to overcome in English quarters.[2] As this plan seems to me so reasonable and naturally suitable, I dare to request that you consider this proposal yourself and in case of approval apply your great authority in support of it. That the plan would become reality in case of an agreement between French and English statesmen can scarcely be doubted in the current situation; and it would be too great a pity if energetic steps for reform would be postponed so long as their effectiveness would be seriously jeopardized through persistent and substantial damage to German production capabilities. In asking you to see that my name not be made public in connection with this matter, I am, with cordial regards to you and Miss Haldane,[3] yours very sincerely,

A. Einstein.

P.S. If you wish, I can procure for you the other publications by Mr. Rechberg on this matter.[4] There is naturally no need to reply to this letter; however, I am gladly at your disposal in case you deem me somehow useful as a middleman.

335. To Paul Painlevé

Berlin, 30 August 1922

Dear Mr. Painlevé,

I am sending you herewith an article by Mr. Rechberg[1] that contains nothing new, of course, but in my opinion marks the only really tractable route. Would you not be inclined to exert your full influence in this same direction? I am also writing Lord Haldane for the same reason.[2] If France and England could concur on this program and take energetic steps along this route, in a short time not only could current obstacles be removed but also tolerable conditions would be secured for the future. Pardon me please for taking the liberty of writing to you about political things as a layman; and please arrange that my name not be mentioned in connection with all these public affairs.

Cordial regards, yours.

336. From Richard B. Haldane

Cloan, Auchterarder, Perthshire [after 30 August 1922]

[Not selected for translation.]

337. To Carl Speyer[1]

Berlin, 31 August 1922

Dear Mr. Speyer,

To my knowledge the University of Jerusalem is initially only envisioning establishing institutes for biology and chemistry, so your intention appears to be premature.[2] ⟨I unfortunately cannot offer you any recommendation, because as a layman I am incapable of assessing your scientific proficiency. Expressing my pleasure at thus having come into contact with you again, I am⟩

In any event, though, I want to inform the people in charge about your willingness; one cannot know how quickly things will develop in Palestine.[3]

With kind regards.

338. To Paul Dienes[1]

[Berlin, 31 August or later, 1922][2]

Dear Sir,

Your attack on the mathematical theories by Weyl and Eddington only touches the formulation, not ⟨however⟩ the content.[3] Weyl had treated the second-order quantities ⟨loosely⟩ negligently but in a manner that is easily demonstrated to be innocuous. I am sending you a small book in which you will find on pages 48–49 ⟨a proof⟩ Levi-Civita's and Weyl's train of thought sketched a little more precisely,[4] ⟨against which there ought to be no objection, in which he avoids these inaccuracies⟩ so such objections do not arise anymore.

In utm. resp.

A. E.

I enclose for you two printed sheets.[5]

339. "On Anisotropic Pressure Forces in Gases with Heat Flow"

September 1922

[p. 1]   On Anisotropic Pressure Forces in Gases with Heat Flow.[1]

Maxwell already discovered that in gases with heat flow anisotropic pressure forces occur that depend linearly on the second derivatives of the temperature against the coordinates, i.e., on the first derivatives of the heat flow against the coordinates $\left(\dfrac{\partial f_\mu}{\partial x_\nu}\right)$.[2] For reasons of symmetry one is also inclined to presume the existence of tensorial pressure forces in gases that depend quadratically on the components f_μ of the heat flow itself. That these really do exist and how they can be calculated in a simple way will be demonstrated in the following. A more detailed presentation can be found in the Zurich dissertation by Miss E. Einstein (1922).[3]

We assume mechanical equilibrium of the flowing gas: If such isotropic pressure forces exist, then—as is gathered from purely formal considerations—they must be of the form

$$p_{\mu\nu} = p\delta_{\mu\nu} + z\,f_\mu\,f_\nu + z'(\sum f_\alpha^2)\delta_{\mu\nu}\,, \qquad \ldots (1)$$

where p signifies the location-independent part of the gas pressure, f_μ the components of the heat flow (on the part appertaining to the molecule's progressive motion), $\delta_{\mu\nu} = 1$ or $= 0$, resp., depending on whether $\mu = \nu$ or $\mu \neq \nu$; z and z' are constants, depending on the state of the gas at the position under consideration. In the following we want to treat them as if they were independent of the coordinates, which however would imply an approximation.

It is known that the heat flow fulfills ⟨both⟩ the relations

$$\sum_\mu \frac{\partial f_\mu}{\partial x_\mu} = 0 \qquad \ldots (2)$$

$$\frac{\partial f_\mu}{\partial x_\nu} - \frac{\partial f_\nu}{\partial x_\mu} = 0. \qquad \ldots (3)$$

Furthermore the equilibrium condition of the gas is

$$\sum_\nu \frac{\partial p_{\mu\nu}}{\partial x_\nu} = 0 \qquad (\mu = 1, 2, 3). \qquad \ldots (4)$$

If one inserts in this ⟨in (1)⟩ the expressions (1), then one obtains with respect to (2) and (3), neglecting the spatial variability of the constants z and z' , the result

$$z' = -\frac{1}{2}z\,,$$

so that one obtains instead of (1):

$$p_{\mu\nu} = p\delta_{\mu\nu} + z\left(f_\mu\,f_\nu - \frac{1}{2}\delta_{\mu\nu}\sum_\alpha f_\alpha^2\right). \qquad \ldots (1a)$$

There now remains the solution of the main aim of finding out z from a consideration in gas theory. We ask about the most probable distribution of velocities at one location of the gas in which a heat flow occurs in the direction of the x axis without a current of the molecules (motion).[4] We look for an extremum of the Boltzmann integral,[5]

$$\int f \lg f \, d\tau,$$

where f means a sought function of the velocity components ξ, η, ζ and $d\tau$ means the element $d\xi \, d\eta \, d\zeta$ of the velocity space. The secondary conditions apply:[6]

[p. 2]

$$\int f d\tau = 1 \quad \ldots (5) \text{ (per the definition of the probability function } f)$$

$$\int \frac{m}{2}(\xi^2 + \eta^2 + \zeta^2) f d\tau [\ ] = L = \text{const.} \quad \ldots (6)$$

$$\int \xi f d\tau [\ ] = 0 \quad \ldots (7) \text{ (condition of the freedom of flow)}$$

$$n \int \frac{m}{2}(\xi^2 + \eta^2 + \zeta^2)\xi f d\tau = f_x \quad \ldots (8) \text{ (given heat flow)}$$

m means the mass of the molecule, n the number of molecules per unit volume. Executing the variation yields

$$f = C e^{-h(\xi^2 + \eta^2 + \zeta^2) + A\xi + B\xi(\xi^2 + \eta^2 + \zeta^2)}, \qquad \ldots (9)$$

where C, h, A, B are independent of ξ, η, ζ. This solution cannot apply to arbitrarily large positive $\xi \, \eta \, \zeta$. Nevertheless, for a small A and B it is still useful in velocity ranges, at the limit of which the factor $e^{-h(\xi^2 + \eta^2 + \zeta^2)}$ practically vanishes. We can, of course, confine our entire consideration to this range of the velocity space.

For finding the constants C, h, A, B we note that we may substitute (9) to second approximation with

$$f = C e^{-h(\xi^2 + \eta^2 + \zeta^2)}[1 + A\xi + B\xi(\xi^2 + \eta^2 + \zeta^2)]. \qquad \ldots (9a)$$

With this taken into account, (5) yields

$$C = h^{3/2} \pi^{-\frac{3}{2}}. \qquad \ldots (10)$$

Furthermore, from (9a) and (7) we get

$$2Ah + 5B = 0. \qquad \ldots (11)$$

Finally, from (8) follows:

$$[\ ] mnAh^{-2} = f_x. \qquad \ldots (12)$$

To first approximation it is also known that[7]

$$h = \frac{M}{2RT}, \qquad \ldots (13)$$

where M denotes the molecular weight with reference to the mole, R the gas constant, T the absolute temperature.

Now after function f of the velocity distribution has been obtained, the pressure components can be calculated by means of the formulas:

$$p_{xx} = mn \int \xi^2 f d\tau$$
$$p_{yx} = mn \int \xi \eta\, f d\tau$$
$$\text{etc.}$$

$$(14)$$

For this calculation the second approximation given by (9a) does not suffice, however; in the expansion of (9), the terms in A and B of second degree must rather be taken into account. If one inserts this third approximation of function f in (14), then one obtains:

$$p_{xx} - p_{yy} = \langle \frac{9}{50} \frac{M}{pRT} f_x^2 \rangle \; \frac{18}{25} \frac{M}{pRT} f_x^2 \; . ^{[8]}$$

The factor f_x^2 calculated for this special case is none other than the sought factor z [p. 3] in equation (1a), so we can summarize our result in the equation:[9]

$$p_{\mu\nu} = p\delta_{\mu\nu} + \langle \frac{9}{50} \rangle \frac{18}{25} \frac{M}{pRT} \left(f_\mu f_\nu - \frac{1}{2} \delta_{\mu\nu} \sum f_\alpha^2 \right). \qquad \ldots (1b)$$

In a gas with heat flow, pressure forces do hence occur that, in the direction of the heat flow, try to dilate the gas and, crosswise to the heat flow, to contract it. The calculation shows that the magnitude ⟨of these forces⟩ of factor z comfortably allows for a test of the result by experiment.[10]

A. Einstein.

340. "Comment on A. Friedmann's Paper: 'On the Curvature of Space'"(1)[1]

[*Einstein 1922p*]

DATED September 1922
RECEIVED 18 September 1922
PUBLISHED 19 December 1922

IN: *Zeitschrift für Physik* 11 (1922): 326.

The results obtained in the cited paper regarding a nonstationary universe seemed   [p. 326] suspect to me. In fact, it turns out that the solution given does not agree with the field equations (A).[2] As we know, from these field equations it follows that the divergence of tensor T_{ik} of matter vanishes. In the case of the assumption characterized by (C) and (D_3),[3] this leads to the relation

$$\frac{\partial \rho}{\partial x_4} = 0 \,,$$

which together with (8) requires the temporal constancy of the world radius R. The significance of this paper therefore lies precisely in that it proves this constancy.[4]

Berlin, September 1922.[5]

(1) *Z[eit]s[chrift] für Phys.* **10**, 377–386, 1922.

341. To Chaim Weizmann

[Berlin, after 2 September 1922]

Dear Mr. Weizmann,

I send you herewith the application of a geologist of about my age for the University of Jerusalem.[1] I know that a geology institute is initially not being planned but I do think that under all conditions such an application should be put into the files.

I am glad that [you] have now reached your grand goal and can take a breather. It must have been hard![2] With cordial regards to you and your wife,[3] yours,

A. Einstein.

342. From Fritz Haber

Oberstdorf, [3] September 1922

Now I've been hiking for almost the whole day
To see the Sturmanns Cave and the gorge,[2]
And could not walk through either one,
Bodily strength fell short of the desire.
This is the torment, this is the inner anguish
Will you make it through what's newly begun?
When life's tempests blow autumnally around you
and cold mist wraps around your step!
Tenacious will is missing in younger years
The colorfulness of life chases from goal to goal!

Overbrimming energy strains against your persistence!
Then your nervous upheaval makes you into a fool
Whatever new thing you set out to do is too much
You must steadfastly trudge along the beaten path.

Greetings,

F. Haber.

343. From Henry N. Brailsford

London, S. W. 1, 67 St. George's Square, 4 September 1922

Highly esteemed Professor,

It is my opinion that an open letter by you during this dreadful crisis in the lives of the German people could possibly exert a great influence on public opinion, indeed, perhaps a decisive one, in accelerating a return to reason.[1] Not just that you may be the only citizen of the German republic whose name inspires gratitude and respect among all, even among those who still uphold their political prejudices. The fact that you are devoted to scientific truth ought immediately to raise above the political debate any statement you should make about the affairs of the present day.

I believe that the questions to which English readers would particularly like to have an answer are the following:–

(1) Some of us know that the economic crisis that Germany has been experiencing since the armistice has had the worst consequences on science and culture in general. But a relatively limited group comprehends this fact; and it knows nothing more about it, either. Could you tell us from your own experience and your own knowledge about the conditions at German universities how the general impoverishment has affected them? Is it exaggerated to assert that these circumstances, if they were to persist for a longer period, pose a threat to the future of European civilization as a whole?

(2) Does your experience confirm the impression I gained during a recent visit in Germany that the material and cultural standards of living for the middle class have suffered even more than for the working class? And do you believe from your experience that students and teachers are particularly grievously affected by this new impoverishment?[2]

(3) Can you recount from personal experience or from that of closer acquaintances how the new conditions influence the real income, health, and physical and mental working efficiency of manual laborers as well as intellectual workers?

(4) Does the devastating tension noticeable in German political life and engendering the many political assassinations[3] arise from the desperation caused by this new pauperization?

(5) Some of us have constantly suspected since the German revolution that the policy by the Entente created an intolerably heavy toll on the republic. Is this your personal diagnosis of the problems troubling the republic?

(6) I encountered the greatest fear among prudent Germans of the consequences of any sudden stabilization of the mark, regarding joblessness and commerce. Are you of the opinion that this obvious danger would be reduced if German statesmen and German experts were appointed as equals to all future European consultations about economic and political issues of concern to us all?

(7) *The New Leader* supports the belief that social service rather than the desire for profit and gain has to be the motivation of all productive work if culture is to be saved. It seems to us that the mentality and experience of scientists must spur us on in this. It would be extremely stimulating for us to learn about your views on this important issue.

I believe I may say in the name of all those who will probably read *The New Leader* that we, indeed, a much broader public in the entire civilized world, will feel gratefully obliged if you were so generous as to find the time to answer these very important questions.

In utmost respect, yours truly,

H. N. Brailsford
Publisher.

344. From Jacques Loeb

The Rockefeller Institute for Medical Research, New York 66th Street and Avenue A
Marine Laboratory, Woods Hole, Mass. September 4, 1922

My dear Professor Einstein:

It was extremely kind of you to take the trouble of reading my book and to write to me in such an appreciative way.[1] The reason why the protein molecule only acts as acid or as base is difficult to explain, except that it is a fact as far as gelatin and proteins in general are concerned. I have ventured to explain it in analogy with the experiences on indicators which have shown that the chromophore group of indicators undergoes a tautomeric change at a definite hydrogen ion concentration. As a preliminary suggestion, I offered the idea that an isomeric change may also occur in the protein molecule at a critical hydrogen ion concentration.

We were all very much worried and shocked at the turn of events in Germany. I think it is an excellent idea for you to follow the invitation which takes you across the seas, and I only wish that you would accept my suggestion that I made to you a year ago, that you spend several years in the United States. If you would consider this possibility, I think arrangements could be made whereby you could have full independence and facilities to work. I do not think that matters will be very much different in Germany in the next few years. Will you let me know how you feel about this matter?

I have had to abandon my trip to Germany for the reason that Mrs. Loeb was not very well and that it looked for a time as if she had to undergo a serious operation. I could not make up my mind to go to Germany with the burden of that worry resting on my shoulders, and it seemed only fair to me that I should inform the committee of this situation. It would have been a great satisfaction to me if I had had a chance to bring some of my results before the physical chemists, since I have a very limited audience in this country and since in Germany they are not familiar with the literature published during the year.[2] In addition, the opposition on the part of the majority of the biologists to any application of physical chemistry to biological problems, and the opposition of the old-fashioned type of colloid chemists like Wolfgang Ostwald,[3] prevent the dissemination of these facts in German literature.

I hope it will be possible, finally, to get support in this country for those German scientists who are able to do the right kind of work; if not, I am afraid there will be added to the brutality following this war the cessation of scientific activity which can only result in keeping the forces of brutality longer in power.

I remain with kindest regards from all of us to yourself and Mrs. Einstein, Yours very sincerely,

Jacques Loeb

P. S.— I hope you will forgive me my dictating this letter in English. My handwriting has become so illegible—probably on account of general weariness—that it will be easier for you to read this dictated letter in English than my own writing in German.

Translator's note: Original written in English.

345. From Albert Karr-Krüsi

Zurich 1, Etzelstrasse [--], 6. Sept. 1922.

[Not selected for translation.]

346. From Helene Stöcker

7 September 1922

Esteemed Professor,

In accordance with our arrangement, I forwarded to Mr. Brailsford your view that you cannot resolve to issue a general appeal on your own but that you would certainly be prepared to respond as best you can to specifically posed questions.[1]

Thereafter, yesterday, I received the following lines from Mr. Brailsford, which I am forwarding to you in German and English.[2]

As a breathing space has luckily set in, it will also suffice if Brailsford receives your reply by the 25th/26th of September, in time for the October issue to appear.[3] However, it would be *much more preferable* to him if he could have the reply a week earlier, hence by the *17th/18th* of September. He would then be able to see about even greater possibilities for its broadest publicity.

I am convinced that any such account will surely be useful for a fair assessment and an interest and comprehension of your conditions and therefore hope that your scientific duties will leave you a few hours of leisure to reply to Brailsford's questions.

Brailsford's letter is supposed to be printed as the inquiring letter, as a *preamble* to your response. He leaves entirely to you whether you wish to answer all these questions or omit a few, change a few things, or add some to discuss subjects that appear to you to be of particular importance. In other words, he asks that you also edit his letter accordingly to meet your personal wishes.

He thinks that although there is a brief respite right now, the problems have not been diminishing, even with the rise of the mark, and that it is consequently the more necessary to make clear to the rest of the world the ordeals of living in Germany.

Brailsford would like to know your opinion on what would be preferable to you: Whether he should submit the letter to the *New Yorker World* or the *Hearst Papers*, both of which have a large number of affiliated newspapers in the United States, would secure the widest dissemination, and would also pay appropriately? Or whether you prefer he submit it to the *Associated Press*, which would perhaps secure even broader publication but not be able to pay in the same proportion?

In France, dissemination would be a little more difficult; however, all the papers of parties not linked to the nationalistic block would probably also issue it.

I would be very grateful for a short note about when Brailsford might count on your material and am meanwhile, with my regards, sincerely yours,

Helene Stöcker.

347. "The Peril to German Civilisation"

[*Einstein 1922m*]

SUBMITTED 11 September 1922
PUBLISHED 6 October 1922

IN: *The New Leader* 1 (1922): 11.

The Questions[1]

Dear Professor Einstein,

I believe that at this terrible crisis in the life of the German people a letter from you for publication might have a great, perhaps a decisive, influence in hastening the return of public opinion to sanity.

I think that the questions to which English readers would particularly like to have an answer are chiefly these:

(1) What has been the effect on Universities of the impoverishment since the Armistice?

(2) Is it, true that the standards of living of the middle class have been lowered?

(3) Have the new conditions affected the real income and health of the workers?

(4) Have the many political murders in Germany a connection with the new poverty?

(5) Is the Allied policy an aggravation of the internal difficulties of the Republic?

(6) Do you share the common fear of the consequences of a sudden stabilization of the mark?

(7) May we draw from the experience of scientific workers hope for a society based on social service instead of acquisitive gain?

I think I may say on behalf of all who read the *New Leader* that we will feel ourselves your debtors, if you will be so generous as to spare time to answer these questions.

H. N. Brailsford.

The Answer[2]

Berlin, September 11, 1922.

Dear Mr. Brailsford,

You have been so kind as to address one or two questions to me regarding economic conditions in Germany. You tell me that you want an objective statement about a state of affairs which ought, in the interests of a return to healthy political relations, to be known to the English public. I thank you for the confidence shown to me by your questions, and will endeavour to confine myself to what I can state with full conviction and certainty. I will proceed straightway to answer the questions one by one.

(1) The salaries of scholars and teachers, expressed in kind, have been continuously reduced as a result of the situation created by the war and the Peace Treaty. At present they amount at best to 20 per cent of their former value, in many cases to far less. This estimate is much too high for brain workers without fixed appointments. Undernourishment is almost universal among brain workers and students, and in addition books have become so dear that the intellectual life and development of the rising generation suffers seriously. The very existence of scientific and artistic activities, especially theatres and journals, is more and more endangered, and some have gone under. The struggle for existence among independent artists, musicians, and writers is desperate. Such conditions, and especially the perpetual consciousness of the insecurity of the individual's material existence, inevitably result in a marked lowering of the estimation in which the public holds professional work and intellectual achievements. I am firmly convinced that, if the present material conditions continue or even become worse, large sections of the so-called middle class, which have hitherto been the principal source and preserver of our intellectual heritage, will sink to the level of the submerged masses.

(2) It is plain that in hard times that work will be relatively best paid which is essential to carry on the economic activities of the moment, but that work which is directed only to the continuation and development of economic activities, and even more to purely cultural purposes, will suffer seriously under the prevailing conditions. Almost all intellectual work falls under the latter head. A colleague assured me on one occasion that scientific meetings are now held far less often than formerly, because those who would attend them must avoid the expense of tram fares. The great majority of students are so far dependent on their earnings that study can only be a secondary occupation. As regards teachers, what I have already said about brain workers in general applies to them.

(3) I know that there are general complaints regarding the reduced productive power of manual and brain workers, but I do not think that I am competent to say how far this is the result of undernourishment or of fear of inability to obtain food, and how far of purely psychological factors. There can be no doubt that people's energy is sapped by the consciousness that under present conditions it is impossible to provide for the future, partly because of the exceedingly heavy burden of taxation, which increases perpetually.

(4) It is a fact that many of the political murders have been committed by people who have lost the means of support as a result of present conditions, but I should not venture to say whether unfavourable economic conditions are *alone* responsible for the lamentable deterioration of political morals. The political intolerance of the supporters of the old regime is doubtless partly due to tradition.

(5) It must be admitted that the policy of the Allies has greatly augmented the difficulties of the Republican Government; in particular it has undermined the prestige of the Government by repeated humiliations, in the face of the whole people. Moreover, everyone here knows that the financial obligations laid upon the country cannot be fulfilled at their present figure, even with the utmost exertion. All this has bred in us the conviction that there is no hope of working our way out of our present serfdom by legitimate means. This paralyses economic activities and drives people to evade taxation, and try to remove their capital from the country.

(6) Even if we admit that the stabilisation of the mark might involve certain temporary difficulties, it can scarcely be doubted that we must try to attain stabilisation in any event and at the earliest possible moment. Without that it is impossible to reach stable economic conditions. The participation of German statesmen and experts in consultations upon international economic relations would certainly be desirable, if not absolutely necessary.

(7) It is only as a layman that I can answer your last question, and, further, with the utmost hesitation. I must admit with regret that I do not see how the hope of individual gain and the fear of want could be dispensed with as motives for productive work. In my opinion the community can mitigate the economic struggle of the individual, but cannot do away with it.

I trust that I have understood all your questions rightly and have answered them. I remain, yours faithfully,

A. Einstein.

Translator's note: Originally published in English. Is a translation of the following document.

348. To Henry N. Brailsford

Berlin, 11 September 1922

[Not selected for translation. Is the German-language draft of the preceding document.]

349. To Richard B. Haldane

Berlin, 11 September 1922

Highly esteemed Lord Haldane,

I thank you cordially for your favorable reply and for your very valuable support.[1] I have nothing against my name being mentioned to those to whom you forward the article. The only thing I would like to avoid is that my name appear in the German papers in this connection via British newspapers because that would embroil me in difficulties in an unuseful way. I would just like to add that I heard that German businesses are currently falling under the influence of foreign [cap]ital at a rapidly rising rate, which naturally means a steadily growing obstacle to a large-scale solution to the problem along Rechberg's lines. This seems to be but one more reason to act promptly.

I also received the Bergson book and read part of it but have not yet been able to make up my mind about it finally. [2]

In repeating my thanks for your kind support, I am, with all due respect, sincerely yours,

A. Einstein.

350. From Franz Selety

Vienna I, 11 Zedlitz Street, 11 September 1922

Esteemed Professor,

Along with this letter I am sending you, esteemed Professor, the offprint of my paper, which just recently appeared in the *Ann. d. Ph[ysik]*, on the possibility of an infinite world,[1] about which I already wrote you in January. It is naturally of greatest interest to me to become acquainted with your assessment. I hope very much that I shall have a chance to learn of it.

For the great generosity and kindness you have hitherto always shown toward me, I am ever very grateful to you.

Hoping that you will continue to preserve your kind favor toward me, I am sincerely,

Dr. Franz Selety.

351. To Max Wertheimer

12 September 1922

Dear Mr. Wertheimer,

You know that I am on a committee of the League of Nations for the international organization of intellectual endeavors.[1] Now Becker of the Ministry of Culture[2] recently summoned me to ask me to have someone substitute for me during my absence in East Asia so that someone from Germany be available. Becker mentioned some names (e.g., Troeltsch!),[3] whom I don't know well enough, however, nor can rely upon sufficiently. I just know a sole individual upon whom and upon whose free and objective way of thinking I can rely in every way, *and you are he*.[4] My request is that you be willing to represent me in Geneva at the meetings of this committee (to which Mme. Curie and Bergson also belong).[5] I was explicitly told there that this substitution is a purely personal affair of mine and that *my personal trust alone should be decisive in my choice*.[6]

Don't turn me down! Your travel will be paid for you, you will make the acquaintance of exceptional people, and have a chance to do some good. The Ministry of Culture wants to be informed about the happenings there. But what you wish or wish not to report to them (orally) upon your return from Geneva is left entirely to your own judgment. The committee just had a session[7] and you will be immediately informed about it. Ilse will show you everything to orient yourself.

Do write me very soon, as I have to direct an official application to the League of Nations in which I would like to suggest you.

With best wishes for the vacation, and regards, yours,

A. Einstein.

The next session will perhaps take place in late fall. I am elected there not as Germany's representative but simply personally. This is how you should regard the situation yourself. Main agenda: International scientific exchange. Organization of intellectual workers. Counteracting the shortfalls in economically afflicted countries. We have strong prejudices against the League of Nations over here. However, I believe there is much goodwill and future potential in it.

352. To Alfred L. Berthoud[1]

Berlin, 14 September 1922

Esteemed Colleague,

Please excuse me for answering your letter of 28 Dec. of last year only today;[2] far too much correspondence is to blame. It just came to hand as I was looking through the accumulated mail and I hurry to pick up the arrears.

The correct mass is the one calculated out of the energy. The electromagnetic mass has to be increased by the additional mass that the moving body possesses because it is subjected to mechanical tensile forces by the action of the electric charges. This follows from the energy's tensor character in relativity theory. It should generally be noted that an energetically negligible frame as carrier of electric masses cannot exist in the interpretation of relativity theory. Bibliographic reference, e.g., von Laue, *Relativitäts-Theorie*, 1st volume.[3]

Very respectfully.

353. To Thorvald Madsen[1]

Berlin, 14 September 1922

Esteemed Professor,

Prof. Einstein asked me to thank you for your kind message and to tell you that he will be glad to keep noon of the 22nd [inst.] open for an interview with you.[2] Mr. Comert[3] already had the kindness to inform Prof. Einstein about the intended election of a German medical doctor to the "Committee on Intellectual Cooperation." Prof. Einstein would now like to know whether all of you have already identified a suitable person; he himself was thinking of Priv. Councillor Prof. Kraus[4] but has not taken any steps yet. If you think it useful that Prof. Einstein do something to prepare the matter before your arrival in Berlin, please have a pertinent message sent out to him.

In utmost respect,

The secretary.

354. From Chugi (Tadayoshi) Akita

Berlin, 72 Brandenburgische St., 15 September 1922

Esteemed Professor,

From our Japanese *Kaizo* magazine company I received two telegrams in succession with the urgent request to ask you, esteemed Professor, for a foreword for the Japanese translation of your collected works.[1] Although there are many demands on your time now, I nevertheless would courteously like to request, if possible, that you write the foreword before your departure.– Five professors of our Tochoku University, under the supervision of your pupil Prof. Ishiwara,[2] are currently working on the translation.[3] They recently finished the first part and hope to have completed the whole of it soon.

Gratefully looking forward to your kind attention, I remain, with amicable regards, also to your gracious wife, very sincerely yours.

355. From Raymond de Rienzi

Paris, 15 September 1922

[Not selected for translation.]

356. To Chugi (Tadayoshi) Akita

Berlin, [15 September 1922 or later]

I am quite willing to write a foreword[1] ⟨would in the meantime I abstained from doing so yet because the publishing house has not had any contract forwarded to me yet⟩ [as] soon as I have concluded a contract with the Japanese publisher.[2] The composition of this foreword is more difficult because I do not know exactly which publications are supposed to be appearing. I hope to be able to settle the whole affair during m[y] presence in Japan.

357. To Tullio Levi-Civita[1]

[Berlin, on or after 15 September 1922][2]

Dearest Colleague,

It would be a great pleasure if we could have a contribution from you. It seems to me a fine idea to create a publication for all Jews who work in the sciences.[3]

Many greetings from your

Albert Einstein.

358. To Arnold Sommerfeld

[Berlin,] 16 September 1922

Dear Sommerfeld,

The last time I was in Leyden[1] I noticed that Ehrenfest was quite unhappy that you had denied him authorship of the adiabatic hypothesis in the last edition of your book.[2] In his last letter to me he informed me of details about this, on the impression that I would be seeing you in Leipzig. You might change your opinion if you read his details in this regard in the enclosed letter; so the passage might be altered in the English edition and eventual later ones.[3] It would please me very much, as I found him quite depressed about the matter.[4]

Amicable greetings, yours,

A. Einstein.

359. From Svante Arrhenius[1]

Leipzig, 3 Dittnitzer St. c/o Dr. Leo Jolowitz [before or on 17 September 1922][2]

Esteemed Colleague,

As I was passing through Berlin yesterday, I heard from Mrs. Margarathe Hamburger[3] that you are in the process of leaving Germany in order to go to Japan. I very much regret that external conditions are so bad that you reached such a decision.[4]

It will probably be very desirable that you come to Stockholm in December and if you are in Japan then that will surely be impossible.[5] The invitation to Stockholm cannot be sent out to you before the middle of November, however. Could you not stay in Europe until then? If you do so, you will presumably also visit Stockholm soon afterwards, unless some "superior force" prevents you.

I was in Paris shortly after you had left that fine city. Everyone was speaking about you, mostly with the greatest admiration, which pleased me very much.

I take the liberty of sending you my cordial regards as well as compliments to your esteemed wife.

In live admiration, I remain yours very truly,

Svante Arrhenius.

360. From Max Wertheimer

Prague, 6 Poric, 17 September 1922

Dear Mr. Einstein,

Thank you very much for your letter![1] and for the fine offer in it. I tried to think everything over. First of all, isn't there some mistake? The meeting in the fall you wrote about, isn't it taking place right now and is surely over by now? Before your letter arrived I read in the paper about the meeting of the Committee, Bergson's speech, etc. So isn't it already too late?[2]

It is a first-rate and fine responsibility and, dear Mr. Einstein, as much as I was surprised by your proposal—it is not easy for me to say no to your suggestion, but many things do speak for that. If only because I do not speak French adequately, or rather little; I can speak English (albeit in this, in some sense in any case, not uncomplicated situation, it might sometimes depend very much on complete comprehension of what is being said at the moment), I am a Jew and, as a nat[ive] German Bohemian, now a Czechoslovak—. And all those reasons that made the matter so difficult for you during our conversation that time and made even you waver about your participation, loom before me, indeed affecting me so much more forcefully: for, if you take part, you are that Einstein, you know! whom the whole world knows, also as a person, and therefore all the obstacles are very much smaller, innocuous, even; whereas I would be exposed to many, not unserious potential mishaps. That might perhaps not be so very bad if I went, even officially, only to listen in and draw up reports for you and for the Ministry,—but even then, that I do not have the sufficient linguistic comprehension, (nor could this be properly remedied by constant interpreting).

For reasons of the political moment concerning the Entente *and* in view of the *Germans*, wouldn't the following solution perhaps be the best, after all? You could send as your personal substitute, also officially, *just* to listen in and draw up reports, a noted, Christian, and linguistically very gifted scholar? Would this be possible? This would appear to me to be very practicable in some respects, also for the sake of the German situation, which depends on many things, of course. The choice might thus be easier, as the person would only be going to listen in and make reports.

Do you think that Brinkmann[3] might perhaps come into consideration? I think he could do a good job of it.

Best regards, dear Mr. Einstein, and again, proper thanks! And I hope you are not cross about this letter of mine—am I not right?

M. Wertheimer

P.S. When are you departing? Couldn't I perhaps still speak with you in Berlin in October?

361. To Swiss Embassy, Berlin

Berlin, 18 September 1922

Highly esteemed Sir,[1]

I permit myself to approach you today with the following request: At the end of this mo. I shall be embarking on a voyage to Japan, China, the Dutch Indies, and Spain, to follow invitations to some universities in these countries. My wife accompanies me. As holders of Swiss passports, we are required to pay the fees normally charged to the Swiss for all visas. As these fees are extremely high for me as an earner inside Germany, the passport office of the Foreign Office kindly issued my exit visas free of charge in various instances last year. It was pointed out to me, however, that this procedure involved quite a lot of trouble each time and that it would be immediately eliminated if I owned a diplomatic passport.

That is why I am so bold as to petition you, highly esteemed Sir, to kindly issue a diplomatic passport for me and my wife. I would be most particularly grateful to you for satisfying this wish, as this would mean not just major monetary savings but a great simplification of my voyage as well. Much looking forward to your valued response, perhaps by telephone (Nollendorf 2807), I ask you please to forgive me for this trouble.

In utmost respect, yours truly.

362. To Max Wertheimer

Berlin, 18 September 1922

Dear Wertheimer,

Your suggestion of Brinkmann is essentially acceptable,[1] but I would definitely prefer to have you there because I know you very much better. You would be there exclusively as my personal substitute; this business has nothing to do with the Ministry, officially. To what extent and whether you would like to or should inform the Ministry depends entirely on your decision. What is now taking place in Geneva is not a meeting of our Committee but a session of the Council of the League of Nations.[2] But it is possible that a meeting will take place in the late fall for about one week; if not, then probably during the first quarter of next year. The fact that you are a Czechoslovakian and a Jew is completely insignificant; it is already a nice gesture to the Ministry that I choose a substitute at all. With Brinkmann there is the complicating factor that he was employed (perhaps he still is) for many years in the Foreign Office. He cuts more of a figure as a diplomat than a scholar. The language difficulty is reduced in that Mme. Curie and the secretary of the Committee can speak German properly.[3] You do not need to present yourself actively at the meetings but can contact the individuals instead.

Dear Wertheimer, write me point-blank, yes or no, hopefully yes; and write me immediately on the same day that you receive this letter; otherwise I am not going to be able to do anything anymore, because I am departing on the 29th of this mo.

Cordial regards,[4] yours,

A. Einstein.

363. From Max von Laue

Leipzig, 18 September 1922

Dear Einstein,

According to reliable news I received yesterday, events could be taking place in November that would make your presence in Europe in December desirable.[1] Do reconsider whether you want to travel to Japan anyway.

Cordial regards, yours,

M. Laue.

364. From Max Wertheimer

[Prague, 19 September 1922]

[Not selected for translation.]

365. To Svante Arrhenius

Berlin, 20 September 1922

Esteemed Colleague,

I thank you cordially for your letter.[1] I am taking the trip to Japan on the basis of an invitation issued from there. No unpleasant circumstances prompted me to accept this invitation, rather the persuasion not to let this opportunity to acquaint myself with such interesting countries slip by. As I am contractually bound with Japan, it is completely impossible for me to postpone the voyage further. I am going to be back in Europe in March. Hoping that thus my proposed invitation to Sweden would only be delayed, not canceled,[2] I am, with best wishes and greetings, very sincerely yours,

A. Einstein.

P.S.[3] I would be pleased if I could still shake your hand before my departure at the beginning of next week. It would be best if you could visit me on the occasion of your passing through.[4] I would, however, also gladly meet you wherever is convenient to you, if I can still somehow manage to arrange it.

366. To Hans Reichenbach

Wednesday [20 September 1922]

Dear Mr. Reichenbach,

Although I am at the point of leaving, I would be very pleased if you told me a little about the progress you are making in the axiomatic analysis.[1] On Saturday would be most preferable to me. On Monday I am not sure anymore whether I shall still be here.

Best regards, yours,

A. Einstein.

367. To Carl Beck

[Berlin,] 22 September 1922

Dear Dr. Beck,

Unfortunately the rescue of the Fatherland once again has come to naught. 300 billion Goldmark equals 70 *billion* dollars.[1] It would therefore require 70,000 photographs, each of which is worth $1,000,000! We, I am sorry to say, only made the mistake of 3 decimal points, which shows in what an enjoyable mood the champagne put us.

A cheery goodbye. With friendliest greetings, yours,

A. Einstein.

368. To Jacques Loeb

[Berlin,] 22 September 1922

Highly esteemed Prof. Loeb,

I think I now understand the either/or behavior (acidic or basic) very simply according to your theory.[1] As colloids are supposed to be very weak acids and bases, the molecules can only separate a considerable number of H^+'s when the H^+ concentration in the solution is substantially lower than in pure water. Likewise, the OH^- separation will be connected with a minuscule concentration of OH^- in the solution. Both conditions cannot be satisfied at the same time because of the equilibrium condition

$$C_{H^+} + C_{OH^-} = \text{const.}$$

This interpretation still has its problems, though, because one does not quite understand why [both] these colloids should work in pure water.

Circumstances here are indeed very unpleasant because a spirit of intolerance and close-mindedness dominates among local intellectuals. Yet I am not suffering under it because I am largely unattached to people and think nothing of their opinions and conduct. Even materially, I am largely independent in that the compensation I receive from the Academy is practically insignificant, so I can dispense with it without any upset to my equanimity. My personal endangerment is probably not as bad as the papers described it. So—excepting my extremely interesting voyage to Japan, which I am embarking on in a few days—I am considering staying calmly put. I thank you wholeheartedly for your repeated generous offer;[2] if really serious problems should arise, I will gratefully and hopefully return to it.

I hope your wife is well again. I deeply regret that you could not come to Leipzig for that reason but believe that for the rest not much good has been lost by it. The strength of your evidence will suffice in asserting your fine ideas about colloids. Truth does not need to be defended, because it is strong enough on its own!

If you wish to donate something to science in Germany, be sure not to give it to organizations that reach their decisions according to the principle of least odium, rather than according to the free assessment of a few competent people.

With best wishes to you and your family and cordial regards, I am yours,

A. Einstein.

P.S. If it should be a matter of support for academic institutions, I would like to draw your attention to the Jewish university fund in Kaunas [*Kowno*], which venture to assist Eastern Jewish students is waging a truly heroic battle for survival.[3]

369. From Michele Besso

Bern, 24 September 1922

Dear Albert,

I don't know anything about your travel itinerary; these lines will perhaps still reach you in Berlin to bring you my farewells for the long journey. If all goes well, these two-times-six weeks' sea voyages will be good periods of rest for you to enjoy to the fullest.

The last I heard of you reached me through my cousin Arrigo Cantoni,[1] who was very pleased about your friendly welcome. I haven't heard directly from you for ages now; you didn't receive my last two letters[2] or you thought that it ought to have already been clear to me that I should have left you alone about the related content because it lies beyond your field of activity.

In connection with the familiar paradox (which forgets the basic assumption of the special theory of relativity) according to which an observer, who is moving away from us and back again at quasi-velocity of light, has *our* clocks necessarily seem to have come to a virtual standstill, just as *his* clocks do for us, Vero raised the question of what the situation is for an observer, moving against us at quasi-velocity of light, who makes a tour around the spherical world: "*To him* we would be aging noticeably during the world tour, just as much as he would *to us*." This paradox is similar to the one I already presented to you of the Lorentz contraction of two rings moving most rectilinearly against each other and is solved in the same way. Nevertheless, this case does seem interesting to me didactically, as homework for a sufficiently developed beginner to think through *exactly*. (It is a fact that the real distribution of masses that results for one observer of a spherical world yields a *different* world for the other observer. What kind of a difference?)

So, warm greetings and bon voyage! And if these lines only reach you among the cherry-blossom worshippers, may they hail you over there!

In all of our names, yours,

Michele.

How's that Pole's six-dimensional cylinder world doing?[3] And the hyperbolic possibilities that I only became acquainted with through Jean Becquerel's pretty little essay?[4]

I myself am at the point of freezing stiff. With the exceptions of a few specialties completely foreign to me and a couple of theologians, there's pitifully little going on over here! Vero, with wife & Marco, in Zurich since 6 weeks ago.[5]

370. "Comment on Franz Selety's Paper: 'Contributions to the Cosmological System'"[1]

[*Einstein 1922q*]

Dated [12–25] September 1922
Received 25 September 1922
Published 19 December 1922

In: *Annalen der Physik* 69 (1922): 436–438.

From the standpoint of Newton's theory there is, it must be conceded, something [p. 436] to the hypothesis of a "molecular-hierarchical" structure for the universe of stars, even though the hypothesized equivalence of spiral nebulae with the Milky Way should be regarded as refuted by recent observations.[2] This hypothesis unassailably explains the non-[luminosity] of the background sky and avoids Seeliger's conflict with Newton's law without conceiving matter as islands in empty space.[3]

Even from the point of view of the general theory of relativity, the hypothesis of a molecular-hierarchical structure of the universe is *possible*. Nevertheless, from the standpoint of this theory the hypothesis should be regarded as unsatisfactory. This will be briefly examined once again in the following.[4] If the geometrical and inertial properties of space are influenced or partly determined by matter, then the compelling view would be that this determinedness was fully conditional, as is the case according to the general theory of relativity if the mean density of the matter is finite and the universe is spatially closed. I would like to attempt to illustrate this with a more simple, fictitious—if imperfect—case.[5]

It is assumed that one could only know gravitation through a close study of the mechanics of masses that are available to us in laboratory experiments. The spherical shape of the Earth is unknown to us. Then the following theory could be postulated. A vertical "cosmic" gravitational field primarily exists that extends everywhere into infinity. The Earth extends downwards into infinity. Its gravita-

[p. 437] tional influence is negligible against the cosmic gravitational field.[1] The cosmic gravitational field is modified by gravitational effects from masses empirically accessible to experience on the Earth's surface.

Although the assumed cosmic gravitational field agrees with Poisson's equation just as do the gravitational fields of the experimentally accessible masses on the Earth's surface, this interpretation would be unsatisfactory because the cosmic field itself had been assumed to be without any material cause. The idea that the gravitational field, which mainly determines the fall of bodies on the Earth's surface, did not exist independently but was caused by the Earth's body, would certainly be received as a major advance.

Today the need to attribute the world's metric and inertial field to physical causes is not being demanded with similar intensity only because this latter field is not being perceived so clearly as a physical reality as was the physical reality of the "cosmic gravitational field" in the above example. To a later generation, however, this contentedness will seem incomprehensible.

Mach's postulate, according to which the inertial effect of an individual body should be determined by the totality of all the others in the same way as its gravitational force, is as little satisfied by the "molecular-hierarchical universe" as by the "island world." I find it hard to understand how Mr. Selety could have allowed this fault in his system to escape him. This fault is particularly serious as, even without considerations of a cosmological nature, the general theory of relativity can show to first approximation that the bodies behave as should be expected according to Mach's thought. In this regard I refer to the fourth of my "Four Lectures on the Theory of Relativity" (delivered in May 1921 at Princeton University).[6]

Finally, one more point is mentioned that causes confusion not just in Selety's

[p. 438] article but in many relevant publications. The theory of relativity states: The natural laws should be formulated independent of any special choice of coordinates, as the system of coordinates does not correspond to anything real; the simplicity of a hypothetical law is only assessed by its generally covariant formulation. From this

[p. 437]   [1] Please excuse that this hypothesis does not fit with Newton's law.

it does not follow, however, that one is not allowed to facilitate the description by a suitably chosen frame of reference without violating the relativity postulate. If, for ex., I approximate the real world by a "cylinder world" with evenly distributed matter and choose the time axis to be parallel to the one generated by the "cylinder," this does not mean the introduction of an "absolute time." Either way, there is no such system of coordinates in the universe in which the formulation of the natural laws would be preferred. As regards the real world, an exact definition of such a system of coordinates would be impossible anyway, even if the real world could be roughly approximated *by* that cylinder world. The principle of relativity does not contend that the world is describable in an equally simple or even in the same way as all systems of coordinates, only that the *general laws* of nature are the same with respect to all systems (more precisely: that the hypothetically possible natural laws should only be weighed against each other, as regards their simplicity, in their generally covariant formulations).

371. To Franz Selety

Berlin, 25 September 1922

Esteemed Doctor,

Thank you very much for sending me your paper[1] and for your letter.[2] I sent a comment on the paper to the *Annalen* and hope that the editors will send you a correction proof when it has been typeset.[3] I have nothing substantial to object to in the first part of your paper;[4] however, I see that you have not completely comprehended the arguments in the general theory of relativity about the closed space connected with Mach's thought.[5] They are presented in a booklet that recently appeared in Braunschweig by Vieweg: *Four Lectures on the Theory of Relativity*;[6] I will ask the publishing house to have a copy sent to you. With kind regards.

372. To Edgar Zilsel[1]

Berlin, 25 September 1922

Esteemed Colleague,

I have not been able to put as much time into the study of your paper[2] as would have been desirable.[3] But as far as I can judge, the main problem remains unsolved, which seems to me to consist in the following: statistical mechanics assumes the validity of the following statement without proof.

A closed mechanical system originating from any given initial state describes a path of the following character along its energy hyperspace. Given a segment of the hypersurface, there is a temporal limit of the ratio: lingering period/total time, which can be written in the form [   ][4] where [   ][4] is a continuous function on the surface independent of the special initial state. *A priori* it would be possible that such a continuous function did not exist at all; if it could be proved that it does exist, its independence from the chosen initial state would probably also be provable. But this first statement—the validity of which I personally do not doubt for the systems coming statistically into consideration—is, to my knowledge, still unproven. I cannot see that you have proved it anywhere, either. As long as it is not proven, an important term is missing in the chain of proofs for statistical mechanics. (That some of the consequences of the law partly are in part not valid in nature, in my opinion can only prove the non-validity of the mechanical equations.)

Regards from your colleague.

373. To Michele Besso

[Berlin,] 26 September 1922

Dear Michele,

Somewhere around the 3rd or 4th of October I'll be visiting you in Bern in transit to Japan. Then we can talk about everything that I, the wretch, did not write you in response.[1]

Warm regards, yours,

Albert

P.S. Zangger is here. I met with him today.

374. To Eberhard Zschimmer

Berlin, 27 September 1922

Esteemed Doctor,

The observations in your essay appear right to me, at least from the aspect of physics, as the only one I can judge with certainty.[1] However, in my view the important question for a comparison between relativity theory and Kantian philosophy does not stand out prominently enough: Are the spatio-temporal forms, etc., that [also] underpin relativity theory "a priori" just suitable means of description, to be viewed merely as conventions? Or are they necessary, individually unalterable givens by the very nature of human thought? I personally back the former standpoint also advocated, e.g., by Helmholtz and Poincaré, whereas it appears to me that Kant's standpoint was more the latter.[2]

In great respect.

375. To Romain Rolland

[Berlin, on or before 30 September 1922][1]

Dear Romain Rolland,

Zangger and I are happily sitting together and thinking of you. I am feeling well. I have the chance to travel to Japan in a few days. Hoping to see you soon, happy and healthy, I am, with cordial greetings, yours,

A. Einstein.

376. To Pierre Comert

Berlin, 1 October 1922

Dear Mr. Comert,

I took a lot of trouble to find someone to represent me but a series of unhappy circumstances prevented me from bringing the business to a good conclusion. Initially I had the intention of having myself represented by my friend the psychologist Wertheimer, professor at the University of Berlin, but he was on vacation in Prague; and he is so unsociable that he hesitated a long time to accept.[1] He let me wait like this so long and so thoroughly that in the end it was too late. Then I sought out Mr. Troeltsch, but he, too, was away on vacation, so today I have to leave without having terminated this affair.[2]

Please do not attribute this failure to ill intentions on the part of my friends here but to the fact that none of the persons I had envisaged were present in Berlin; and I am persuaded all ⟨were⟩ would have been appointable to replace me.

With amicable regards, yours,

A. Einstein.

377. To Michele Besso

[Zurich, 4 October 1922][1]

Dear Michele,

I'm only coming on Friday, 10 o'clock,[2] because I had to postpone the departure by one day, that is, together with Vero.[3]

To a happy reunion, yours,

Albert.

378. Poem to Albert and Luise Karr-Krüsi

[Hof Mayenbühl, on or before 6 October 1922][1]

[Not selected for translation.]

379. Travel Diary Japan, Palestine, Spain

[6 October 1922–12 March 1923]

[p. 1]   6th October. Night trip in overfilled train after reunion with Besso and Chavan.[1] Lost my wife at border.

7th Oct. Sunrise shortly before arrival in Marseille. Silhouettes of austere flat houses surrounded by pines. Marseille, narrow alleyways. Well-endowed women. Vegetative living. We were taken in tow by honest-looking youth, let down in ghastly inn by the station. Beetles in the morning coffee. Errand to the shipping office and to the old harbor near the old city quarter. By the ship,[2] energetic dispatch of the scamp, who drove away offended after a jarring ride to harbor on luggage cart over dreadfully bumpy pavement of Marseille. There, just verbal baggage check. Friendly welcome by ship officer. Comfortably settled in cabin. Made acquaintance of young Japanese physician whom a Munich medical doctor had thrown out with an inflammatory ultimatum to scholars of the Entente countries.[3]

8th Oct. Leisurely morning in the harbor. Joyous greeting by rotund Russian Jewess, who recognizes my being a Jew. Noon casting off in bright sunshine. Virtually only English and Japanese on the ship. Quiet, fine company. After exiting the
[p. 1v]   harbor, wonderful view of Marseille and its framing hills. Then by dazzling, craggy

chalk cliffs. Coast slowly recedes away to the left. Conversation with Europeanized Japanese physician Miyake from Fukuoka.[4]

Afternoon, 4 o'clock, safety drill. All passengers—wearing the life-belts stored in their cabins—must report for review at the spot where the lifeboat designated to them for manning in the event of danger is located. Crew (all Japanese) friendly, precise without pedantry, without individualistic stamp. He (the Japanese) is unproblematic, impersonal, cheerfully fulfills the social function falling to his lot without pretension, but proud ⟨about⟩ of his community and nation. The abandoning of his traditional ways in favor of European ones does not gnaw away at his national pride. He is impersonal but not really reserved; for as a predominantly social being, he seems not to possess anything personally that he could have the need to be taciturn or secretive about.

9th Oct. 4 o'clock in the morning, major racket. Cause: scrub-down of vessel. Great cleanliness of people and things. The ship is as if licked clean. It is already becoming significantly warmer. The sun assuages me and removes the gulf [p. 2] between "ego" and "id." I begin reading Kretschmer's *Physique and Character*.[5] Wonderful description of temperaments and their physical character. I can thus treat objectively many of my fellow beings but not myself, because my type is a hopeless mixture. Yesterday I looked into Bergson's book on relativity and time.[6] Strange that time alone is problematic to him but not space. He strikes me as having more linguistic skill than psychological depth. He is not very scrupulous about the objective treatment of psychic factors. But he does seem to grasp the substance of relativity theory and doesn't set himself in opposition to it. Philosophers constantly dance around the dichotomy: the psychologically real and the physically real, and differ only in evaluations in this regard. Either the former appears as a "mere individual experience" or the latter as a "mere construct of thought." Bergson belongs to the latter kind but objectifies in *his* way without noticing.

I've been thinking about the gravitation-electricity problem again. I find that [p. 2v] Weyl is right that a $g_{\mu\nu}$ field, or an invariant ds disconnected from the electric one, has no reality, therefore cannot be mathematically objectified either. But I do think that the final solution is further away from Riemann than for Weyl and also think that ⟨disconnected from the electromagnetic⟩ nothing directly corresponds to the elementary law of vectorial parallel displacement and that this formalism has no objective legitimacy as the basis of the theory beyond Riemann. However, it does seem to me possible that the field theory will be retainable; whether the expression of the natural laws by differential equations will also be retainable appears doubtful.[7]

10th Oct. This morning sailed past the Stromboli on the left-hand side. Clouds of steam.[8] Magnificent in the morning sun.

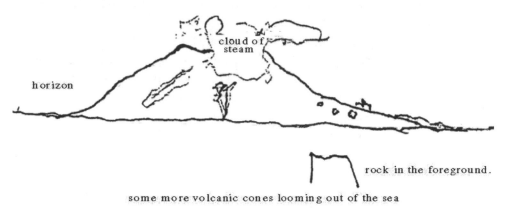

horizon

cloud of steam

rock in the foreground.

some more volcanic cones looming out of the sea

[p. 3]   Mild, intoxicating air. Steel-colored sea. Italian suggestion of solid ground diffusely cloudy. Japanese women crawling about with children. They look rosy and bedazzled, almost as if ⟨schematic⟩ stylized. Black-eyed, black-haired, large-headed, pattering.

Yesterday, was still shaken by reading Kretschmer. Felt as if grabbed between thongs. Hypersensitivity transformed into indifference. During youth, inwardly inhibited and alienated. Pane of glass between subject and other people. Unmotivated mistrust. Paper alternate world. Ascetic impulses. I am so thankful to Oppenheim for the book.—[9]

Splendid passage through the Strait of Messina at midday. Bald, forbidding mountain landscape on both sides. Towns likewise forbidding, predominance of the horizontal. Low, flat white houses. Overall impression: oriental. Temperature relentlessly rising. I'm convinced that the Greeks and Jews of classical antiquity lived in a less sluggish atmosphere. It is no coincidence that the zone of active intel-

[p. 3v] lectual life has since moved northward. The pleasanter for vegetating. The quest for contentment is here more easily satisfied, because it is already almost too hot even for lively wishing.

11th Oct. Sunny day. Whitish sky. Sea somewhat restless. I now think that seasickness is based on orientational dizziness, not directly on the apparent gravitational changes in direction and magnitude.

12th Oct. Radiant day. Sea quiet, almost windless. Atmosphere completely transparent. Distinct horizon. Almost calm. $\frac{1}{2}$ past 7 o'clock in the morning, rocky mountains of Crete visible, falling steeply into the sea. In the evening, wonderful sunset—purple with finely illuminated narrow wind-swept clouds. Then, sparkling starry sky with prominent Milky Way in mild air. ⟨Noon arrival Port Said.⟩ In morning, conversation with [Ishii] and Japanese lawyer.[10] Both strongly Europeanized—down to facial expressions. The former is very cautious and realistic. In the evening, conversation with exporter from Hamburg betw. 60 & 70, shrewd and level-headed.

13th. Midday, 15:00 hours, Port Said. Water green before the coast was visible. In the Mediterranean, deep blue. Long artificial dams with [irregular hewn blocks?]. Houses, as far as visible from the sea, European style. Picturesque Egyptian sailboat of the type   [p. 4]

In the harbor, a swarm of rowboats with screaming and gesticulating Levantines of every shade, who lunge for our ship. Deafening clamor, as if spewed from hell. The upper deck transformed into a bazaar, but nobody buys anything. Just a few pretty, athletic young fortunetellers are successful. Bandit-like filthy Levantines, handsome and graceful to look at. Sunset, sky *locally* and very intensively reddened, as if flaming. On the facing side of walls and buildings, one of those garish colors often illustrated on tropical pictures. In the evening, conversation with Fren[ch] civil servant from Siam. Encounter with Japanese sister steamship. Nationalistic raptures by the crews. Japanese in love with their country and nation.

14th. Awakening during canal passage through desert, cool temperature.[11] Palms, camels. Piercing yellow color. Frequent views of immense surfaces of sand, intermittently broken by tufts of vegetation. Steamer encounter in canal. Wide desert landscapes. Green Bitter Lake.[12] Hilly shore in brilliant lighting. Wonderfully   [p. 4v] clear air. Then, last part of the Suez Canal. At mouth, city of Suez, villa quarter of the canal administrators very pretty. Homes with verandas and palms. Welcome sight after so much desert. Barren rocky mountains on either side. Forerunners of the Sinai Arabian small merchants sail up. They are handsome sons of the desert, full-grown, flashing black eyes, and better mannered than in Port Said.

Gulf opens out after departure. Sun vanishes after 6 over the Egyptian desert mountains. Local reddish violet tint with magnificent ⟨relief⟩ silhouette of the Ütliberg-like mountain range.[13] Additionally, reddish yellow reflection in the east. Then, wonderful star-lit night. Never saw the Milky Way so fine. Spots with distinct borders. Oblong spots clearly leaving the disk edgewise. At night, naked with fan. Not troublesome.

15th. In Red Sea under overcast skies without view to the shore. Lightly cloudy skies. Brief rain shower in the morning. Greenhouse temperature. At noon passed by two small flat coral islands.

[p. 5]   16th. Saw two sharks next to steamer with huge dorsal and tail fins. Flying fishes, too. Temperature rising but very bearable. Wonderful sunset. Reflection in the east with reddish surface of the sea. Eastern sky blue-gray, reddish further up. Venus sparkles beautifully and is mirrored in the sea.

17th. In evening, mighty distant storm on the Arabian side under bright sky. A stroke of lightning of small angular area almost every second. Thereupon a strong wind rose with swelling seas.

18th. Reached the exit of the Red Sea in rough waters. Arabian mountain range visible; lone-standing mountains looming out of the sea, glowing red in the morning sun.

19th. Enteritis with ghastly hemorrhoids.

Japanese professor comes to the rescue.[14]

20th. More or less shipshape again. Steamed past the Somali coast (cape).[15] Splendidly illuminated mountains in the afternoon sun. Brisk breeze with moderate swell.

25th. Passed by coral islands with palm groves. At night, tropical rain. Now, ½ past 9, distant tropical rain visible (with variably cloudy skies).

[p. 5v]

On approaching the equator, clouds from the Arabian Gulf gathered. In October the most illuminated ⟨layer⟩ part of the Earth's surface lies at the equator. There moisture-saturated air rises, thereby precipitating water. Air flows in afterwards from both (north and south) sides (winds of the subtropical zone, diverted by the Earth's rotation). Seasons shift the zone of maximum warming and hence the entire complex of phenomena northward or southward. Additionally, *land* intensifies the temperature maximum as opposed to the sea, because it heats up to a greater degree.

The strong rising currents of air also encourage storminess. We often saw sheet lightning without any or with only weak cloud formation.

In the Arabian Gulf many sharks and flying fishes. Nothing of the kind was to be seen on the open sea, which is several thousands of meters deep. Little light penetrates to the sea floor, so weak plant growth at the bottom, scant fauna below, the less so above.

[p. 6]

During the night, ship's siren. Thought it was an accident. But was just acoustic signals owing to intransparency of the air in heavy rainfall, in case of encounters with vessels. Temperature very tolerable, only inside the cabin very hot (between sunbathed wall of ship and a corridor bordering on the engine room. Often feel unwell; Japanese doctor always helpful.[16] On the ship I was ⟨very⟩ frequently photographed, with and without people, mainly by Japanese.

Yesterday I recalculated the electromagnetic equations in a vacuum

$$\left(\frac{\partial \varphi^{\mu\nu} \sqrt{-g}}{\partial x^{\nu}} = 0 \right)$$ according to Weyl's geometry in hope of finding an expression

for the current density. But a useless result $\varphi^{\mu\alpha} \varphi_{\alpha}$ comes out.[17]

28th. Yesterday evening we approached ⟨Honk⟩ Colombo with considerable delay. Before the coast came in sight we got caught in a severe tropical storm with a cloudburst, forcing the ship to stop. When it brightened up around 9 o'clock, it turned out that we were near the harbor. A pilot came up in a rowboat and we soon docked next to another Japanese steamship. We saw here for the first time an elderly Indian, fine, distinguished face with gray beard, who brought us two telegrams and—entreated us for a tip. We saw other Indians as well, brown to black sinewy figures with expressive faces and bodies and humble demeanor. They look like nobles turned into beggars. Much pride and depressiveness are indescribably united there.

[p. 6v]

This morning at 7 o'clock we went on land and together with the Du Plâtre couple viewed the Hindu quarter of Colombo and a Buddhist temple.[18] We drove in individual little carts that were drawn on the double by herculean and yet fine people. I was very much ashamed of myself for being a part of such despicable treatment of human beings but couldn't change anything. These beggars of majestic proportions descend in droves on any stranger until he has capitulated before them. They know how to entreat and to beg until one's heart is wrung out. On the streets of the native quarter one can see how these fine people spend their primitive lives.[19] For all their fineness, they give the impression that the climate prevented them from thinking back or ahead by more than a quarter of an hour. They live in great filth and considerable stench down to the ground, do little, and need little. Simple economic cycle of life. Far too packed together to allow any special

[p. 7]

existence for the individual. Half-naked, they show their fine and yet powerful bod-
[p. 7v] ies and their fine, patient faces. Nowhere shouting like the Levantines in Port Said.
No brutality, no liveliness of the market place, but quiet, drifting along, albeit not
lacking in a certain lightheartedness. Once you take a proper look at these people
you can hardly appreciate Europeans anymore, because they are softened and more
brutal and so much rougher and more covetous—and therein unfortunately lies
their practical superiority, their ability to take on big things and carry them out. In
this climate, wouldn't we, too, become like the Indians?

In the harbor, lively bustling. Herculean laborers with shiny black bodies take
care of the cargo. Divers perform their neck-breaking craft. Always that smile and
self-effacement for filthy money and satisfied people who are mean enough to be
able to enjoy it. At ½ past 12 we set out into the rainy desert of water. Ceylon is a
plant's paradise and yet a stage of pathetic human existence.

31st. Yesterday it was the mikado's birthday.[20] Celebration on the upper deck
[p. 8] before noon. Banzai and singing of the national anthem, which sounds very alien
and is strangely structured.[21] Japanese very devout. Unsettling fellows whose
nation is at the same time their religion. Weather clearer. Traveling along the coast,
first of Sumatra now along the mainland. Interesting refractive effects at the edge
of the horizon, because of the temperature or humidity gradients, ships seem to be
floating in the air, likewise the far shore. Yesterday evening, spontaneous show by
the Japanese. One man sang and wailed like a tomcat whose tail has been stepped
on, in accompaniment from time to time with a wild gesture, he coaxed a tone out
of a short-fretted guitar-like instrument that seemed not to have anything to do with
the sing-song.[22] Slender, distinguished young Japanese (botanist) performs astonishing magic tricks, mainly with three red balls that he makes disappear and reap-
pear again. Yesterday, saw navigational instruments and was acquainted with the
[p. 8v] standard methods for determining position. Compass very primitive, with balanced
inertial moment. Sextant, clock. Speedometer, the propeller of which is towed
behind on a long line. Tomorrow morning Singapore. Mrs. Ishii turns out to be the
aunt of Sakuma's young bride.[23] Weather is brightening up.

2nd November. 7 o'clock in the morning, through narrow passage between small
green islands. Arrival in the harbor of ⟨Colombo⟩ Singapore. There Zionists were
waiting for us and gave us a warm welcome. Mr. and Mrs. Montor (he, a brother of
the Hamburg actor, himself also theatrically talented, *she*, of the true Viennese type
but having grown up in Singapore) brought us to their spacious home.[24] Drive
through the wonderful zoological garden through various parts of town, in not too
hot weather. I found out that the indefatigable Weizmann had decided to make use
of my voyage for the Zionists' benefit.[25] Having arrived at the house, I immediately had to compose an answer to a welcoming address that Mr. Montor had writ-

ten with a friend in the ⟨service⟩ at the behest of the Zionist Association in Singapore. The address that was supposed to be ceremoniously handed over to me was on silk and in a valuable silver case (Siamese ornamental relief), was quite [p. 9] clever in content but far too solemn and—as their father himself with a grin let on—concocted with the aid of *Meyer's Convers. Lexikon*.[26] At 11 o'clock a journalist joined us and I had to tell Mr. Montor, his friends, and the newspaper man, among other things, the beetle story about spherical space. After the splendid midday meal our friendly hosts took us to a guestroom where we were allowed to sleep under a mosquito cage. Next door was a lavatory with chamber pots and a big washing tub. Elsa was appalled, I a little also, because of such unaccustomed installations. Well rested, we drove in Montor's car at 4 o'clock to Meyer, the Jewish Croesus of Singapore.[27] His palatial house with Moorish-like halls is situated on the peak of a hill with a view of city and sea; directly below it, a sumptuous synagogue essentially for contact between Croesus and Jehovah, built by the former. Croesus is a still quite upright, slim, strong-willed ancient of 81 years. Gray goatee, narrow [p. 9v] flushed face, thin Jewish hooked nose, clever, somewhat sly eyes, a black skull cap on his well-arched brow.[28] Similar to Lorentz, just with his shining kindly eyes exchanged for cautiously shrewd ones and the facial expression speaking more of planned order and work than—with Lorentz—philanthropy and civic sense.[29] It was the fortress that, according to Weizmann's designs, I was to conquer for the benefit of the Jerusalem University. His daughter, narrow ⟨black⟩, pale noble face, is one of the finest Jewish female figures I have ever seen. ⟨She proves⟩ Looking at her, one is tempted to take seriously the joke of the "oldest aristocracy."[30]

So when we arrived up there, first came the business of picture taking. Croesus next to me, surrounded by his family and many Jewish couples, a group photo.[31] After this enormously important act came filing into a big oriental refreshment hall. Malaysian band played Viennese and Negro music in European schmaltzy coffee-house manner. I sat with Croesus and the—archbishop (beside spouse), a seasoned, [p. 10] exclusively English-speaking slim, big-nosed English nobleman, who not in vain flirts with Croesus's money, without the least weight on his conscience.[32] Dreadful linguistic calamity with tasty cake. Then Montor brought us (Croesus and me) to two raised [seats] with lectern set up at the end of the hall, sat down to one side, delivered his address with a convincing voice, with my reliably translated reply safely in his pocket. I responded freely, during which he simulated taking notes, in order to be able thereupon to read out my answer, composed that morning and translated by him and his friend, which he gave off as an improvised translation of my free address, the slick operator. Then endless handshaking, reminiscent of America. Genuine cordiality among Jews everywhere. When I managed to get away, already quite seasick, and the sun had already hidden itself, we quickly drove

home, where I had to provide autographs for a small batch of albums. Then we
[p. 10v] drove through the Chinese quarter (awful thronging but there wasn't enough time
to look, only to smell) to Croesus's dinner, a pompous meal in an open hall for
about 80 people. The meal was wholesome and endless. I had to get up at last
because I couldn't even look at food anymore, let alone eat it. Then the familiar
band returned and merrily went at it again, ⟨The ancient⟩ and everyone danced. Not
even Croesus spurned doing so, after having shown how mightily fit his 80-year-
old stomach still was. At last the well-planned assault by Weizmann on Croesus
took place (about contribution to Jerusalem University); I still do not know whether
despite much effort one of my bullets could penetrate Croesus's thick outer skin.[33]
Then we drove home and (after finishing off a few more albums) slipped into the
well-earned mosquito cage. At night, enormously hefty tropical downpour with
thunderstorm. I quickly secured all the window shutters, which nevertheless only
partially helped against so much water. Throughout the stay, the temperature was
[p. 11] not particularly elevated, but the great humidity makes one think of a greenhouse.
There is something orientally intoxicating about it.

3rd November. After breakfast, wonderful drive over rubber plantation hills to
the harbor. Splendid vegetation; cheerful-looking Chinese villas. View over sea
with islets. Another fine group of trusty Jews come onto the ship. Cast off only
around 11 o'clock in a scenic trip between green islands.

The Chinese may well surpass every other nation in diligence, frugality, and
progeny. Singapore is almost completely in their hands. They enjoy great respect
as merchants, far more than the Japanese, who are deemed unreliable. It may be
hard to understand them psychologically, I hesitate to try ever since the Japanese
singing remained so entirely incomprehensible to me. Yesterday I heard another
one singing away again to the point of making me dizzy.

[p. 11v] 7th Nov. In the interim, rainy weather in greenhouse air. In Singapore, addition
to the company on the steamer of two old congenial Swiss officers and a young
German salesman. From evening of 5th to evening of 6th, typhoon with huge
waves, wind, and much hefty rain. The ship danced about mightily. Wonderful
show at the prow. Many flying fish startled by the steamer; powerful vertical
motion. Today the sea still shows reminiscences in its motion. Fatigue from eternal
balancing. Women much more seasick than men.

10th. On the morning of the 9th we arrived in Hong Kong. It is the prettiest landscape I've seen up to now on the entire voyage. Mountainous, stretched-out island
next to the similarly mountainous mainland. Between the two, the harbor. Many
small steep islets. The whole thing like a half-drowned area from the Alpine foot
hills. The city lies terrace-like at the foot of a gently sloping, roughly 500-m high
mountain. Air pleasantly cool.[34]

Declined the reception by the Jewish community with thanks.[35] But two Jewish [p. 12]
businessmen spent the entire day with us.[36] In the morning, started tour around
the island by automobile. Views onto sea, fjordlike bays, and mountainsides of
inexhaustible variety and magnificence. Along the way we ate lunch in an Ameri-
can-type luxury hotel, where both our guides not only spoke animatedly with me
about the country and science but also revealed a great affinity for worldly
pleasures.[37] On the homeward trip we saw a Chinese fishing village composed of
sailing barks, a seemingly very cheery Chinese funeral and—tormented people,
men and women, who have to hammer stones and carry stones for 5 cents per day.
Thus the Chinese are severely punished for their fecundity by the heartless economic machine. I think they hardly notice it in their lethargy, but it is sad to see.
Incidentally, they supposedly carried out a successful wage strike with remarkably [p. 12v]
good organization a while ago.[38] In the afternoon we visited the Jewish clubhouse, which is set in a lush garden at quite a high elevation and has a magnificent
view of city and harbor. There are supposedly only 120 Jews there, mostly Arab,
whose religiosity seems to have frozen more into formality than is the case with our
Russo-Europeans.[39] In the clubhouse two women socialized with us, the wife of
one of our hosts and her sister. I am now convinced that the Jewish race has maintained itself quite purely in the last 1,500 years, as the Jews from the lands of the
Euphrates and the Tigris are very similar to ours. The feeling of belonging together
is also quite strong. We also all drove to the summit of the mountain together, at the
foot of which the city lies (by cable railway; Chinese and Europeans separated). At
the top, grandiose view onto harbor, island mountains, and sea. The sight of the
many small islets looming steeply out of the sea reminds one of the sea of mist in
the Alpine foothills. In the evening there was a sudden storm that snatched away [p. 13]
my hat into the street so I had to run after it with all my might to get it back.

This morning I visited the Chinese quarter on the mainland side with Elsa.
Industrious, dirty, numbed people. Houses very uniform, verandas ⟨arranged⟩ like
beehive-cells, everything built close together and monotonous. Behind the harbor,
one eating place after the next in front of which Chinese don't sit but squat on
benches as they eat, as Europeans do when they answer Nature's call out in the
leafy woods. Quiet and civility in all doings. Even the children are spiritless and
look numbed. It would be a pity if these Chinese would push out all other races. For
the likes of us the mere thought is unspeakably boring. Yesterday evening three
Portuguese middle-school teachers visited me, who claimed that the Chinese could
not be trained to think logically and specifically had no talent for mathematics. I
noticed how little difference there is between men and women; I don't understand
what kind of charms Chinese women have to enthrall the corresponding men so [p. 13v]

decisively that they are so unable to defend themselves against the formidable blessing of children. At 11 o'clock the Kitano Maru left through the shining green sea between green island mountains that were delightful in shape and color but bald, i.e., without tree growth. The current lush flora on Hong Kong is supposed to have all been planted by the English. They have an admirable understanding of governance. The policing is done by foreign-born black Indians of awe-inspiring stature, Chinese are never used. For the latter the English have built a proper university in order to captivate those Chinese who have prospered.[40] Who can match that? Poor Continental Europeans, you don't understand how to take the bite out of nationalistic opposition through tolerance.

11th. At night, wonderful marine phosphorescence. The crests of the sea's waves glowed bluish as far as one could see.

[p. 14]   14th. On the 13th about 10 o'clock in the morning, arrival in Shanghai. Travel along flat, picturesque, yellowish-green illuminated shores upriver.[41] Departure of the two Swiss officers, the one from Bern who so kindly mended my little pipe, as well as the chauvinistic but otherwise well-meaning young German former officer. In Shanghai, welcomed on the ship by Inagaki and spouse, our dear escorts Shanghai-Kobe[42], and by the German consul,[43] Mr. and Mrs. Pfister.[44] First, journalists, a respectable group of Japanese and American ones who asked their usual questions. Then, with Inagakis and two Chinese (a journalist and secretary of Christian Chinese federation), led into a Chinese restaurant.[45] During the meal we observed through the window a noisy, colorful Chinese funeral—a ⟨somewhat⟩— for our taste—barbaric, almost comical affair. The food, extremely refined, interminable. One constantly fishes with sticks from common little bowls set out on the table in great numbers. My innards reacted quite temperamentally so it was high

[p. 14v]   time when I landed (literally speaking) around 5 o'clock in the haven of the friendly Pfister couple. After eating, walk through the Chinese quarter in splendid weather. Streets becoming ever narrower, teeming with pedestrians, rickshaws, caked with dirt of every kind, in the air a stench of never-ending, ever-changing variety. Impression of grim fight for survival by the meek and mostly apathetic-looking, mostly neglected people. Beyond the street, loud open workshops and shops, great racket but nowhere any quarreling. We visited a theater, on every floor a separate show by comedians.[46] Public always appreciative, very entertained, the most diverse people with small children. Really filthy everywhere. Inside and outside in the terrific bustle, quite happy faces. Even those reduced to workhorses never give the impression of conscious misery. Peculiar herding nation, often a respectable paunch, always sound nerves, often resembling automatons more than humans. Sometimes curiosity with grinning. With European visitors like us, comical reciprocal staring—Else particularly imposing with seemingly aggressive lorgnette.

Then, drive to Pfisters' spacious country house with already praised safe [p. 15]
haven.[47] Pleasant tea. Then came a deputation of about 8 Jewish dignitaries with
venerable rabbi and very difficult communication.[48] Then, drive with Inagakis
through dark alleyways to a wealthy Chinese painter for a Chinese meal in the evening. House dark outside with cold high wall. Inside, festively lit halls around a
romantic courtyard furnished with a picturesque pond and garden. The halls
adorned with the host's magnificent authentic Chinese pictures and with lovingly
collected old pieces of art. Before the meal, the entire dinner party including the
host, us, Inagakis, Pfisters, a German-speaking Chinese, a couple related to the host
with a trusting, pretty little daughter, about 10 years old, most sweetly reciting in
German and Chinese, the rector of Shanghai University and a few teachers from
this institution.[49] Endless, extremely fine fare, inconceivable to a European,
downright sinful indulgence with schmaltzy speeches translated this way and that
by Inagaki, one of them by me.[50] The host had an uncommonly fine face, similar [p. 15v]
to Haldane.[51] On the wall hung a wonderful lapidary self-portrait of him. The
mother of the little reciting daughter played hostess and entertained quite amus-
ingly and ably in German. At ½ past 9 departure with the Inagakis to the Japanese
Club, where we were greeted by about a hundred mostly young Japanese in a pleasantly informal, modest and cheerful way.[52] Same casual welcome and response,
translated by Inagaki. Then return to the ship. There, a visit by an interesting and
sympathetic English engineer. Finally, bed.

Today after breakfast, trip by car to interesting temple with many courtyards and
with a gorgeous Chinese tower, presently being used as barracks. Next door, a
highly enjoyable village, entirely Chinese with very narrow alleyways and little
houses opening toward the front, small shops or workshops everywhere. Reciprocal staring even more impressive than in the city. Children teeter between curiosity
and fear. Almost throughout, happy impression apart from filth and stench; I shall
remember it often and with pleasure. We had a close look at the temple. The neigh- [p. 16]
boring people seem to be indifferent toward its beauty. Architecture and interior
decor (larger than life-sized Buddhas and other figures) work strangely together to
form a great artistic total impression. Lofty Buddhist thought amidst baroquelike
figures of abstruse superstition (half symbolic).

Afternoon, 3 o'clock departure.

16th and 17th. Passage through Japanese straits with countless green islets.
Delightful, ever-changing fjord landscapes. 17th afternoon, arrival in Kobe. Welcomed by Nagaoka, Ishiwara, Kuwaki, Mr. Nagaoka with dainty wife, German
consul and German Club, Zionists.[53] Great hubbub. Masses of journalists on
board the ship. Half-hour interview in the salon. Landing with great participation
by the crowd. Brief nap in the hotel next to the wharf.[54] ⟨under large [. . .]⟩ In the

evening, 2 hour train ride with the professors. Light carriages. Public sits in two long rows along the windows. In Kyoto, magically lit streets, cute little houses. Drive to hotel somewhat higher up.[55] The city below, like a sea of lights. Grand impression. Graceful little people patter about, clip-clop, on the streets. Hotel, a [p. 16v] large wooden structure. Communal meal; dainty waitresses in small separate room. Japanese modest, fine, altogether very appealing. In the evening, scientific conversation. All in all, highly strenuous.

18th. In the morning, car tour around Kyoto. Temple. Great gardens, old palace surrounded by wall and moats, wonderful ancient Japanese architecture (derived from Chinese, lighter, less baroque).[56] On streets the dearest of schoolchildren. From 9 o'clock in the morning until 7 o'clock in the evening, train trip to Tokyo in touring car under cloudless skies. Trip along lakes and sea coves. Over the Fujiyama mountain pass,[57] snowcapped, gleams far out over the land. Unmatched sunset near [Mt.] Fuji. Magnificent silhouettes of woods and hills. Villages quaint and clean. Nice schools. Country carefully built up. After sunset, journalists in the train. Usual daft questions as always. Arrival in Tokyo! Milling crowds and photographers with flashbulbs. Were completely blinded by countless flashes & magnesium ones. Brief respite in lobby in train station hotel.[58] Reception by academy, [p. 17] German, and club deputations.[59] Arrival at hotel, quite exhausted among gigantic flowery wreaths and bouquets. Visits by Berliners[60] and burial alive still to come.

19th. From ½ past 1–4 and 5–7 public lecture in university auditorium in small segments, translated by Ishiwara.[61] The latter in decorative Japanese clothing. Looks like a cross between penitent and priest.

20th. Nagaoka picked us up for the academy dinner in the botanical gardens. The academicians were very cordial. Nagaoka collected us and brought us home. Following lecture, an address was delivered to me, for which I gave brief thanks.[62] Supper in the hotel together with Yamamoto[63] and *Kaizo* employees. Then Japanese theater with song and dance.[64] Female roles performed by men. Audience sits on the floor in separate little boxes with bag and baggage & actively participates. Access to stage through parterre, passages of which also belong to the stage. [p. 17v] Roles strongly stylized. Chorus of 3 men sing incessantly, similar to priests at mass. Orchestra in a kind of cage backstage. Picturesque staging. Music provides rhythm and emotional expression, comparable to bird chatter, lacking symphonic logic and unity. Actors intent on pathos and pictorial effects. Then, with Inagaki and Yamamoto pair, saunter through shopping street with booths of all sorts of pretty trinkets for children and adults. Bright lighting everywhere, but few people, owing to very cold weather. Large business thoroughfare as good as empty at 10 o'clock. We went into a charming little restaurant and chatted. Then, at home with fruits and cigars touchingly provided by Y.

21st. Chrysanthemum festival in the imperial palace garden.[65] Great difficulty in procuring a fitting frock coat along with top hat. The former from unknown donor via Mr. Bärwald, who brought it personally;[66] the latter from Mr. Yamamoto; far too small, so I had to carry it around in my hand the whole afternoon long. We were with the foreign diplomats, who were arranged in a semicircle. Met and accompanied by the German embassy. Jap. empress stepped around the inside of the semicircle and spoke a few words with husbands and wives from the embassies, with me a few kind words in French.[67] The refreshments in the garden at tables, where I was introduced to infinitely many people. Garden marvelous, artificial hill, water, picturesque autumn foliage. Chrysanthemums in booths properly lined up like soldiers. The hanging chrysanthemums are the most beautiful. In the evening, comfortable evening at the Berliners' charming Japanese home. *He*, an intelligent political economist, *she*, a gracious, intelligent woman, true native of Berlin. Lazing about under such conditions is more tiring than working, but the Inagakis help us with touching solicitude. [p. 18]

22nd. Around ½ past 10, picked up by Gakis and Y. for Kaizo at the publishing house.[68] Employees were ceremoniously waiting for us in front of the door. Comradely mood unmistakable. Y. beamed with his childlike eyes under great big horn-rimmed glasses. Everyone from the publishing house photographed together in the alleyway.[69] Curious crowd of onlookers with many children. Then we drove to the magnificent Budd. temple.[70] Glimpse into the monks' dining hall. Wonderful building with splendid carvings. Monks very friendly, gave us a magnificent book with illustrations of art work. In the courtyards, usual photographing, partly together with giggling schoolgirls from Osaka, who were also just visiting the temple. Then lunch in Yamamoto's charming house. A splendid person. In addition to wife and children,[71] he houses in his little home three maids and a servant, four students. How peaceful and undemanding these people must be! In the afternoon we saw a farmhouse and other very simple Japanese cottages, everything spotless and welcoming. Many well-cared-for, boisterous little children, inured to the cold. Visit to the chairman of the academy. One of his sons turns out to be a student at the Zurich Polytechnic and pupil of H. F. Weber.[72] Upset about reception at Tujisawa's.[73] In evening, grand reception in the German-East Asian Club.[74] Conversations with many Germans and Japanese after dinner. My head is spinning, but learned much and encountered enthusiastic friendliness. Japanese scholars have much sympathy for Germany. Japanese erected independent optical workshops with the help of German engineers.[75] [p. 18v] [p. 19]

23rd. From 9 to 11 o'clock, studied files on the persecution of the employee Schulz's wife by the embassy. Poor woman who is supposed to be sacrificed in a scandal coverup.[76] ⟨11–10½⟩ 12–1 o'clock banter with Japanese journalists about

Japanese impressions.[77] Sitting in a circle, question-and-answer game. Then, communal Lucullan meal. From 2–4, Japanese concert at the music school, with lamenting flutes in unison, highly ornamented without proper melody.[78] Harp with few strings and mandoline-like instrument with litany-like singing. There was also a piece for plucked instruments mimicking the natural lute, graceful, melodi-

[p. 19v]   cally weak. Not a trace of proper structure and harmonics. Evening meal at Fuji-sawa's with Nagaoka and a couple of others (someone from the German embassy among them as well). Dainty little daughter of the host. Pleasant company with harmless chitchat. Beforehand visits with Bärwald and Prof. Nagai (chemist).[79]

24th. Morning walk on foot with Inagaki. Meal in a Japanese inn.[80] Sitting on the floor difficult. Roast lobsters; poor creatures. Charming establishment. Fine, quiet manners among the customers. Afternoon visit in artistic Japanese private home (Nezu).[81] Wonderful traditional Japanese pictures of most refined rhythm and color. Beautiful evidence of the Japanese psyche. Chinese-Buddhist influence not in accord with the national soul; appears baroque compared to the country's own original artwork. From ½ past 5–7 and 8–10, second public lecture with Ishi[wara]. Immense interest by the public.[82] 24th–1st Dec. scientif. talks about relativity at the university.

25th. Visit by crazy American woman who believes she can cure other nuts. 2 o'clock, visit at the physics institute. Then my first scientific lecture.[83] ½ past 3,

[p. 20]   reception by the students in the auditorium. Profound impact; I spoke about science as an international benefit.[84] 6–8, Japanese theater. Again chorus with string and percussion instruments. Ballet-style treatment of familiar children's fairytales. Very interesting game of gestures—in part very strange.[85] In the evening from ½ past 8–10, dinner in a Japanese restaurant, invited by the journalists as a group, with a number of geishas. The latter performed charming dances with music. Dancers very young; older geishas with very expressively sensual faces; unforgettable. Question-and-answer in chest voice. Dolls' meal for people. Then we were politely dismissed so that the more relaxed second part could begin.[86] Private conversation with short Ina[87] about geishas, morality, etc.

26th (Sunday). Okura Museum with splendid Chinese and Japanese statues, paintings, reliefs, and gorgeous hilly planted garden.[88] Noh plays in the afternoon. Antique drama with Japanese chorus. Very slow movements and masks. Highly dramatic effect.[89] Then, visit to a gigantic bookstore. All 250 employees present.[90] Curious reception, meal in room. Business with Yamamoto.

[p. 20v]   27th. Visit by Nagaoka's son-in-law with charming little wife. Meal at Nagaoka's with conversation about endless Japanese univ. examinations.[91] The lecture on tensor theory.[92] Then, supper at Tokugawa's.[93] 2 pieces of music, vocal, 2 plucked instruments, Tokug. flute. Content: landscape impressions. Pantomime dances by Tok.'s little children. Then masterful pantomime dances with accompaniment by two singing and plucking women, the latter also with little children. Supper ostentatious, Japanese, cooked by the devil's grandmother. Then, violin music by me (Gluck, Hauser, Bach) and Japanese lady (Wieniawski).[94]

28th. Reception at University of Commerce.[95] Grand speech by one student in German. I replied about the originality of Japanese cultural values. Deep impact. University meal. Festive, without speech. Sat next to university president.[96] Afternoon lecture and physics seminar (talk about Kármán problem).[97] Kaizo Publishers employees communal dinner in train station restaurant. Speech ([by me], as the publisher's youngest employee!). Ceremonial welcoming handshake by the actual youngest employee.[98] Also dictated a short article about Japanese music.[99]

29th. While in shirtsleeves received a card from pastor Steinichen announcing [p. 21] his visit about Frau Schulze to keep me informed. Changing and getting dressed fast, half-way in his presence. Then the English physician Gordon-Munroe, who was attending the Schulzes.[100] Found out that wife's psychosis is due to husband's maltreatment (employee at the German Embassy). ½ past 10, tea ceremony in a fine Japanese home. Exactly prescribed ceremony for a meal to celebrate friendship. Glimpse into the contemplative cultural life of the Japanese. The host has written 4 thick volumes on the ceremony, which he proudly showed us.[101] Then, reception by 10,000 students of Waseda University, founded by Okuma (?) in the democratic spirit, with addresses.[102] Lunch at hotel. Then lecture. Viewing of the institute. Interesting communication about electric-arc line-shift.[103] At ½ past 6 reception by pedagogical societies. During farewells, greeting by female seminar participants outside. Sweet, cheerful vision of crowds in semi-darkness.[104] Too much love and spoiling for one mortal. Arrival home dead tired.

30th. With wife to the tourist information at train station. Sole excursion on our [p. 21v] own. Funny difficulties in communicating. Inagaki seeks and finds us along the way by car and guides me to imperial ensemble to listen to ancient Chinese music (½ past 10–½ past 1), which will only survive there by tradition.[105] Common Indian roots for Byzantine and Sino-Japanese music. Chorale-like. Wonderful tone painting. Flute, plucked instruments, reed instruments, also for very high pitches, producing a silvery tone. Talk at univ. Discussion: Tamaru elucidation of inconsistency of Doi's theoretical proof. Festive address by student delegations (in total ca. 20,000) of the universities of Tokyo.[106] Dinner at the embassy. Diplomats and other big fish. Gorgeous music. But otherwise insipid and stiff. I bungled through some violin as well; very badly due to fatigue and lack of practice. Danish ambassador brought us home; pleasant married couple.[107]

1st December. Meal with Mr. and Mrs. Witt, who met us yesterday at the train station. Information about captivity in Canada. Last lecture on cosmological [p. 22] problem. Thanks from registered students.[108] Huge dinner at hotel. Entire intellectual elite in attendance. After the meal I had to deliver an address (after Yamamoto) and—play violin (Kreuzer sonata). (In the morning, visit by chemist Tamaru in hotel).[109]

2nd. Visit in the School of Technology. Student reception. Address Takeuchi.[110] Drive to Sendai (1–9). Honda & Aichi drive 4 hours to join us.[111] Arrival. Fellow academics, university president at train station, botanist Molisch as well. Life-threatening crowds on the way to facing hotel. There reception by authorities.[112]

3rd. In morning, lecture ½ past 9 until 12 & 1–½ past 2.[113] Trip with Yam., the caricat. painter Okamoto to pine island.[114] Spectacular coastal landscape. Stop in Japanese restaurant, Japanese-style. Dinner with physicists at the hotel. Acquaintance of the poet Tsuchii. Gave me sketchbook by Hokusai and a book of Italian poems he had written himself.[115] In the evening, moving reception at the university. Student assembly. Then with professors. Sat beside Molisch and dean of the med. faculty. Had to write name and date in ink on wall.[116]

4th) Departure for ⟨Sendai⟩ Nikko with In., Yam., und Okamoto.[117] Honda came along for 1 hour.[118] Superb persons. Humorous, modest lovers of nature and [p. 22v] art. Unforgettable. Splendid mountainous landscapes seen from train. Yesterday and today everywhere special favors by railway officials. Women lost along the way, because they missed the train in Tokyo.[119] Picturesque trip. Conversation with semi-Americanized German-American silk-stocking manufacturer. Through the village Nikko with Inag. and Okamoto by foot to hotel. The latter sketched a number of very charming caricatures that same evening.[120]

5th. Hard to get Inag. out of bed because reunited with his wife.[121] Around ½ past 9 set out for the temple lake lying at 1,300-m altitude.[122] Car ride up to actual climb. Then hike through magnificent forests with splendid views onto mountains, narrow valley, and plateaus. At top, hefty snow storm in raw cold that loyally accompanied us to the bottom. Okamoto, poor blighter, in straw sandals, but always full of humor and mischief. In the evening xth telegram from the German society in Kobe. At least in Japan, I much prefer dealing with Japanese. They are more similar to Italians in temperament, but even more refined, still entirely drenched in [p. 23] their artistic tradition, not nervous, full of humor. Along the way conversations about Buddhist religion. Educated Japanese flirt with primitive Christianity. Additionally, conversation about Japanese worldview prior to contact with Europe. It seems that the Japanese never thought about why it is hotter on their southern islands than on their northern islands. Nor do they seem to have become aware that the height of the sun is dependent on the north-south position. Intellectual demands of this nation seem to have been weaker than its artistic ones: natural predisposition?[123]

6th. Visited Temple system at Nikko. Nature and architecture magnificently united. Cedar avenue. Enhancement through system of courtyards. Central building marvelously decorated with colorful carvings. Somewhat overdone.[124] Joy in representing nature outweighs the architectonic and even more so the religious. Long talk by priest about historical matters—not about experience. Wonderful stone stairway under cedars to the grave of the eldest Tokugawa.[125] Afternoon, Brother Beck with daughter here. Walking path to [train] station with these mountains fabulously illuminated by the setting sun. Trip to Tokyo.[126] There, great scurry of [p. 23v] packing in the hotel.

7th. Even greater scurry with noise, suitcase shutting. Bärwald also there. Trip to train station. Final departure from Tokyo. Travel with Ishiw., Inagakis, Yamamoto & wife to Nagoya. I busy with article writing, very hurriedly, about Japanese impressions.[127] Arrival, met by great crowd of students and pupils. Pleasant supper with the entire company with four people from Kaizo's publishing house in tavern room decorated with maple.[128] Upstairs in Inagakis' room. Met Michaelis in the hotel.[129]

⟨8th⟩ 9th. Morning walk along the main street of Nagoya up to train station; futile attempt to buy pipe tobacco. With Yam., Inag., & Ishiwara, visit of Shintoist temple. Large grove. In it, temple complex according to courtyard system. Elegant, smooth wooden structures. Undecorated. Empty little houses for souls.[130] Must have come from the south. Characteristic roof projection.7

Natural religion, utilized by the state. Much god worship. Ancestors & emperor [p. 24] cult. Trees the main thing in the temple complex. Trip to Kyoto following great farewells at station by students and teachers. Kyoto, friendly welcome by univ. physicists and students.[131]

8th).[132] Visit of the imperial palace with magnificent fortified building (tower-like). Inside the palace, splendid paintings of nature on the walls and doors. Tigers, winter room, plants and birds. Courtly scenes.[133] Afternoon, music making with Michaelis, major talk in the circus with Ishiwara.[134]

⟨10⟩ 11th.[135] Trip to Osaka (large factory and commercial town). Met by mayor & students at train station. In hotel ½ past 11, important dignitaries. Introduction with substantial hand-pressing. Banquet by Prof. Sata in great hall. Military trumpet music, immense meal. Speeches with much pathos, also by me.[136] (Japanese scaled-down America.) From ½ past 5–7 and 8 until ½ past 9, talks.[137] The whole thing not so terribly stressful because everyone considerate and modest. Upon return home, great scene by left-at-home wife.

[p. 24v]   10th. ½ past 10–12 and 1–3, talk in Kyoto. Very cold in splendid hall.[138] Then visit of imperial garden and coronation palace.[139] The inner palace courtyard is among the finest architecture I have ever seen.[140] Completely surrounded by buildings. Audience room and coronation hall open out on sand-covered courtyard. Emperor's demeanor is that of a god; for him very uncomfortable. Inside the hall, where the coronation thrones are visible from the courtyard, portraits of about 40— Chinese—statesmen in acknowledgment of the cultural influence Japan has received from China.[141] This veneration of foreign teachers still lives on today among the Japanese. Moving recognition among many Japanese who had studied in Germany for their German teachers. There is supposedly even a temple in memory of the bacteriologist Koch.[142] Ernest respect without a trace of cynicism or even skepticism is characteristic of Japanese. Pure souls as nowhere else among people. One has to love and admire this country.

[p. 25]   12th. Morning, nap. 2 o'clock, old Tokugawa palace with beautiful landscape paintings (clouds, trees, birds, funny tiger, painted on gold background). Painting interrupted by beams makes walls seem to vanish and the interior to extend into fabulously colorful outdoors.[143] With Ishiwara, calculation on energy tensor of the electromagnetic field in isotropic ponderable substances for joint article in Japanese academy's reports.[144]

13th. Trip to Kobe. Lunch with ⟨Okamot⟩ Yamamoto and the important young social politician   in fishing village near Kobe (tourist sight).[145] Talk with Ishiwara from ½ past 5 until 8 o'clock, dinner at the consul's, Trautmann. Then reception at the German Club.[146] Trip home alone with wife on local train (arrival Kyoto, 1 o'clock in the morning).

14th. Festive noon meal with professors from the university. Large assembly of students. Address by university president and representative of the student association in impeccable German (very cordial).[147] Then talk by me about the coming into being of the theory of relativity (by request).[148] Visit in the phys. institute. (Highly interesting, particularly Kimura's investigations on broadening of spectral lines.)[149]

[p. 25v]   In the evening, Nagaoka arrives from Tokyo with a suitcase full of magnificent presents from the University of Tokyo.[150]

15th. Departure from Nagaoka. Visit to splendid Buddhist temple, memorial mass for the dead. Friendly welcome by monks. Viewing of the big bell; striker hor-

izontal and outside. Blossoming cherry tree in front of the temple.[151] Picture taking at the univ. phys. institute with small reception and colloquium. Dusk, visit of
a Shintoist temple standing on high stilts against the side of a mountain. Then visit
of festively lighted street with luxury shops and much goings-on. Inexpressibly
cheerful scene, like *Oktoberfest*. Swarm of lanterns and little flags. Street surface
extremely dirty, everything else sparkling clean and brightly colored.[152]

16th. ⟨Morning⟩ Early visit to the Buddhist temple at the foot of the ⟨mountain⟩
hill next to the hotel. Wonderful architecture, subtrop. lush vegetation.[153]
Morn[ing], western temple with magnificent paintings. Also harmonious treatment
of human figures in the landscape, reminiscent of Ital. Renaissance. Never portraits
or group compositions.[154] Afternoon, Lake Biwa with wonderfully situated and
architect[urally] very perfect old rockside temple.[155] Evening ⟨many⟩ wrote a   [p. 26]
number of letters.

⟨1[8?]th⟩ 17th. Visit with wife to silk store. Splendid landscape & animal
embroidery.[156] Afternoon, climbed hill alone at sunset. Japanese wood (maple)
and light effects unrivaled. In the evening, traveled to Nara. With Gaki on foot to
hotel. Very tasteful, semi-Japanese style and excellent.[157]

18th. Tour around the temple territory. Tame deer milling about and snuffling
about one all over the countryside. Temple architecturally magnificent. Temple
with large Buddha figure especially majestic (more than 1,000 years old); the figure quite raw.[158] Much superstition. Lanterns, memorial stones. Bits of paper on
trees and by temples. In afternoon, state museum of old sculptures.[159] Some
charmingly pretty things from the period 700–1200. Profound impression of Japanese art of caricature.

⟨20⟩ 19th. With Inagaki climbed up bald hill (of young grass) that sets a Japanese
in raptures as it embodies the delights of spring.[160] ⟨Evening⟩ Afternoon, letters
& postcards, also calculation for joint paper with Ishiwara. From 6 o'clock in ev.
until 6 o'clock in the morning, voyage to Miyajima.[161]

⟨21⟩ 20th. Arrival in the dark of night. Bathed and into bed and slept until 10.   [p. 26v]
From 11–12, enthralling walk along the coast to the temple with graceful pagoda
built in the water (flood region).[162] Afternoon, with Gaki tour to the peak of the
mountain lending the island its main form. Wonderful view over the Japanese
Inland Sea.[163] Subtlest of colors. Along the way, countless small temples, dedicated to natural deities. Stone figures often delightful. The entire path of steps hewn
into granite rocks (height around 700 m). Memorial to Japanese love of nature and
all sorts of nice superstitions.— Noon telegram from Solf about denial of Harden's
assertion that I had to escape to safety in Japan. My answer: Affair too complicated
for telegram; letter to follow. I wrote the latter that evening, in accordance with the
truth.[164]

21st. Coastal walk in brilliant sunshine. Telegram to the Community of Shanghai. Okamoto along. Afternoon, walk in woods and along coast. Jellyfish hunt with rocks (Gaki and Okamoto).

22nd. Yamamoto came, shorter walks. Wood blocks puzzle. Great difficulty solving it. Minor poisoning from open coal fire (hard coals among them!) in room; the women were particularly affected.

[p. 27]   23rd. Trip to Moji. Copious reception. Had to pass examination with journalists upon arrival in Shimonoseki. In evening, princely accommodations in Mitsui Club outside of Moji.[165]

24th. Photographed for the 10,000th time. Then, trip to Fukuoka immediately before talk, which lasted from 1–3 and 4–6; Ishiwara, suffering from cold, had to translate. Student reception called off because the people, as in Sendai, generally count on lecture free of charge and poor Yamamoto has been duped.[166] After lecture, Kaishosha dinner that almost lasts forever. A large part of the company and another group of high-school teachers at the next table, properly tipsy from rice wine and very funny.[167] Prof. Miyake thoughtfully accompanies me everywhere, in the end to a Japanese hotel, where the lady of the inn is deeply thrilled and, on her knees, bows her head to the ground around 100 times. Rooms in extremely good taste. Living room and 2 adjoining rooms provisionally outfitted with Euro-

[p. 27v]   pean seating, everything separated by paper sliding doors that a little finger could easily shift aside.[168]

25th. Wild day! Around 9 o'clock Kuwaki shows up, Inagaki also there. Then the funny innkeeper. She comes with about 6 pieces of silk about ¾ [m] in length and a bundle of brushes and Japanese ink, and I am supposed to paint it all up with my name.[169] Discussion with Kuwaki about epistemol. questions in relativity. Then 11 o'clock picked up women (Elsa & Mrs. Inagaki). Trip to Japanese hotel. Touring of the city and many shops.[170] At 1 o'clock to university dinner in rooms of the medical faculty. Shaking of hands with very many professors. Viewing of gall stones, microscopic preparations on Weil's disease, fish representing the result of crossbreeding, all specially set up in reception hall. Afternoon meal, talk by president and by me. Received many gifts.[171] Visit to a temple, the physics and solid mineralogy institute.[172] [Then] visit with Myake, who has four of the dearest

[p. 28]   children and finally to a trade exhibition in the town hall where the provincial governor in attendance had specially arranged to have wonderful paintings hung.[173] Then departure, to which *all* came, including the friendliest of all innkeepers. But I was dead, and my corpse drove back to Moji where it was carried on to a children's Christmas and had to play violin for the children. At last, 10 o'clock return, supper, many letters from home, bed.[174]

26th. Morning and evening climb up a hill with beautiful view onto mountains and sea. Yamamoto arrived. Embarrassed, because he has to relocate us as the Mitsui Club is baring its claws and demanding horrendous sums. Wrote introduction to lecture. Idea about $\left(\dfrac{\partial}{\partial x_\alpha}\right)(R^\alpha{}_{k,\,lm}) = 0$. In the evening a Japanese comes by with a stack of paper and wants me to communicate to him my impressions on the top of the hill![175]

27th. Tour aboard Mitsui steamboat on China Sea. Walk through Shimonoseki, supper with Yamamoto & Watanabe. Afterwards Nagai arrived. Poem with drawing for Mrs. Yamamoto and inked up much silk.[176] Most animated spirits. So much for peace!

28th. Rainy day. Evening invitation by the Commercial Club of Moji.[177] I play the fiddle, the Japanese sing one at a time in Japanese. Hard finance. Shrewd and not as refined as the professors, but more like European analogues, after all. Here, too, modesty of form. [p. 28v]

29th. Touching farewells.[178] Mr. & Mrs. Yamamoto, Mr. & Mrs. Inagaki, Kuwaki (with little son), Ishiwara, Miyake, as well as gentlemen of the Mitsui company. All boarded ship.[179] Marvelous present, poem, and letter from Zuckii (Sendai, poet) arrived.[180] Sailing around 4 o'clock. Ship large and comfortable. Found electrodynamics energy tensor and wrote Ishiwara.[181]

30th. Peaceful trip. Idea for development of Weyl-Eddington theory.[182] Letters to Yamamoto and Zuchii. Read issues of *Frankfurter Zeitung* that Prof. Berliner's wife sent from Tokyo; miserable Europe![183]

31st. Arrival in Shanghai in glorious weather. At noon picked up by De Jong (engineer) and Mr. Gaton (parvenu). ⟨Supper⟩ Staying with the latter snob, but good piano. New Year's Eve there; I sat next to fine Viennese lady, otherwise noisy and, for me, sad.[184]

1st Jan. Shanghai unpleasant. European crowd Chinese servants, are lazy, conceited and shallow. Noon meal at de Jong's. Friendly internationally minded Englishman. Afternoon "reception" in Gaton's home, flocks of Jewish and other [p. 29] schmaltzy, clawing bourgeois, usual hand clasping and speeches—disgusting.[185] Then discussion in "Question Club" (comedy with dumb questions).[186] In the evening also visited Chinese popular entertainment establishment. Picturesque life. Chinese indiscriminately accept all European music for any occasion (party, wedding, funeral), never mind if funeral march or waltz, so long as there is plenty of trumpet blowing. There was also a small temple, in the midst of the worldly hustle and bustle. In the morning short car drive to the city environs; everything full of burial mounds and coffins or little coffin houses which cannot be removed. Chinese dirty, tormented, blunted, kindly, steady, gentle and—healthy. All are unanimous in praising the Chinese but also regarding the intellectual inferiority in business skills; best evidence: he obtains 10 times lower wages in an equivalent position, and the European can still compete successfully as a business employee.

2nd January. Noon weighing of anchor. Gloomy, windy weather. I enjoy the tranquillity beyond words.

[p. 29v]   3rd January. Cold, windy weather. Contemplative, enviable existence on board, where I, in order to maintain this state, fearfully avoid making acquaintances. Pondering and calculations on Eddington's theory.[187] (Attempt at a completion of the latter.) Improvement of the variation method in regular general relativity.

⟨4⟩ 5th January. Arrival in Hong Kong. 7 o'clock in the morning. In hope of being left in peace for once, we secretly go ashore at ½ past 9, initially to do a few errands. Met at Nippon Yusen Kaisha a fellow by the name of Gobin, who had already shown us Hong Kong the last time. He informed us of the community's reception that afternoon. We also went to see the French consul and quickly said our good-byes.[188] We drove up to the Peak again; I climbed up to the top. Resplendent panorama of harbor, sea, and island. Quite hot up there. We down to the town, downhill for about one hour, tropical woods all the way. The entire path was taken up by Chinese men, women, and children, groaning while hauling bricks uphill. Most pitiful of people on Earth, cruelly mistreated and worked to death in reward

[p. 30]   for modesty, gentleness and frugality. Then drove to ship. Barely arrived, again collected by that fellow, driven to Jewish clubhouse beside synagogue. Despite his busy effort, practically nobody came to the "reception," which was very funny. Then we went, perforce, to his family to eat. Friday evening prayers, then long, terribly spicy meal, an interesting young Russian Jew also there. Finally, thank God, back on board ship.

6th. Wonderfully cloudless crisp morning in the harbor. 11 o'clock departure. Brilliant sunshine. Sat long in the sun in the afternoon, which (with hat) was still just barely tolerable. Voyage through islets and bustling with numerous Chinese sailboats that were dancing on the waves. New idea for the electromagnetic problem of the gen. th. of rel.

7th & 8th. Somber and humid with temperature on the rise. Thinking about gen. relativity and electricity.

9th. Writing of paper on gravitation and electricity.[189]

10th. Letters to Arrhenius, Planck, Bohr.[190] Evening mooring before Singapore.

11th. 6 o'clock entry into harbor. Oppressive, dark atmosphere, frequent rain. Montor there with letter from Java, Voute. Declined by telegraph and wrote.[191] Trip into jungle preserve in light rain. Unrivaled impression of wild, lavish plant [p. 30v] growth, swampy, impenetrable. Midday meal at Montor's. Toward evening, trip to Fränkel's cultivated palms.[192] Trees wonderful, people banal. At night on board quite oppressive.

12th. Visit to Mayer-Croesus with noble daughter ("Portia").[193] Then to fat Weil with fancy wife, real Jewish gypsies. Tropical house with splendid view of city and water. Afternoon, in pouring rain, back onto ship. Departure 5 o'clock between glaringly green velveteen islets.

13th. Arrival near Malacca in the early morning, where the ship stood in the open sea until 3 o'clock in the afternoon. We visited Malacca in the morning. Portuguese church & other buildings. Vibrant mix of Indians, Malays, and Chinese. Two-wheeled carts with straw roofs, drawn by long-horned oxen. Beating tropical sun but less humid than in Singapore. Discovered fly in my electricity ointment in the afternoon. A pity. True tropical heat.

14th. Noon arrival in Penang. Scorching heat on the ship, which remained in the bay quite far away from the city. A lot of Indians were brought on board, handsome [p. 31] tall-statured men and women. We went ashore around 3 o'clock and walked, with rickshaws in pursuit, around town, which is very interesting. Boats, houses, people, they all have style. Heat downtown quite tolerable. We saw Buddhist temples with mysterious, terrifying, colorful decor, also a mosque with bath inside where the men were lounging about, elegant Arabic structure with slender towers whitish in color. Beautiful, pushy beggar woman. Together with Japanese traveled back to ship in skiff in pitching waves. Else very frightened but had enough energy left to grumble. But the haggard Indian with flaming black eyes ⟨land⟩ calmly rowed his long oar steadily on and delivered us safe and sound to the Haruna. Sweltering heat until midnight.

15th. Cruising in pleasant breeze. New ideas about the electrical problem. Evening conversation with Indian school teacher about land and life in Ceylon. Splendid star-studded night sky. Enviable way of life.

16th–18th. Disciplined toiling on the problem despite the heat. Moving forward with many setbacks.

19th. Colombo. In morning, futile attempt to assemble a car party. Interesting [p. 31v] finding about ⟨Japanese⟩ captain's business practice, for easily surmised reasons, not letting the passengers know about cheaper wireless telegrams offered by communication firms from Colombo. Morning excursion by tram & on foot. Left train station in the company of pushy natives. 1 h 25–3 h 5, trip to Negombo, a little town with no Europeans somewhat northward along the coast.[194] We took 2 rickshaw men, one of whom was an absolutely naked primitive, the other a former Hagenbeck zoo elephant-keeper, who enthused about Hamburg.[195] They drove us through the little town, its main row of houses like single houses ducked under palm groves. We were gawked at everywhere just as at home the Sinhalese are. Then, visit to a fishing village. Children stark naked, men loincloth. Handsome people. Fishing boats made of two rigidly attached narrow parallel pieces. Great speed but uncomfortable sitting. Boat came home with many fishes and a swarm of en-

[p. 32] vious crows. Then past bay and rivulet, before which we saw at a distance of 12 m a large crocodile lying in the grass that, after being pelted with stones and screamed at by many natives, very slowly waddled into the water. Then, European restaurant and train. The English administer faultlessly without needless chicanery. From no one did I hear words of discontent against them, not even from a Sin-

halese teacher who traveled along as a steerage passenger from Penang to Colombo. The rickshaw man was so delighted by us or, resp., our 5 rupees, that he brought bananas to us at the station in farewell. At the station we also made the acquaintance of a pretty-as-a-picture fine young Sinhalese woman with sister and mother, village aristocrats. But they had a Dutch great-grandfather. Rarely have I ever seen anything so fine. Particularly magnificent starry sky. Homeward trip still hot and

[p. 32v] teeming with mosquitoes in the carriage, which in this ⟨swampy⟩ rice region was quite unsettling. Upon arrival back in Colombo the rickshaw coolies descended upon us. We surrendered ourselves to them after long useless resistance. They regarded it as an insult for a European to travel on foot. Then, trip by rowboat to ship in rough waters with Else's deathly fear and severe remonstrances. Afterwards on board, quenching of dreadful thirst. At night almost no cooling down. Daytime temp. about 29 degrees in the steamship in the well-ventilated section; unusually cool for this region.

22nd. Final draft of paper on gravitation and electricity.[196] Sea voyage extremely pleasant without noteworthy experiences. Japanese meal with captain. Wonderful star-spangled clear evenings. Temperature drops slowly with distance from the equator. Telegrams about the French marching into the Ruhr region, no wiser than 100 years ago.[197]

31st. On the Red Sea, first a constant 28–29 degrees with almost clear skies. Fabulous sunsets with yellowish red to purplish red sky and jagged islets

[p. 33] glaringly lit up or in sharp dark silhouette. Today arrival in Suez. Deeply blue, remarkably transparent sea. Sky lightly veiled in mist. Silvery dull colors, picturesque sailboats, yellow shoreline. Passage through saline lakes with barren, abandoned shores. Pale silvery tints, sky gloomy. Air quite cool.

On the last hot day there was a masquerade party for the passengers, the day before, for the stewards. Japanese are virtuosos in this art. Recently made some nice acquaintances. Greek envoy, who is returning home from Japan, likeable English widow, who contributes one pound for the Jer. University despite my protests, not to forget the Og[a]tas, a fine, nice Japanese merchant couple, with whom we chatted a lot on the ship.

1st February. Arrival in Port Said in the early hours. Greek envoy facilitated our landing and customs. Young Jew ⟨Goldstein⟩ Cantor appears at customs house with telegram from Jerusalem to assist us. City a real meeting place of foreigners with [p. 33v] accompanying rabble. Visit to the Local Council (Palestinians). 6 o'clock in the evening, train to Qantarah at the Suez Canal.[198] Cantor and his companion Goldstein accompany us there and by ferry across the canal, sojourn from 8–1 o'clock at night.

2nd. The departure to Palestine eased[199] by young Jewish conductor, who had seen me in Berlin at a meeting—not completely delighted with his fellow Jews, but a reliable, good person. Trip first through desert, then from about ½ past 7 through Palestine in quite dreary weather and frequent showers.

Trip first through flat terrain with very sparse vegetation, interspersed with Arab villages and Jewish colonies, olive trees, cactuses, orange trees. At a junction not far from Jerusalem, welcomed by Ussishkin, Mossinson, and a few others of ours.[200] Traveling past colonies through wonderful rock valley up to Jerusalem.[201] There Ginzberg, happy reunion. In automobile with officer to governor's palace, formerly a possession of Kaiser Wilhelm, very Wilhelminian. Acquaintance with Herbert Samuel. English formality. Fine, well-rounded educa- [p. 34] tion. Lofty view on life tempered by humor. Unassuming, fine son, cheerful, robust daughter-in-law with nice little son. Day rainy, yet hint of the magnificent view of city, hills, Dead Sea, and Transjordanian mountains.[202]

3. Walked with S. Samuel into the city (Sabbath!) on footpath past the city walls to picturesque old gate, walk into town in sunshine. Stern bald hilly landscape with white, often domed white stone houses and blue sky, stunningly beautiful, likewise the city crowded inside the square walls.[203] Further on in the city with Ginzberg. Through bazaar alleyways and other narrow streets to the large mosque on a splendid wide raised square, where Solomon's temple stood. Similar to Byzantine church, polygonal with central dome supported by pillars.[204] On the other side of the square, a basilica-like mosque of mediocre taste.[205] Then downwards to the temple wall (Wailing Wall), where dull ethnic brethren, with their faces turned to [p. 34v] the wall, bend their bodies to and fro in a swaying motion.[206] Pitiful sight of

people with a past but without a present. Then diagonally through the (very dirty) city teeming with the most disparate of religions and races, noisy, and orientally alien. Gorgeous walk over the accessible part of the city wall.[207] Then to Ginzberg-Ruppin for midday meal with cheerful and serious conversations.[208] Stay owing to heavy rain. Visit to the Bukharian Jewish quarter and to dark synagogue where pious, grimy Jews await the end of the Sabbath in prayer.[209] Visit with Bergmann, the serious holy man from Prague who is setting up the library with too little space and money.[210] Atrocious rain with even messier streets. Homeward drive with Ginzberg and Bergmann by car.

4th. Drive with Ginzberg and S. Samuel's capable and earthy cheerful daughter-in-law[211] over wonderful bald rolling hills and deep valleys in beaming sunshine and a fresh wind, downhill to Jericho and its ancient ruins. Splendid tropical oasis [p. 35] in desert region. Meal in Jericho Hotel. Then drive through wide Jordanian valley up to the Jordan bridge on horrendously marshy road, where we saw magnificent Bedouins.[212] Then in dazzling sunshine homeward again. At home, in piercing sunshine, resplendent view of Dead Sea and Transjordanian hills from S. Diez's official residence, where we drank tea.[213] Then in the darkening room, interesting conversation about religion and nationality with S. Diez. Evening, pleasant chat with S. Samuel & daughter-in-law. A day of unforgettable magnificence. Extraordinary enchantment of this severe, monumental landscape with its dark, elegant Arabian sons in their rags. There were many four-legged camels and donkeys to be seen.

5th. Tour of two Jewish colonial settlements west of Jerusalem, belonging to the city. The construction is carried out by Jewish workers' cooperative in which the leaders are elected. The workers arrive without specialized knowledge or expertise [p. 35v] and achieve excellence after a short time. The leaders are paid no more than the workers.[214] Visit to the Jewish library. Bergmann from Prague functions there ably but without a sense of humor.[215] A local mathematician (high-school teacher) shows me some of his really interesting investigations regarding continuous matrices and their operations.[216] In the evening music making with officer in S. Samuel's apartment, far too long, because starved of music.[217]

5th.[218] Visit of Jewish school of art. Splendid work under difficult conditions. Revival of antique Jewish ornaments.[219] Afternoon welcoming reception by Jewish schoolchildren forming a ⟨ceremonial⟩ passageway amongst them and then Jewish citizens in general in school hall. Speeches by Ussishkin and Yellin and presentation of Hebrew address.[220] In the evening, invited for music at Bentwitch's. Highly musical family. We played Mozart quintet.[221]

6th.[222] Church of the Holy Sepulcher. Via Dolorosa.[223] Afternoon talk [p. 36] (French) in university building *in spe*.[224] I have to start with greeting in Hebrew which I read out with difficulty. Afterwards, words of thanks (quite witty) by Herb. Samuel & walk back and forth over the hilly road. Philosophical conversations. Evening, invitation from major dignitaries with scholarly and other conversations.[225] That evening thoroughly contented with all these farces!

8th. Drive to Tel Aviv ½ past 9–12 in automobile. Reception in high school. Visit of many hours. Light gymnastics by pupils. Short thanks to them.[226] Reception in city hall; named honorary citizen. Moving speech.[227] After lunch, visit to the electrical Ruthenberg center under construction, the municipal central power station, the quarantine camp, the sand-brick factory.[228] Then major public address before the high school with speech by Mossinson and me,[229] visit of agricultural experiment station, scientific night courses (Czerniawski) and the engineers' association, where I received a diploma and a splendid silver box.[230] Supper at Talkowski's. Evening, cultivated gathering with talk by me.[231] The accomplishments by the Jews in but a few years in this city excite the highest admiration. Modern Hebrew   [p. 36v] city with busy economic and intellectual life shoots up from the bare ground. What an incredibly active people our Jews are!

9th. Morning workers' assembly. Very impressive.[232] Visit to agricultural school. Mikve[233] Large wine cellars. Eggs being artificially incubated must be cooled once a day. And the Jewish Rothschild colony, Rishon le Zion.[234] Both ⟨establ⟩ enterprises already 50 years old. Old man delivered a welcoming address in the village. School lesson, children in the garden. Happy impression of healthy life, but economically still not yet quite independent. Train ride to Jaffa together with Joffe (physician and cousin of the Russian) through plains with gradually approaching mountains. Arab and Jewish settlements. Jewish salt works station before Jaffa. Workers came to train station to greet me.[235] Arrival in Haifa after the beginning of the Sabbath, despite Mr. Struck's forewarning. Walk through enormous filth with Ginzberg and the physicist Czerniawski to his brother-in-law Pefzner. Wife delicate and of acute intelligence.[236] Comfortable room upstairs. German maids. In the evening, a lot of unimportant people out of curiosity, but also Struck and wife.[237]   [p. 37]

10th. Viewed technical high school (Sabbath). Prussified but capable director Biram.[238] Struck's apartment. Lunch with him there with pleasant conversations. Visit with Weizmann's mother, surrounded by x sons, daughters, etc.[239] Walk up [Mt.] Carmel with Struck. Encountered Jewish woman laborer. Climbed onto pastor's roof with magnificent panorama of Haifa and sea.[240] A Jewish pioneer [Chaluz] accompanied us down the sloping street to the apartment of an Arab friend. The common folk know no nationalism. Visit to an Arab writer with German wife.[241] Evening party at the technical college. Again speeches, remarkable ones by Czerniawski and Auerbach. Psalms and Eastern Jewish songs by candlelight.[242]

11th. Visit to the technical college workshops. Then Rothschild mill and oil factory. The former almost finished. Incredibly sophisticated, almost entirely automated installation.[243] Afternoon trip across Israel's plains, ⟨Bethlehem⟩ Nazareth, to Lake of Tiberias. On the way, visit to the Nahalal colony, which is being con-

[p. 37v] structed according to Kauffmann's design. Virtually all Russians. Village with private properties. Construction work cooperative.[244] To Nazareth, reachable after a scenic mountain drive, in pouring rain. Drive into the night up to the Migdal farm. For the last leg, the car had to be dragged through deep mud by mules up to the Migdal farmhouse. Cozy evening huddle with opulent meal. Farm is supposed to be divided up into garden plots. Our host is a strapping sedentary former gypsy. His family was not able to stay put there and is presently living in Germany.[245] Hilarious pilgrimages with a big lantern to the outhouse. Torrential rain during the night.

12th. Driving tour down to Lake Tiberias: palm and pine avenue. Landscape resembles Lake Geneva. Sun emerges. Lush region but malaria-infested. Pretty-as-a-picture young Jewess and interesting educated worker in the farmhouse. After lunch via picturesque Tiberias to the Communist settlement Degania at the outlet

[p. 38] of the Jordan from Lake Tiberias, first passing by Magdala, Mary's home town where Arab archaeologists sold land at horrendous prices.[246] Colonists extremely likable, mostly Russians. Grimy but of earnest will and with tenacity and love in pursuit of their ideal, carrying on the fight against malaria, hunger, and debt. This communism will not last forever but will raise people of integrity. After detailed conversation and viewing, drive up to Nazareth in good weather. Along the way, magnificent view on the lake, rocky hills, and finally the picturesque little town of Nazareth. In the evening at home in a German inn.[247] Pouring rain anew.

13th. Drive from terraced, very scenic Nazareth across the Israeli plain, Nablus, to Jerusalem. Quite hot at departure, then chilling cold with pelting rain. En route, road blocked by a truck sunk in the mud. People and car take separate detours over

[p. 38v] ditch and field. Cars get heavily battered about in this country. In the evening, talk in German in Jerusalem in a packed hall with inevitable speeches and presentation of diploma by ⟨Pal⟩ Jewish medical doctors, the speaker scared stiff and tongue-tied.[248] Thank heavens that there are some with less self-confidence among us Jews. I am absolutely wanted in Jerusalem and am being assailed on all fronts in this regard. My heart says yes but my reason says no. ⟨My wife⟩ Else seriously feverish on the night before our voyage.[249]

14th. 6 ¾ departure with Hadassa to railway station. ½ past 7 departure after farewells by the tracks. Hadassa travels with us until Lod.[250] Changing of trains.

Wife feeling worse up to Qantarah, where she completely collapses. Kind Arab conductor. In Qantarah acquaintance with a few officials who give my wife eggs and a place to lie down. Layover ½ past 5–10 o'clock. Onward travels very arduous. Arrival in Port Said. Safely housed in Mr. Muschli's beautiful house.[251] All will be well.

15th. Walk to Lesseps monument.[252] Cubist colorful picture of rectangular bath [p. 39] house and larger houses along the beach. Radiant sunshine. Sense of liberation. Wife better, is being devotedly cared for by Mrs. Muschli.[253] Visit with governor (broad-faced Orientals) and a few consuls.

16th. Morning departure on "Ormuz" (Oriental line).[254] Bad food. Almost exclusively English colonialists on board, acquaintance with Jewish businessman Haye from Australia and a few Americans.

17th, 18th, 19th. Indigestion from bad food. High seas and rain. 19th in the morning, Stromboli nicely in view. Afternoon 6 o'clock, Naples. Vesuvius in gray clouds, overcast sky. So cold and unpleasant that one is happy to be able to stay on the ship. Englishman from Australia turns out to be from Mecklenburg. News of railway strike in France and ever more reprisals along the Ruhr; how will things go on?[255] In Toulon people friendly, in Marseille, dangerous to speak German. Director of freight depot refuses to dispatch our luggage to Berlin, or even to Zurich.[256]

22nd–28th. Stop in Barcelona. Very tiring, but dear people. (Terradas, Campalans, Lana, Tirpitz's daughter.) Folk songs, dances. Refectory.[257] How nice it was![258]

[p. 39v]

1st March. Arrival in Madrid. Departure from Barcelona, cordial farewells. Ter- [p. 40] radas, German consul with Tirpitz's daughter, etc.[259]

3rd March. First lecture at university.[260]

4th March. Drive with Kocherthalers. Reply to Cabreras. Wrote academy [p. 40v] speech. Afternoon academy session chaired by king. Wonderful speech by the acad. president.[261] Then tea at arist. lady socialite's.[262] Evening at home but very catholic.

5th. Morning. Mathematical society honorary membership. Discussion about general relativity. Meal at Kuno's. Visit with Kuchal. Wonderful old thinker. Seriously ill. Talk. Evening dinner invitation by Mr. Vogel.[263] Kind-hearted, humorous pessimist.

6th. Excursion to Toledo concealed through many lies. One of the finest days of my life. Beaming skies. Toledo like a fairytale. An enthusiastic old man, who had supposedly written something of import about ⟨Gra⟩ El Greco, guides us. Streets and market place, views onto the city, the Tagus with stone bridges, stone-covered hills, charming level cathedral, synagogue, sunset on the homeward trip with glowing colors. Little gardens with views near synagogue. Magnificent fresco by El Greco in small church (burial of a nobleman) is among the profoundest images I have ever seen.[264] Wonderful day.

[p. 41]   7th. 12 o'clock audience with the King and Queen Mother. The latter demonstrates her science. One notices that no one tells her what they think. The king, simple and worthy, I admire him for his manner. Afternoon, third university lecture; devout public that could surely understand virtually nothing because the latest problems were being discussed. In the evening, great reception at the German envoy's. Envoy & family magnificent, modest people.[265] Socializing as punishing as always.

8th. Honorary doctorate. Truly Spanish speeches with associated Bengali fire. Long but good in content by the envoy about Germano-Spanish relations; ⟨but in⟩ genuine German. Nothing rhetorical. ⟨Evening⟩ Then visit with the techn. students. Speeches and nothing but speeches, but meant well. Evening talk. Then music making at Kuno's. A professional (director of the conservatory) Poras played violin exquisitely.[266]

9th. Excursion to mountains and Escorial. Glorious day. Evening reception in the student residence with talks by Ortega and me.[267]

10th. Prado (mainly looked at paintings by Vélasquez and El Greco). Good-bye [p. 41v] visits. Meal at the German envoy's. Evening with Lina and the Ullmanns in primitive little dancing establishment.[268] Merry evening.

11th. Prado (splendid masterpieces by Goya, Raphael, Fra Angelico).[269]

12th. Trip to Zaragoza.[270]

380. From Chaim Weizmann

London, W. C. 1, 77 Great Russell Street, 6 October 1922

Dear Professor Einstein,

I am sorry to trouble you again about *The New Palestine*, but I find the book is now going to press. All the contributions, with the exception of two, have been received, and I should be very grateful if you would be kind enough to give me some idea when it would be possible for you to let me have yours,[1] so that I may inform the publishers. I need hardly say how much I appreciate your kindness in contributing.

With kind regards, yours very sincerely,

Ch. Weizmann

Dearest Professor,[2] send us the article if ever you can. I hear that you are soon leaving Europe for a longer period.[3] Perhaps you will also come by Palestine and could make a detour from Port Said *on the return trip*. I am going to Palestine on 3 Nov. From there, to America—again [. . .]. It's a heavy lot.

Warm greetings to you and your d[ear] family, yours truly,

Ch. W.

Translator's note: Original written in English with the exception of the postscript.

381. To Hans Albert and Eduard Einstein

Marseilles, 7 October [1922]

Dear Boys,

Sunny like in Bologna.[1] At 2 o'clock boarding the vessel[2] (to Tokyo). Like sardines here in the train. Besso and Chavan send their greetings.[3] Both were extremely happy, also the person (Schenk from the patent office).[4] Here, narrow streets, slender houses, two-wheeled carts, throngs of colored people, lots of little donkeys, too.

Warm greetings, yours,

Albert.

382. To Marcel Grossmann[1]

[Marseille, 7 October 1922][2]

Dear Grossmann,

Seek him not in the college course,
Seek him by a glass of Tokai-er,
Seek him not in Hedwig's Church,
Seek him by Mademoiselle Maier–

This is Heine.[3] But I sought you in vain at the Poly[technic] and did not have anymore time to look for you at your roost. So, here's another hearty greeting in case I drown or am wrung out to the point of disintegration from all that yapping.[4] Yours,

Albert

Congratulations on junior [filius]![5]

[. . .][6]

383. From Richard B. Haldane

Westminster, 23 Queen Anne's Gate, 23 October 1922

Dear and esteemed Professor,

Negotiations with your government need time. But now I have personally visited the Foreign Office and also the Ministry of Finance.[1]

The idea of a plan analogous to the one by Mr. Arnold Rechberg is not entirely [new]. Cecil Rhodes presented the concept to the Kaiser many years before the war. It has its merits. But problems stand in the way of a rapid advance along such lines. [Something] quicker [would be] necessary.

I submit to you for private use a letter I received from Sir Basil Blackett. He is the "permanent adviser about the Reparations procedure" of the Treasury here. You shall see best what his opinion is when you read [his] own words.[2]

In any event, it is only good that the idea be before the government in London by virtue of the importance of your name.

I shall disregard the further developments.

Most devotedly yours,

Haldane.

384. From Christopher Aurivillius[1]

Stockholm, 10 November [1922]

nobel prize for physics granted you more by letter[2]

Aurivillius.

385. From Christopher Aurivillius

Stockholm, 10 November 1922

Highly esteemed Professor,

As I already informed you by telegram,[1] during the meeting held yesterday the Roy. [Swedish] Academy of Sciences chose to award you with the previous year's

Nobel Prize in physics (for 1921)[2] in reward for your research in theoretical physics and specifically for your discovery of the law of the photoelectric effect but without taking into consideration your theories of relativity and gravitation pending future confirmation of their due merit.[3]

On the 10th of December, during the formal plenary session, the diplomas will be awarded to the prize winners along with the gold medals.

In the name of the Academy of Sciences, I therefore invite you to attend this meeting, in order to receive the prize in person.

According to § 9 of the Statutes, it is incumbent upon you to deliver a public lecture in Stockholm referring to the prize-winning text. If you do come to Stockholm, it would be decidedly best if you held your lecture on one of the days following the prize awards.

Hoping that the Academy will have the pleasure of seeing you here in Stockholm, I am, in utmost respect, sincerely yours,

Chr. Aurivillius
Secretary

386. From Niels Bohr

[Copenhagen,] 11 November 1922[1]

Dear Professor Einstein,

I would like to congratulate you most warmly on the award of the Nobel Prize. This public acknowledgment cannot mean anything to you, of course, but the associated funds might perhaps bring about some relief in your working conditions.

For me it was the greatest honor and joy I could possibly get through external circumstances that I should be considered for the prize award at the same time as you.[2] I know how little I deserve this, but I would like to say that I perceived it as a great good fortune that—quite apart from your great engagement in the human world of ideas—the fundamental contribution made by you to the more specialized field in which I work, as well as the contributions by Rutherford and Planck, should also be formally acknowledged before I should be considered for such an honor.[3]

With my warmest regards to you and your wife from my wife and me, yours truly.

387. "Comment on E. Trefftz's[1] Paper: 'The Static Gravitational Field of Two Mass Points in Einstein's Theory'"(1)

[*Einstein 1922r*]

PRESENTED 23 November 1922
PUBLISHED 21 December 1922

IN: *Preußische Akademie der Wissenschaften* (Berlin). *Sitzungsberichte* (1922): 448–449.

[p. 448] The author grounds his analysis on the field equations in vacuo,

$$R_{ik} - \frac{1}{4}g_{ik}R = 0,\qquad(1)$$

which are equivalent to the equations:

$$\left(R_{ik} - \frac{1}{2}g_{ik}R\right) - \lambda g_{ik} = 0,\qquad(1a)$$

as is easily proved by reducing (1a).[2] The author believes he has found a solution that has a spherical connection in space and except for the two masses no singularity, also not containing any other masses.[3]

In view of the importance of the problem to the cosmological issue, i.e., the question of the large-scale geometrical structure of the universe, I was interested to know whether the equations really did yield as a physical possibility a static universe whose material mass was concentrated in just two celestial bodies.[4] It became apparent, however, that Trefftz's solution does not permit this physical interpretation at all. This will be demonstrated in the following.

Mr. Trefftz sets out on the assumption for the (four-dimensional) line element:

$$ds^2 = f_4(x)dt^2 - [dx^2 + f_2(x)(d\vartheta^2 + \sin^2\vartheta d\phi^2)].\qquad(2)$$

This assumption corresponds to a space of spherical symmetry around the origin. The special case $f_4 = \text{const}; f_2 = x^2$ would correspond to the Euclidean-Galilean isotropic and homogeneous space.

(1)Mathem. Ann. 26, 317, 1922.

In (2), x signifies the radial naturally measured distance from one of the two mass points (up to an additive constant, $\sqrt{f_x(x)}$),[5] the naturally measured circumference divided by 2π of a sphere having a constant value x that separates and centrally surrounds each of the two masses. The surfaces of both spherically shaped masses would be expressed by two equations $x = X_1$ and $x = X_2$, between which $(X_1 < x < X_2)$ there is empty space.

Mr. Trefftz gives as a general solution to the problem: [p. 449]

$$\left.\begin{array}{c} x = \displaystyle\int \frac{dw}{\sqrt{1 + \dfrac{A}{w} + Bw^2}} \\[3em] f_2 = w^2 \\[1em] f_4 = C^2\left(1 + \dfrac{A}{w} + Bw^2\right), \end{array}\right\} \tag{3}$$

where initially C can be set equal to 1 without limiting the generality. According to (2), one can hence set

$$ds^2 = \left(1 + \frac{A}{w} + Bw^2\right)dt^2 - \frac{dw^2}{1 + \dfrac{A}{w} + Bw^2} - w^2(d\vartheta^2 + \sin^2\vartheta d\phi^2).$$

For negative A and vanishing B this yields the well-known Schwarzschild solution for the field of a material point. Thus the constant A will have to be chosen as negative here as well, in accordance with the fact that only positive gravitating masses exist. The constant B corresponds to the λ-term of equation (1a). A positive λ corresponds to a negative B and vice versa.

If equation system (3) really represents the field of two spherical masses, then this world obviously has to behave metrically as follows. From the first sphere $x = X_1$, the circumference divided by 2π of concentric spheres $x = $ const, which are expressed by $w(=\sqrt{f_2})$, must initially grow, then, upon approaching the second sphere, diminish, provided we are dealing with a closed world with the topology of a spherical world.[6] Therefore, somewhere in the empty space between the two material spheres[7]

$$\frac{dw}{dx} = \sqrt{1 + \frac{A}{w} + Bw^2} = 0$$

must be valid. But there, according to (3), f_4 would vanish. According to (1), $\sqrt{f_4}$ is the running speed of a standard clock that is positioned at rest at that location.

The vanishing of f_4 hence signifies a true singularity of the field. That the f_4's are not allowed to disappear is also shown in that the logarithmic derivatives f'_4/f_4 occur in the differential equations. Thus it is demonstrated that solution (3) should not be continued up to that point. In actual fact, it presumes the existence of other masses distributed in spherically symmetric extension, as H. Weyl has already shown.[8]

388. "Einstein's Opinion on the Investigation of Responsibilities for the War"

[*Einstein 1922n*]

PUBLISHED 25 November 1922

IN: *Les cahiers des droits de l'homme* 22 (1922): 547.

An Opinion by Einstein[1]

I expect nothing from this delving into the sorry past and all the discussions aimed at the moral cleansing of our two nations. Joint efforts by Germany and France toward the reconstruction of the ravaged territories seem to me to be of much more importance. Working together is fertile, it engenders trust.

389. From Hantaro Nagaoka et al.

Physics Institute of Tokyo Imperial University, 1 December 1922

[Not selected for translation.]

390. From Alexander Friedmann[1].

Petrograd, Central Physics Observatory, Vassili OstroV, 23rd Line, 6 December 1922

Esteemed Professor,

In a letter from one of my friends who is currently abroad, I had the honor of learning that a brief comment by you had been submitted to press in the 11th volume of *Zeitschrift für Physik*,[2] which points out that from assumptions (D_3) and (*C*) of my article on "The Curvature of Space" [*Die Krümmung des Raumes*] and from the cosmological equations you had postulated it should follow that the radius of the curvature of the universe was a time-independent quantity. You obtain this result by exploiting the circumstance that one obtains as a necessary consequence of the universal equations a vanishing of the divergence of tensor T_{ik}. From the vanishing of the divergence of tensor T_{ik}, you obtained the relation:

$$(*) \qquad \frac{\partial \rho}{\partial x_4} = 0.$$

Such a relation naturally demonstrates the continuity of the radius of curvature R, however, and consequently also the incorrectness of my article.

I did not, of course, manage to obtain relation (*) from the vanishing of tensor T_{ik}; the result I obtained does *not contradict* the case of a nonstationary world.[3] In view of the definite interest attached to the problem of a possible existence of a nonstationary world, I permit myself to submit to you the calculations of the divergence of tensor T_{ik} I carried out for your evaluation and consideration. Let Q_k be the kth component of the contragradient tensor that represents the divergence T_{ik} and then, according to the formula, we shall have for the divergence:

$$Q_k = \frac{1}{\sqrt{g}} \frac{\partial \sqrt{g} g^{\alpha\sigma} T_{\alpha k}}{\partial x_\sigma} - \begin{Bmatrix} k\sigma \\ s \end{Bmatrix} g^{\alpha\sigma} T_{\alpha s};$$

Q_4 interests us, as Q_1, Q_2, Q_3 turn into zero and specifically as a consequence of the circumstance that we have the unstationary world expressed in my article by the formula (D_3) under the conditions (*C*):

$$\begin{Bmatrix} 41 \\ 4 \end{Bmatrix} = 0, \qquad \begin{Bmatrix} 42 \\ 4 \end{Bmatrix} = 0, \qquad \begin{Bmatrix} 43 \\ 4 \end{Bmatrix} = 0, \qquad \begin{Bmatrix} 44 \\ 4 \end{Bmatrix} = 0.$$

For Q_4 we shall have:

$$Q_4 = \frac{1}{\sqrt{g}} \frac{\partial \sqrt{g} g^{\alpha\sigma} T_{\alpha 4}}{\partial x_\sigma} - \left\{ \begin{matrix} 4\sigma \\ s \end{matrix} \right\} g^{\alpha\sigma} T_{\alpha s} =$$

$$= \frac{1}{\sqrt{g}} \frac{\partial \sqrt{g} g^{\alpha\sigma} T_{44}}{\partial x_\sigma} - \left\{ \begin{matrix} 4\sigma \\ 4 \end{matrix} \right\} g^{4\sigma} T_{44},$$

where all the T_{ik}'s, with the exceptions of the T_{44}'s on the basis of conditions (C) of my article, equal zero. Under the conditions (D_3), $g^{4\sigma} = 0$ for all σ's with the exception of $\sigma = 4$, however, the preceding formula is rewritten in this way as follows:

$$Q_4 = \frac{1}{\sqrt{g}} \frac{\partial \sqrt{g} g^{44} T_{44}}{\partial x_4},$$

as $g^{44} = \dfrac{1}{g_{44}} = 1$ (in our case equal to the interval set by formula (D_3)), but T_{44} is equal to $c^2 \rho g_{44}$, i.e., $T_{44} = c^2 \rho$; hence, in this way Q_4 is written in the form of the following formula:

$$Q_4 = \frac{1}{\sqrt{g}} \frac{\partial \sqrt{g} c^2 \rho}{\partial x_4}.$$

By our setting Q_4 equal to zero, which follows from your universal equations, of course, we do *not* have the equation indicated by you and appearing in your article, but the following instead[4]:

$$(**) \qquad\qquad \frac{\partial \sqrt{g} \rho}{\partial x_4} = 0;$$

in this way, one arrives at the necessity that $\sqrt{g} \rho$ be independent of x_4 but

$$g = -\frac{1}{c^6} R(x_v)^6 \sin^4 x_1 \sin^2 x_2, \qquad \sqrt{g} = \frac{1}{c^3} \sqrt{-1} R(x_v)^3 \sin^2 x_1 \sin x_2 ;$$

however, on the basis of formula (8) of my article, ρ is expressed as follows:

$$\rho = \frac{3A}{\frac{1}{2} R(x_4)^3},$$

whence

$$\sqrt{g} \rho = \frac{3A}{c^3 \kappa} \sqrt{-1} \sin^2 x_1 \sin x_2,$$

and really is independent of x_4, which is also being sought.

Do not, highly esteemed Professor, deny me notification about whether my calculations discussed in the present letter are right. I recently examined the case of a world of constant and variable (in the temporal sense)[5] negative curvature. There it was, of course, necessary (in order to obtain the only interesting ⟨material⟩ real[6] world, from the point of view of physics and geometry) to use a different expression for the interval, in that I took it (according to Bianchi, *Lezioni di geometria differenziale*, volume 1) in the following form:

$$d\tau^2 = -\frac{R(x_4)^2}{x_3^2}(dx_1^2 + dx_2^2 + dx_3^2) + M^2 dx_4^2 \; ;$$

the result of the computations showed that in this case not only a world of constant (already negative) curvature can exist but also a world of (in the temporal sense)[7] variable curvature. The existence of the possibility of obtaining from your universal equations a world of constant negative curvature is of very special interest to me, which is why I earnestly ask you not to deny me an answer to this letter of mine, although I know how very busy you are going to be.

In the event that you find the calculations discussed in this letter correct, please do not reject my letting the editorial office of the *Zeitschrift für Physik* know about it; in this case you might perhaps publish a correction or make possible a printing of an excerpt from the present letter.[8]

Your sincere admirer,[9]

A. Friedmann.

At head of the document: "Physical Central Observatory, Petrograd, Wassili Ostrov, 23 Linie, 2nd professor."

391. "Musings on My Impressions in Japan"

[*Einstein 1923b*]

Manuscript completed on or after 7 December 1922.

PUBLISHED January 1923[1]

IN: *Kaizo* 5 (1923): 338–343.

I have been traveling much around the world in the last few years, actually more [p. 343] than befits a scholar. The likes of me really ought to stay quietly put in his study and ponder. For my earlier trips there was always an excuse that could easily pacify my not very susceptible conscience. But when Yamamoto's[2] invitation to Japan arrived, I immediately resolved to go on such a great voyage that must demand months, even though I am unable to offer any excuse other than[3] that I would never have been able to forgive myself for letting a chance to see Japan with my own eyes go by unheeded.

Never in my life have I been more envied in Berlin, and genuinely[4] so, than the moment it became known that I was invited to Japan. For in our country this land is shrouded more than any other in a veil of mystery. Among us we see many Japanese, living a lonely existence, studying diligently, smiling in a friendly manner. No one can fathom the feelings concealed behind this guarded smile. And yet it is known that behind it lies[5] a soul different from ours that reveals itself in the Japanese style, as is manifest in numerous small Japanese products and Japanese-influenced literature coming into fashion from time to time. All the things I knew about Japan could not give me a clear picture. My curiosity was in utmost suspense when, on board the Kitanu Maru, I passed through the Japanese channel and saw the countless delicate green islets glowing in the morning sun. But glowing most of all were the faces of all the Japanese passengers[6] and the ship's entire crew.

[p. 342] Many a tender young lady, who otherwise would never[7] be seen before breakfast time, was roaming restlessly and blissfully about on deck at six o'clock in the morning, heedless of the[8] raw morning wind, in order to catch the first possible glimpse of her native soil. I was moved[9] to see how overcome they all were with deep emotion. A Japanese loves his country and his nation most of all; and despite his linguistic proficiency and great curiosity about everything foreign, away from home he still does feel more alien than anyone else. How is this explained?

I have been in Japan for two weeks now,[10] and yet so much is still as mysterious to me as on the first day. Some things I did learn to understand, though, most of all the shyness that a Japanese feels in the company of Europeans and Americans. At home our entire upbringing is tuned[11] toward our being able to tackle life's struggles as single beings under the best possible circumstances. Particularly in cities, individualism in the extreme, cutthroat competition drawing on our utmost energy, feverish laboring to acquire as much luxury and pleasure as possible. Family bonds are loosened, the influence[12] of artistic and moral traditions on daily life is relatively slight. The isolation of the individual is seen as a necessary consequence of the struggle for survival; it robs a person of that carefree happiness that only integration in a community can offer. The predominantly rationalistic education—indispensable for practical living under our conditions—lends this attitude of the individual even more poignancy;[13] it makes the isolation of the individual bear even more keenly on the conscience.[14]

Quite the contrary in Japan. The individual is left far less to his own devices than in Europe or America. Family ties are very much closer than at home, even though they are actually only provided very weak legal protection. But the power of public [p. 341] opinion is much stronger here than at home, assuring that the family fabric is not undone. Published and unpublished Dame Repute helps force to completion what is mostly already sufficiently secured by a Japanese upbringing and an innate kindheartedness.[15]

The cohesion of extended families in material respects, mutual support, is facilitated by an individual complaisance about bed and board.[16] A European can generally[17] accommodate one person in his apartment without there being perceptible disruption[18] to the household order. Thus a European man can mostly only care for his wife and children, all being well. Often wives, even women of higher status, must[19] help earn a living and leave the children's education to the servants. It is even rare that adult siblings, let alone more distant relatives, provide for one another.

But there is a second reason, as well, that makes closer protective ties between individuals easier in this country than where we live. It is the characteristic Japanese tradition of not expressing one's feelings and emotions but staying calm and relaxed in all circumstances. This is the basis upon which many persons, even those not in mental harmony with one another, can live under a single roof without embarrassing frictions and conflicts arising. Herein lies, it appears to me, the deeper sense of the Japanese smile, which is so puzzling to a European.

Does this upbringing to suppress the expression of an individual's feelings lead to an inner loss, a suppression of the individual himself? I do not think so. The development of this tradition was surely facilitated by a refined sensitivity characteristic to this nation and by an intense sense of compassion, which seems to be more potent than for a European. A rough word does not injure a European any less than it does a Japanese. The former immediately counters by stepping forward to the offensive, amply repaying in kind. A Japanese withdraws, wounded, and—   [p. 340] weeps. How often is the Japanese inability to utter sharp words interpreted as falseness and dishonesty!

For a foreigner like me it is not easy to delve deeply into the Japanese mind. Being received everywhere with the greatest attention in festive garb, I hear more carefully weighed words than meaningful ones that inadvertently slip out of the depths of the soul. But what escapes me in direct experience with people is completed by the impressions of art, which is so richly and diversely appreciated in Japan as in no other country. With "art" I mean all things of permanence that human hands create here by aesthetical intent or secondary[20] motivation.

In this regard I never cease to be astonished and amazed. Nature and people seem to have united to bring forth a uniformity in style as nowhere else. Everything that truly originates from this country is delicate and joyful, not abstractly metaphysical but always quite closely connected with what is given in nature. Delicate is the landscape with its small green islets or hills, delicate are the trees, delicate the most carefully farmed land with its precise little parcels, but most especially so the little houses standing on it; and, finally, the people themselves in their speech, their movements, their dress, including all objects they make use of. I took a particular liking to the Japanese house with its very[21] segmented smooth walls, its many little rooms laid out throughout with soft mats. Each little detail has its own sense and import there. To this come the dainty people with their picturesque smiling, bowing, sitting—all[22] things that can only be admired but not imitated.[23] You, oh foreigner, try to do so in vain!—Yet Japan's dainty dishes are indigestible[24] to you; better content yourself with watching.— Compared to our people, the Japanese are cheerful and carefree in their mutual relations,— they live

[p. 339]   not in the future but in the present. This cheerfulness always expresses itself in refined form, never boisterously. Japanese wit is directly comprehensible to us. They too have much sense for the droll, for humor. I was astonished to note that, regarding these psychologically surely deeply lying things, there is no great difference between the Japanese and the European. The soft-heartedness of the Japanese reveals itself here in that this humor does not assume a sarcastic note.

[25]Of greatest interest to me was Japanese music, which had developed partly or entirely independently of ours. Only upon listening to completely strange art does one come closer to the ideal of separating the conventional from the essential conditioned by human nature. The differences between Japanese music and ours are indeed fundamental. Whereas in our European music, chords and architectonic structure appear to be universal and indispensable, they are absent in Japanese music. On the other hand, they have in common the same thirteen tonal steps into which the octaves are divided. Japanese music seems to me to be a kind of emotive painting of inconceivably direct impressions. The exact tonal pitch does not even seem to be so absolutely imperative for the artistic effect. It much rather suggests to me a stylized depiction of the passions expressed by the human voice as well as by such natural sounds that conjure up in the human mind emotive impressions such as birdsong and the beating surf of the sea. This impression is amplified by the important role played by the percussive instruments, which have no specific tonal pitch of their own but are particularly suited for rhythmic characterization. The main attraction to me of Japanese music lies in the extremely refined rhythms. I am fully aware of the circumstance that the most intimate subtleties of this kind of music still elude me. Added to the fact that long experience is always presup-

[p. 338]   posed for distinguishing the purely conventional from a performer's personal expression, the relation to the spoken and sung word, which plays a considerable role in most Japanese pieces of music, also escapes me. As I see it, a characteristic of the artistic approach of the Japanese soul is the unique appearance of the mild flute, not the much more strident wind instruments made out of metal. Here, too, is manifested the particular characteristic preference for the charming and the dainty that is especially prominent in Japanese painting and design of products for every-day life. I was most greatly affected by the music when it served as accompaniment to a theater piece or a mime (dance), particularly in a Noh play. What stands in the way of the development of Japanese music into a major form of high art is, in my opinion, a lack of formal order and architectonic structure.

To me,[26][27] the area of most magnificence in Japanese art is painting and wood carving. Here it is properly revealed that the Japanese is a visual person delighting in form, who untiringly refashions events in artistic form, converted into stylized lines. Copying nature in the sense of our realism is foreign to the Japanese, just as is religious repudiation of the sensual, despite the influence[28] of Buddhism from the Asian continent, which is intrinsically so foreign to the Japanese mind. For him, everything is experienced in form and color, true to Nature and yet foreign to Nature insofar as stylization broadly predominates. He loves clarity and the simple line above all else. A painting is strongly perceived as an integral whole.

But among the great impressions I had during these weeks, so few could I mention,[29] saying nothing about political and social problems. On the exquisiteness of the Japanese woman, this flower-like creature—I was also silent;[30] for here the common mortal must cede the word to the poet. But there is one more thing I have at[31] heart.[32] The Japanese rightfully admires the intellectual achievements of the West and immerses himself successfully and with great idealism in the sciences. But let him not thereby forget to keep pure the great attributes that he possesses over the West: the artful shaping of life, modesty and unpretentiousness in his personal needs, and the purity and calm of the Japanese soul.

392. From Yuanpei Cai

Peking, 8 December 1922

Esteemed Professor Einstein,

The news about your wanderings and activities in Japan are being followed here with great interest; and the whole of China is ready to welcome you with open arms.

You surely do still remember the arrangement you made with us through the intermediary of the Chinese envoy in Berlin, and will also carry it out, as we in joyous anticipation hope.[1]

We would be pleased if you could inform us of the day of your arrival in China. We shall make the necessary preparations to make your travels as untroublesome as possible.

Respectfully,

J. P. Tsai,
rector of the National University of Peking,
and at the same time in the name of the following institutions:[2]

393. To Bansui Tsuchii (Doi)

Nikko, 9 December 1922[1]

Esteemed Mr. Tsuchii,

Through your personality, the wonderful book with the sketches of Hokusai, your Italian poem, and the poem published in the newspaper, you gave me indescribable pleasure.[2] I shall always cherish everything I have by you as a precious souvenir of a high-ranking Japanese artist and will look at them again and again. This visit in your very pleasing country, so highly developed artistically and humanly, is to me one of the finest experiences of my life.

Accept my cordial thanks, and friendly greetings from your

A. Einstein.

394. Preface for the Japanese edition of Georg Nicolai's *Biologie des Krieges*[1]

[Kyoto, 10 December 1922][2]

[Not selected for translation.]

395. To Heinrich Zangger

Nagoya, 11 December [1922]

Dear Zangger,

Here I sit with Hashida in Nagoya,[1] where we are thinking of you. Until you have been in Japan, you will never know where real human beings are to be found and what they look like.

Cordial regards, yours,

A. Einstein.

396. From Henrik Sederholm and Knut A. Posse[1]

Nobel Foundation, Stockholm G, 6 Norlandagatan, 11 December 1922

Highly esteemed Professor,

As the Roy. Academy of Sciences already informed you, it bestowed on you the physics Nobel Prize for 1921, for conferral during the formal award ceremony on 10 December of this yr.[2]

In view of your sojourn in the Far East, through which you were unfortunately prevented from appearing for the awards, we found it appropriate to deposit the prize amount, *121,572.54 Swedish kronor*, at the local Stockholmer Enskilda Bank into an account with a 14-day mutual termination notice, into which account interest at the highest rate valid for deposit accounts is paid.[3]

This interest rate is presently 3% p[er] a[nnum].

In the enclosed we submit to you the bank's receipt for this deposit.[4]

We are storing for you until further notice the prize medal in gold as well as the prize certificate bestowed on you at the same time in your account, and ask you please to inform us when and to which address you wish to have these objects sent.

Looking forward to your kd. acknowledgment of receipt of the bank receipt, we sign in great respect, for the Nobel Foundation,

H. Sederholm
K. Bosse.

397. To Sanehiko Yamamoto

Kyoto, 12 December 1922

Esteemed Mr. Yamamoto,

Under the influence of the great services you are doing by reason of my and my wife's presence in Japan, I consider it an absolute duty to inform you of the following. The ship[1] is leaving from Moji in only 16 days and in the interim I have nothing for you to do. So I feel it is unfair of me to have us impose upon Mr. Inagaki and his wife during this interval.[2] Much as I love them both, I ask you please, in order to relieve my conscience, to leave my wife and me *alone* in Kyoto during this quiet period. You are truly doing enough for us by making possible for us such a long stay in this wonderful city. I also ask you please not to have anyone travel to Fukuoka and Moji *as a pure personal favor to us*.

On this occasion I would like to give expression to my profound need to thank you for giving us a chance to see this wonderful country and for having made life so nice for us throughout in such a generous and caring manner.

Cordial greetings to you, yours,

A. Einstein.

My wife was not in Osaka yesterday, because I had asked her to stay in Kyoto. I did this because it was not known to me in time that an official function was supposed to take place in Osaka. My wife was very unhappy about having been the source of a disruption through no fault of her own.[3]

398. "Answer to Questions on Religion"

[*Einstein 1923c*]

DATED 14 December 1922
PUBLISHED February 1923

IN: *Kaizo* 5 (1922): 194–195 and 197.

[p. 197]   Response to a Question about Religion

When Professor Einstein was asked about his opinion on religion, he supplied the answers below to the following questions:[1]

1)   Do you think that scientific truth and religious truth rest on different viewpoints?

2)   Do they promote each other; that is, can scientific discovery improve religious faith and remove superstition because the religious sense can provide the impetus for scientific discovery?

3)   What concept, Professor, do you have of "God"?

4)   What, Professor, is your opinion of the "Savior"?

1)   It is not easy to attach a clear sense even to the term "scientific truth." For, the meaning of the word "truth" differs, depending on whether it involves a fact of experience, a mathematical theorem, or a scientific theory. Under "religious truth" I cannot conceive of anything clear at all.

2)   Scientific research can reduce superstition by encouraging causal thinking and providing an overview. It is certain that a passionate conviction, related to the

religious sense, about the rationality or intelligibility of the world lies at the basis of all more refined scientific work.[2]

3)   That deeply felt conviction about a superior rationality which manifests itself in the perceptible world forms my concept of God; one can therefore express it with the common designation of "pantheistic" (Spinoza).[3]

4)   I can regard confessional traditions only historically and psychologically; I have no other relationship toward them.[4]

EINSTEIN'S LECTURE AT THE UNIVERSITY OF KYOTO

[See documentary edition for editorial note.]

399. How I Created the Theory of Relativity
(Jun Ishiwara's Notes of Einstein's Lecture at Kyoto University)

PRESENTED 14 December 1922
PUBLISHED 1923

IN: *Kaizo*, Vol. 5, No. 2 (1923): 2–7.

[See documentary edition for English translation.]

400. To Hans Albert and Eduard Einstein

Kyoto, 17 December 1922

Dear Children,

Now you, d[ear] Albert, have been a student for a couple of months already.[1] I often think of that with pride. The voyage is wonderful, even though Japan is quite exhausting. I have already given 13 lectures. I am very glad that I left you, d. Albert, in Zurich; as I would not have been able to pay much attention to you anyway, and your studies mean more to you than any trip, no matter how nice, in which you'd have had to make official appearances in so many instances.[2] The Japanese do appeal to me, by the way, better than all the peoples I've met up to now: quiet, modest, intelligent, an artistic sense, and considerate, putting nothing on appearances, rather everything on substance. So now all of you really will be getting the Nobel Prize.[3] Look into the matter about the house.[4] The rest will be deposited somewhere in your names. Then you'll be so rich that, God knows, I may have to beg off you some day again, depending on how things go. After my return home (end of March or beginning of April), I am going to have to travel to Stockholm in order to accept the prize. When I travel to Geneva then,[5] I shall visit you, of course; I'm already looking forward to that. Then we can also consult about what we'll be doing next summer. I have decided definitely not to ride around the world so much anymore; but am I going to be able to pull that off, too?

You rascals didn't write me at all; now it's too late for Asia. If you want to write me before my return home to Germany, e.g., about the house, then write to Spain (University of Madrid) or—if you want to write fast—to the Zionist Organization in Jerusalem. I am enclosing for you, d. Tete, a few postage stamps collected along the way.

Warm regards to you and Mama from your

⟨Albert⟩ Papa.

401. To Michele Besso and Anna Besso-Winteler

Naru, 19 December 1922

[Not selected for translation.]

402. To Wilhelm Solf

[Miyajima, 20 December 1922][1]

[…]

I hasten to forward to you the more detailed sequel to my reply by telegraph.[2] Harden's statement is certainly awkward for me, in that it makes my situation in Germany more difficult; nor is it completely right, but neither is it completely wrong.[3] For, people who survey the situation in Germany well are indeed of the opinion that a certain threat to my life does exist. Albeit, I did not assess the situation the same way before Rathenau's murder as I did afterwards. A yearning for the

Far East led me, to a large part, to accept the invitation to Japan; another part was the need to get away for a while from the tense atmosphere in our homeland for a period of time, which so often places me in difficult situations. But after the murder of Rathenau I was certainly very relieved to have an opportunity for a long absence from Germany, taking me away from the temporarily heightened danger without my having to do anything that could have been unpleasant for my German friends and colleagues.

403. To Yuanpei Cai

[Miyajima,] 22 December 1922

Highly esteemed Rector,

It pains me greatly that at the present moment, despite every good intention and despite earlier formal promises, it is impossible for me to come to China.[1] When I arrived in Japan I waited five weeks in vain for a message from Peking. When this failed to arrive I was convinced that the University of Peking no longer intended to come back to its earlier invitation. That is why I thought it not tactful to contact you about it. This was made even less feasible as Dr. Pfister in Shanghai—purportedly on your authority—made offers regarding a sojourn in China that were in contradiction to our earlier negotiations.[2] From this also I had to conclude that you were not planning to hold to the latter.

Under these conditions I stayed longer in Japan than would have been compatible with a visit to China and reached all travel decisions under the assumption that my trip to China was canceled.

In possession of your letter, which arrived today, I realize that this was a misunderstanding;[3] however, I do not see any possibility of changing my arrangements retroactively anymore. But I do rely on the pleasant hope that you will understand my way of proceeding, for you can imagine how greatly interested I would have been to visit Peking now. Furthermore, I optimistically hope that this omission, caused by such a sorry misunderstanding, will later be rectified.

In expression of my utmost respect,

A. E.[4]

404. To Max and Hedwig Born

[Miyajima,] 23 December [1922][1]

Dear Borns,

Resplendent sunshine at Christmas. Cheerful, beautiful country with fine, delicate people. On the 29th it's back onto the great water, homeward bound via Java, Palestine, and Spain; it will probably be April by the time we get back.

Until then, warm regards from your

Einstein.

405. To Jun Ishiwara

Moji, [between 23 and 29 December 1922][1]

To my dear colleague Ishiwara as a souvenir, with whom I saw so many fine things, collaborated, and chatted many pleasant hours away. He is one of those few in whose company I would like very much to ponder and work; because, for all the differences in origin and tradition, a mysterious harmony exists between us.

Albert Einstein
Moji 1922.

406. "Preface" to Japanese Collection of Papers

[*Einstein 1923f*]

DATED 27 December 1922
PUBLISHED 1 May 1923

IN: *Einstein 1922–1924*, Vol. 2, pp. [i–ii].

[p. i] On the occasion of my sojourn in Japan, the indefatigable director of Kaizo Publishers[1] prepared a complete collection of my scientific papers to date,[2] which are thus made accessible in convenient form to my Japanese colleagues in the field and to students. It is a pleasant duty of mine to express my deep gratitude to Mr. Yamamoto for this achievement, no less also to my esteemed friend and colleague Ishiwara,[3] who took upon himself the great trouble of translating; his name vouches for a faithful rendition.

Our science advances so rapidly that most original papers very quickly lose their current importance and appear outdated. On the other hand, however, there is always a special charm in retracing the development of theories by means of the original articles; and not seldom such a study lends deeper insight into the material than a smoothed systematic exposition of the finished subject by the labors of many contemporaries. In this sense I hope that the present collection constitutes an enrichment of the professional literature. In particular, I would like to take the liberty of recommending to our younger colleagues the papers on the special and general theory of relativity, the papers on Brownian motion, and the quantum theoretical articles from the years 1905 and 1917, which contain considerations that in my opinion have still not been taken sufficiently into account even today. [p. ii]

This is the first edition of my complete scientific works.[4] That this is coming to pass in the Japanese language is new proof to me of the intensity of scientific life and interest in Japan, which in these past weeks I have learned not only to respect highly as a site of science, but also—what is even more important—to love from the human perspective.

407. To Yoshi Yamamoto

Moji, [27 December] 1922[1]

[Not selected for translation.]

408. From Wilhelm Solf

[Tokyo, 27 December 1922]

[Not selected for translation.]

409."Farewell to Japan"

[*Einstein 1922s*]

DATED 28 December 1922
PUBLISHED 29 December 1922

In: *Fukuoka Nichinichi Shinbun*, p. 2.

[Not selected for translation.]

410. To Ayao Kuwaki

[Moji, 29 December 1922][1]

   To Prof. Kuwaki the physicist and epistemologist, the first Japanese physicist I had the pleasure of getting to know,[2] in friendly remembrance,

Albert Einstein, 1922.

Nature is a prim goddess.

411. To Bansui Tsuchii (Doi)

[On board *S. S. Haruna Maru*,] 30. Dezember 1922

Very esteemed Mr. Tsuchii,

   With great joy and admiration did I read the translation of your thoughtful poem[1] and your very friendly letter. It does not matter that you hugely overrate my accomplishments,[2] because the word simply emanates from the pure soul. The scientific quest really is different from that of an artist. The artist develops with certainty if he has the faculty to see and feel, the power to create, and a perseverance and love of perfect creation. Science, however, is like riddle-guessing or even playing the lottery. It is a rare, chance event if something really valuable is found. Many a highly talented young man labors right into old age without the severe goddess unveiling anything of her deep secrets to him; she is unpredictable and inquires little about merit earned from a devoted search for Truth. And the little she entrusted to me must appear gigantically magnified to the uninformed, who do not know about the achievements of my forerunners and fellow seekers. Be this as it may— I am delighted about your enthusiastic words.

   I found exceptionally fine what you said about your beautiful land and the peculiar transitional state in which it currently finds itself. But I do believe your characterization is a little too harsh. In nurturing Western sciences over the course of these couple of decades, Japan has already climbed to a high level and is tackling the profoundest of problems. Japan is not merely borrowing the superficial elements of civilization![3] A flood of alien culture is dangerous for any land in which its own high values are too easily underrated and forgotten—I mean what I so much admire and love in the artistic, social, and ethical tradition of your country. In such matters the Japanese does not recognize his superiority over the European; it would be of great service to bring this to his consciousness so that he can feel that, with the indiscriminate adoption of European life forms, great values are being jeopardized. Japan may accept the civilizing spirit of Europe and America, but it must know that its soul is far more valuable than these external shimmering little things.

With joy and with a trembling hand I accept the splendid reproductions of Japanese-Chinese works of art which you have given me.[4] They will accompany me on this voyage and they will soften the roughness of the transition to Europe. The Japanese artistic hand is of incomparable delicacy.

My extant collected works will soon appear in Japanese [5] and I will gladly permit myself to have them sent to you, although it will be difficult to insert a dedication. But I will try to do so. I am attaching a small card for your son.[6]

Accept my most cordial greetings and please accept my warmest thanks, from your,

A. Einstein

412. To Eiichi Tsuchii (Doi)[1]

[On board SS *Haruna Maru*, 30 December 1922][2]

He who is familiar with the exertions of pondering scientific problems never feels empty and alone; he also gains a firm foothold against the vicissitudes of fate.

A greeting for the young man E-i-ichi,

Albert Einstein

413. To Sanehiko Yamamoto

[On board SS *Haruna Maru*, 30] December 1922[1]

You, Mr. Yamamoto, are one of those natural leaders who, out of an inner compulsion, devote all their energy to improving the social conditions of their nation. This goal, not the program of any party, is your guiding star. But your ideal extends beyond the shaping of Japan, in that you also devote your unflagging diligence to the greatest political goals of the present day, namely, the creation of an international organization for the prevention of war catastrophes. For this, first mutual understanding between men of the different nations is needed as well as an

350 DOCS. 414, 416 DECEMBER 1922–JANUARY 1923

emphasis on the truly international assets of humanity, among which science perhaps takes first place. Thus I understand the invitation you sent me.[2]

Not only do you love society as a whole but you are also full of benevolence toward the individual. This I experienced of you with joy and gratitude and shall never forget what fine things I learned from you and through you.

Yours,

Albert Einstein.

414. To Yoshi Yamamoto

[On board SS *Haruna Maru*, 30] December 1922[1]

You, Esteemed Mrs. Yamamoto, will always symbolize for me the ideal model of Japanese femininity. Quiet, cheerful, and flowerlike, you are the soul of your home, which looks like a jewelry case, within which, like jewels, lie your darling little children.[2] In you I see the proper soul of your people and the embodiment of its ancient culture, directed primarily toward daintiness and beauty.

Yours,

Albert Einstein.

415. To Charlotte Weigert

[Shanghai, between 31 December 1922 and 2 January 1923]

[Not selected for translation.]

416. From Rafaele Contu

Rovigno (Istria), [8 January 1923][1]

Esteemed Prof. Albert Einstein,

Kindly forgive my long silence. In the meantime, I completed an important work on relativity theory and compiled the reference material for the theory itself in a work that appeared on the 1st of this mo.—Our Hoepli will have one copy of it forwarded to you.[2]

Would you—please—notify me about whether you received the contract for the *Prospettive Relativistiche* from the "Audace" publishers.[3] At the end of March one issue is appearing about your first paper and the "Dialogue about . . . . . ," under the terms you already approved.[4] I hope to arrange everything so that the majority of the whole work will be published over the course of the year.

Scholars and admirers would like to confer upon you a chair at the University of Rome. With this in mind I was asked by many quarters for bibliographic and other information, which I also provided. I am now commissioned, under strict confidentiality, to find out whether you would be inclined to teach higher mathematics and theoretical physics; and would you please have a statement sent out to me in this regard with indications as to whether and to what extent I may make use of it?

I shall, in any case, treat it with the greatest discretion but can already assure you in advance that our wish, which is excellently supported, will be fulfilled, the more so as it meets strict scientific requirements.

Requesting, if possible, a prompt reply, I again ask you please to accept this expression of my highest admiration.

Respectfully,

Rafaele Contu

417. "On the General Theory of Relativity"

[On board SS *Haruna Maru*, ca. 9 January 1923][1]

On the General Theory of Relativity. A. Einstein   [p. 1]

§1. Generalities

⟨The most powerful impulse⟩ Theoretical efforts of the past few years in the field of general relativity theory ⟨spring from⟩ correspond to two ⟨sources⟩ trains of thought. First, the aim was to comprehend the gravitational field and the electromagnetic field ⟨under a single⟩ as a unified entity; second, the ⟨splitting⟩ distinction of the concept of affine connections from the originally purely metric foundation of Riemannian geometry yielded new possibilities or, resp., limitations on the choice of equations to express natural laws.

The basis of Riemannian geometry is the ⟨fundamental⟩ metric invariant

$$ds^2 = g_{\mu\nu}dx_\mu dx_\nu,$$ (1)

from which all the theory's other concepts are derived. Among these derived concepts, the parallel displacement of vectors which determines the affine connection is of primary interest to us here, determined according to the formula

$$\delta A^\mu = -\Gamma^\mu_{\alpha\beta} A^\alpha dx_\beta, \tag{2}$$

where according to Riemannian geometry the $\Gamma^\mu_{\alpha\beta}$'s are given by

$$\Gamma^\mu_{\alpha\beta} = \frac{1}{2} g^{\mu\sigma}\left(\frac{\partial g_{\sigma\alpha}}{\partial x_\beta} + \frac{\partial g_{\sigma\beta}}{\partial x_\alpha} - \frac{\partial g_{\alpha\beta}}{\partial x_\sigma}\right) \tag{3}$$

The $\Gamma^\mu_{\alpha\beta}$ quantities seem to have been first introduced by Christoffel.[2] However, its geometric interpretation as the coefficient of the law of parallel displacement originally goes back to Levi-Civita and Weyl.[3] The fundamental meaning of the Γ's is primarily that they alone determine the Riemannian curvature tensor $R^i_{\kappa, lm}$ as well as its important contraction $R_{\kappa l}$ in gravitational field theory.

The dependence (3) of the parallel displacement law (2) on the fundamental invariant (1) arises in Riemannian geometry, as well as in the original form of the general theory of relativity (with the exception of the independently postulated symmetry condition $\Gamma^\mu_{\alpha\beta} = \Gamma^\mu_{\beta\alpha}$),[4] because the parallel displacement law demands that the quantity $g_{\mu\nu} A^\mu A\nu$ of a contravariant vector not vary upon dis-

[p. 2]   placement according to (2).[5] Analytically this amounts to the condition that the tensor formed out of the fundamental tensor $(g_{\mu\nu})$ or $(g^{\mu\nu})$, resp., through differentiation ("extension")

$$g_{\mu\nu;\sigma} = \frac{\partial g_{\mu\nu}}{\partial x_\sigma} - g_{\alpha\nu}\Gamma^\alpha_{\mu\sigma} - g_{\alpha\mu}\Gamma^\alpha_{\nu\sigma} \tag{4}$$

vanishes identically.[6]

Because the newer theories by Weyl and Eddington[7] modify or eliminate this connection, they arrive at modifications of the general theory of relativity that we will look at briefly. Ultimately, we want to construct a new theory that, although related to Eddington's theory, follows more naturally and simply than it does from the original general theory of relativity.

§2. The Theories by Weyl and Eddington

H. Weyl sets out from the idea that a more fundamental importance be attached to the elementary law of light propagation, or the "light cone"

$$g_{\mu\nu} dx_\mu dx_\nu = 0,$$

than to the ds itself. According to him, only the relations of the $g_{\mu\nu}$'s not of the $g^{[\mu\nu]}$'s have their own real meaning (for an established system of coordinates). In accordance with this interpretation, he requires of the displacement law (2) that it leave unchanged not the amount of *one* vector, but the ratio of the values of two vectors impinging on the same point. Thus he succeeds in having a linear form

$$\varphi_\mu dx_\mu$$

enter into the geometric theory besides the quadratic form (1). By equating these φ_μ's with the electromagnetic potentials, one obtains a mathematical theory with invariants in which the potentials for gravitation and electromagnetism appear in a mutually dependent way.

In my opinion, the great importance of this theory by Weyl is not that it be physically accurate but, much rather, that it first demonstrated the independence of the law (2) from its original metric basis (1). Its weakness, however, lies in its point of departure. The concept of parallel displacement originates, like all concepts of Euclidean geometry, ⟨its emergence⟩ from the consideration of positioning laws or laws of the relative displacements of rigid bodies. That is how the ⟨assertion⟩ stipulation that upon parallel displacement a path does not change its value acquires its cogency. The transition to four-dimensionality does not essentially change anything in this situation. Two (infinitesimally small) bodies of reference originally with equal measuring rods and clocks always measure the same *ds* between two events; the lengths of two measuring rods or, resp., the running speeds of two clocks that are brought together, remain the same if they had ever been the same.   [p. 3]

Now, one certainly can eliminate from the theory those elements that concern the rigid body and the clock. One can also assume that just the equation $ds^2 = 0$ has real meaning. One can introduce a law (2) of affine connection without any physical interpretation by means of the rigid body. But it is then pretty arbitrary to demand that this ⟨affine⟩ law leave unchanged the ratio between the values of two vectors upon displacement ⟨if the interpretation of (2) is⟩.[8]

Eddington consistently continues down the path that Weyl had blazed.[9] Because (1) does not form a sufficient basis for (2), yet the Riemannian tensor so fundamental for the physics is grounded solely on (2), one must try to make do with (2) *alone*. Eddington does not doubt that a metric invariant of type (1) exists in nature; however, his goal is to derive a type-(1) invariant from (2) that has metrically physical meaning. This easily works. The Riemannian curvature tensor $R^i_{\kappa, lm}$ derived just from (2) yields, after contracting *i* and *m*, the symmetric covariant tensor $R_{\kappa, l}$, which with the line element's component dx_κ yields the invariant $R_{\kappa l} dx_\kappa dx_l$. Because this is the simplest invariant that can be attributed to the line element on the basis of (2), it is regarded as the fundamental metric invariant. The weakness in Eddington's theory lies in that it did not lead to all the necessary equations for determining the 40 quantities $\Gamma^\mu_{\alpha\beta}$. Furthermore, Eddington did not succeed in linking it to the existing secure results of the general theory of relativity.   [p. 4]

Eddington also saw no way to bring his theory into the convenient Hamiltonian form.[10]

Because I departed from Eddington's approach to arrive at my new theory, I would like to indicate here the following consideration, which clearly illuminates the connection between my ideas and Eddington's. I first asked myself whether an invariant volume integral existed whose invariance is based only on (2); if yes, then one can require that the first variation of this integral vanish for *all* ⟨possible⟩ variations of the quantities $\Gamma^{\mu}_{\alpha\beta}$. One then obtains 40 differential equations for the 40 $\Gamma^{\mu}_{\alpha\beta}$ -quantities, between which four identical relations hold.

The existence of such invariants can be seen as follows: According to Riemann's theory, within a manifold of 4 dimensions

$$\frac{\delta^{iklm}}{\sqrt{-g}}$$

there is ⟨one tensor, if g is the determinant of the $g_{\mu\nu}$'s and δ^{iklm} is a magnitude that vanishes, if not all four indices⟩ one antisymmetric contravariant tensor of 4th order, all of whose components have the absolute quantity $\dfrac{1}{\sqrt{-g}}$. By multiplying by $\sqrt{-g}$, the tensor density,

$$\delta^{iklm} ,$$

independent of the fundamental tensor, appears. Out of this and the Riemannian curvature tensor, nonvanishing scalar densities are formable, for ex.,

$$R^{i}_{i, \, lm} R^{\iota}_{\iota, \, \lambda\mu} \delta^{lm\lambda\mu}$$

$$R^{i}_{\iota, \, lm} R^{\iota}_{i, \, \lambda\mu} \delta^{lm\lambda\mu} .$$

If one designates $\Im$ as a linear combination of such scalar densities, the Hamiltonian equation

$$\delta \left\{ \int \Im d\tau \right\} = 0$$

then yields a system of 40 differential equations of 2nd order for the $\Gamma^{\mu}_{\alpha\beta}$ -quantities. This system appears interesting because it is connected with manifolds of four dimensions.[11]

[p. 5]   Hence it has indeed proved possible to erect a complete theory in Eddington's sense that is based solely on the assumption of the existence of an affine connection between the vectors or line elements; and it will be necessary to check its agreement with experience by the centrally symmetric case.[12] I do not believe, though, that one will arrive at a theory true to reality along this path; for Eddington's integral invariants have no similarity to the integral invariants

$$\int g^{i\kappa} R_{i\kappa} \sqrt{-g}\, d\tau$$

and

$$\int g^{i\sigma} g^{\kappa\tau} \varphi_{i\kappa} \varphi_{\sigma\tau} \sqrt{-g}\, d\tau \ ,$$

which correctly produced the laws of the gravitational field and the electromagnetic field (outside of electrons),[13] albeit without providing the essential connection between the gravitational field and the electromagnetic field.

In order to establish this link to the forms of the original theory, I thought of the following theory, although its foundation is less uniform than that of Eddington's theory.

§3. The New Theory

I assume a continuum in which a physically meaningful (measurable by clocks and measuring rods) mass-invariant *ds* exists in the sense of equation (1). Furthermore, we presuppose that an affine connection exists between the line elements as expressed by equation (2). But from the very beginning we do not introduce any precondition about a correlation between affine connection and metric. In accordance with this, the $g_{\mu\nu}$'s and $\Gamma^{\mu}_{\alpha\beta}$'s may be made to vary independently from one another.[14]

We start out from a scalar density $\mathfrak{I}$, which depends exclusively on the $g_{\mu\nu}$'s,   [p. 6]

the $\Gamma^{\mu}_{\alpha\beta}$'s, and the first derivatives $\dfrac{\partial \Gamma^{\mu}_{\alpha\beta}}{\partial x_{\sigma}}$ of the latter quantities. The general laws

of nature would then be determined by the Hamiltonian condition

$$\delta \left\{ \int \mathfrak{I}\, d\tau \right\} = 0 \tag{5}$$

whereby the $g_{\mu\nu}$'s and $\Gamma^{\mu}_{\alpha\beta}$'s should be varied independently. The natural laws then explicitly read

$$\left.\begin{aligned}
\mathfrak{I}_{\mu\nu} &= \frac{\partial \mathfrak{I}}{\partial g^{\mu\nu}} = 0 \\[2mm]
\mathfrak{I}^{\alpha\beta}_{\mu} &= \frac{\partial \mathfrak{I}}{\partial \Gamma^{\mu}_{\alpha\beta}} - \frac{\partial}{\partial x_{\sigma}} \left(\frac{\partial \mathfrak{I}}{\partial \Gamma^{\mu}_{\alpha\beta,\sigma}} \right) = 0,
\end{aligned}\right\} \tag{5a}$$

where

$$\frac{\partial \Gamma^{\mu}_{\alpha\beta}}{\partial x_{\sigma}} = \Gamma^{\mu}_{\alpha\beta,\sigma}$$

is set.

On the choice of the Hamiltonian function $\mathfrak{I}$, the following remark: If the Riemannian curvature tensor

$$R^i_{\kappa,\,lm} = -\frac{\partial \Gamma^i_{\kappa l}}{\partial x_m} + \Gamma^i_{\alpha l}\Gamma^\alpha_{\kappa m} + \frac{\partial \Gamma^i_{\kappa m}}{\partial x_l} - \Gamma^i_{\alpha m}\Gamma^\alpha_{\kappa l}$$

is contracted by the indices i and m, one obtains the tensor

$$R_{\kappa l} = -\frac{\partial \Gamma^\alpha_{\kappa l}}{\partial x_\alpha} + \Gamma^\alpha_{\kappa\beta}\Gamma^\beta_{l\alpha} + \frac{\partial \Gamma^\alpha_{\kappa\alpha}}{\partial x_l} - \Gamma^\alpha_{\kappa l}\Gamma^\beta_{\alpha\beta}\,, \tag{6}$$

which in the general theory of relativity together with the Riemann-Christoffel constraint (3) or (4) produced the theory of the pure gravitational field. If the Riemannian relations between the Γ's and g's are given up, the Riemannian tensor has a second, generally vanishing contraction (by the indices i, κ):[15]

$$\varphi_{\mu\nu} = \frac{\partial \Gamma^\alpha_{\mu\alpha}}{\partial x_\nu} - \frac{\partial \Gamma^\alpha_{\nu\alpha}}{\partial x_\mu}\,. \tag{7}$$

[p. 7]   We want to interpret this tensor as the expression of the electromagnetic field, as Eddington has already done.[16] We want to introduce the obvious assumption that the ⟨sought⟩ Hamiltonian function contain the Γ-quantities only in combinations (6) and (7).

§4. Law of the Pure Gravitational Field

In the case of the pure gravitational field, the Hamiltonian function will depend only on (6). One arrives at the goal by assuming the simplest dependence, namely, the linear one:

$$\mathfrak{I}_1 = g^{\kappa l}R_{\kappa l}\sqrt{-g}\,. \tag{8}$$

Equations (5a) then take on the form

$$\frac{1}{\sqrt{-g}}\mathfrak{I}_{1\mu\nu} = R_{\mu\nu} - \frac{1}{2}g_{\mu\nu}R = 0 \tag{9}$$

$$\left.\begin{aligned} \frac{1}{\sqrt{-g}}\mathfrak{I}_1{}^{\alpha\beta}{}_\mu &= \frac{1}{\sqrt{-g}}\left(\frac{\partial g^{\alpha\beta}\sqrt{-g}}{\partial x_\mu} - \frac{1}{2}\frac{\partial g^{\alpha\sigma}\sqrt{-g}}{\partial x_\sigma}\delta^\beta_\mu - \frac{1}{2}\frac{\partial g^{\beta\sigma}\sqrt{-g}}{\partial x_\sigma}\delta^\alpha_\mu\right) \\ &+ \frac{1}{2}g^{\alpha\sigma}\Gamma^\beta_{\mu\sigma} + \frac{1}{2}g^{\beta\sigma}\Gamma^\alpha_{\mu\sigma} - \frac{1}{2}g^{\sigma\tau}\Gamma^\alpha_{\sigma\tau}\delta^\beta_\mu - \frac{1}{2}g^{\sigma\tau}\Gamma^\beta_{\sigma\tau}\delta^\alpha_\mu - g^{\alpha\beta}\Gamma^\sigma_{\mu\sigma} = 0 \end{aligned}\right\} \tag{10}$$

Form (10), which must, of course, have a tensor character—as an Italian mathematician already discovered on another occasion—can be brought into the form

$$\frac{1}{\sqrt{-g}}\mathfrak{I}_1{}^{\alpha\beta}{}_\mu = g^{\alpha\beta}{}_{;\mu} - \frac{1}{2}\delta^\alpha_\mu g^{\beta\sigma}{}_{;\sigma} - \frac{1}{2}\delta^\beta_\mu g^{\alpha\sigma}{}_{;\sigma} - g^{\alpha\beta}g_{\sigma\tau}g^{\sigma\tau}{}_{;\alpha},^{[18]} \tag{11}$$

where $g^{\alpha\beta}{}_{;\mu}$ means the extension

$$\frac{\partial g^{\alpha\beta}}{\partial x_\mu} + g^{\alpha\beta}\Gamma^\alpha_{\mu\sigma} + g^{\alpha\sigma}\Gamma^\beta_{\mu\sigma}.$$

With the aid of (11) it is easy to prove (by forming both contractions) that equation (10) is therefore equivalent to the equation

$$g^{\alpha\beta}{}_{;\mu} = 0. \tag{10a}$$

Equations (3) and (9) together, however, are precisely the equations of the gravita- [p. 8] tional field which the general theory of relativity had reached earlier under the precondition that relations (3) or (4), resp., or (10a) hold a priori.

§5. Law of the Electromagnetic Field

In the general case that an electromagnetic field exists besides the gravitational field, the Hamiltonian function must also depend on tensor (7). Pursuant to the earlier results of the general theory of relativity, one should put

$$\mathfrak{I} = \mathfrak{I}_1 + \mathfrak{I}_2{}^{[20]} \tag{11}$$

where $\mathfrak{I}_1$ is given by (8) and

$$\mathfrak{I}_2 = \frac{1}{2}g^{\mu\sigma}g^{\nu\tau}\left(\frac{\partial\Gamma^\alpha_{\mu\alpha}}{\partial x_\nu} - \frac{\partial\Gamma^\alpha_{\nu\alpha}}{\partial x_\mu}\right)\left(\frac{\partial\Gamma^\beta_{\sigma\beta}}{\partial x_\tau} - \frac{\partial\Gamma^\beta_{\tau\beta}}{\partial x_\sigma}\right)\sqrt{-g} \tag{12}$$

is set. Executing the variation yields

$$\frac{1}{\sqrt{-g}}\mathfrak{I}_{2\mu\nu} = -\frac{1}{4}\varphi_{\sigma\alpha}\varphi_{\tau\beta}g^{\sigma\tau}g^{\alpha\beta}g_{\mu\nu} + \varphi_{\mu\alpha}\varphi_{\nu\beta}g^{\alpha\beta} \tag{13}$$

$$\frac{1}{\sqrt{-g}}\mathfrak{I}_2{}^{\alpha\beta}{}_\mu = -\delta^\alpha_2 i^\beta + \delta^\beta_\mu i^\alpha,^{[21]} \tag{14}$$

where, to abbreviate,

$$i^\alpha\sqrt{-g} = \frac{\partial(\varphi_{\sigma\tau}g^{\sigma\alpha}g^{\tau\beta}\sqrt{-g})}{\partial x_\beta} \tag{15}$$

is set. The first system of field equations (5a) yields the equations of the gravitational field, taking into account the ⟨field excitation⟩ influence of the electromagnetic field. The second system of (5a), however, in combination with (11) and (14), yields a relation between the current density, the metric and gravitational fields. One obtains after a simple transformation:[22]

$$g^{\alpha\beta}{}_{;\mu} = \frac{6}{7}(\delta^{\alpha}_{\mu}i^{\beta} + \delta^{\beta}_{\mu}i^{\alpha}) - \frac{4}{7}g^{\alpha\beta}i_{\mu}. \tag{16}$$

[p. 9] Outside of the charged masses ($i^{\alpha} = 0$), however, equations (3) of Riemannian geometry apply again. But within the charged masses the generally covariant derivative of the metric tensor does not vanish. It is here that this theory differs from the one of the independent electromagnetic field.

It is highly interesting that according to the theory developed here, the two signs of electricity do not appear equivalently.[23] The reason lies in the connection between gravitational field and electromagnetic field in equation (7).[24] We shall only be able to decide whether the two invariants $\Im_1$ and $\Im_2$ will suffice to describe the electron after first calculating out the centrally symmetric static problems;[25] the interesting problem is this: whether such singularity-free solutions exist for both electric signs.

Singapore. January 1923.[26]

418. Calculations on Back Pages of Travel Diary

[on or around 9 January 1923][1]

[Not selected for translation.]

419. Fragment and Calculation on the General Theory of Relativity

[ca. January 1923]

[Not selected for translation.]

420. To Svante Arrhenius[1]

Near Singapore, 10 January 1923

Esteemed Colleague,

The news about the award of the Nobel Prize reached me via telegraph on the Kitano Maru shortly before my arrival in Japan.[2] I am very pleased—among other reasons, because the reproachful question: Why don't you get the Nobel Prize? can no longer be posed to me. (I reply each time: Because *I* am not the one who awards the prize.)

Mrs. Hamburger[3] informed me that you were so kind as to invest the money temporarily.[4] Thank you very much for this kind solicitude. You (and Bohr) also wrote that the official award ceremony has been scheduled for June, which I appreciate very much. I am returning from this wonderful trip at the latest at the beginning of April. I am very enthralled by the land and people of Japan, everything is so subtle and peculiar. And how conducive to thinking and working the long sea voyage is—a paradise without correspondence, visits, meetings, and other inventions of the devil! It is a special pleasure for me that I receive the prize together with my admirable and beloved Bohr.[5]

In the happy prospect of a cheerful reunion, at the latest in Stockholm, I am, with all due respect and kind regards, yours,

A. Einstein.

421. To Niels Bohr

Near Singapore, 10 January 1923

Dear, or rather, beloved Bohr!

Your affectionate letter[1] reached me shortly before my departure from Japan.[2] I can say without exaggeration that it pleased me as much as the Nobel Prize. I find your fear of possibly getting the prize before me especially endearing—that is genuinely Bohr-like. Your new analyses on the atom accompanied me on my journey, and my love for your mind has grown even more.[3] I now believe I finally understand the connection between electricity and gravitation.[4] Eddington came closer to the gist of it than Weyl.

This is a splendid voyage. I am delighted by Japan and the Japanese and am sure that you would be, too. Traveling by sea is a splendid existence for a ponderer anyway—it's like a cloister. Add to that the caressing warmth near the equator. Warm water drips languidly down from the heavens and spreads calm and vegetative inspiration—this little letter is a testimonial.

Warm regards. To a happy reunion, in Stockholm at the latest.

Your admiring

A. Einstein.

422. From Jun Ishiwara

Hota, 12 January 1923

Esteemed Professor,

Your stay in our country was a special pleasure for me that I shall always cherish as such a fine memory. You are probably still happily continuing your journey!

I now tried the problem in the following way:

I assume two six-vectors $\varphi_{\mu\nu}$ and $\psi_{\mu\nu}$, whose components for bodies *at rest* should read:

$$
\begin{array}{ccccccc}
& (23) & (31) & (12) & (14) & (24) & (34) \\
\varphi_{\mu\nu}: & 0, & 0, & 0, & -ie_x, & -ie_y, & -ie_z \\
\psi_{\mu\nu}: & b_x, & b_y, & b_z, & 0, & 0, & 0.
\end{array}
$$

If one transforms these field vectors, e.g., through Lorentz's transformation equations:

$$
\sqrt{1-v^2}\,x'_1 = x_1 - ivx_4 \, , \quad x'_2 = x_2 \, , \quad x'_3 = x_3 \, ,
$$

$$
\sqrt{1-v^2}\,x'_4 = x_4 + ivx_1 \, ,
$$

one obtains (because of the formula: $\varphi_{\mu\nu}' = \dfrac{\partial x'_\mu}{\partial x_\alpha}\dfrac{\partial x'_\nu}{\partial x_\beta}\varphi_{\alpha\beta}$)

$$
\varphi'_{\mu\nu}: \ 0, \quad \frac{-v}{\sqrt{1-v^2}}e_z, \quad \frac{v}{\sqrt{1-v^2}}e_y, \quad -ie_x, \quad \frac{-ie_y}{\sqrt{1-v^2}}, \quad \frac{-ie_z}{\sqrt{1-v^2}}
$$

$$
\psi'_{\mu\nu}: \ b_x, \quad \frac{b_y}{\sqrt{1-v^2}}, \quad \frac{b_z}{\sqrt{1-v^2}}, \quad 0, \quad \frac{-iv}{\sqrt{1-v^2}}b_z, \quad \frac{iv}{\sqrt{1-v^2}}b_y.
$$

At rest, $\mathfrak{d} = \varepsilon\,e$, $\mathfrak{h} = \dfrac{1}{\mu}\,b$, is valid; and generally one also must set for the elec. displacement or, resp., the mag. field strength

$$
\varepsilon\varphi_{\mu\nu} \ , \text{resp.,} \ \frac{1}{\mu}\psi_{\mu\nu}\,.
$$

Furthermore, I assume as the four-element electric current

$$
\mathfrak{T}_\mu = \sigma\varphi_{\mu\nu}u_\nu + \rho u_\nu \, ,
$$

the space and time components of which read

$$\mathfrak{I}_\mu : \frac{\sigma e + \rho v}{\sqrt{1 - v^2}}, \; i\frac{\sigma(ev) + \rho}{\sqrt{1 - v^2}} \; .$$

Then the Maxwell- Lorentz equations can be written thus:

$$\frac{\partial}{\partial x_\nu}\left(\varepsilon\varphi_{\mu\nu} + \frac{1}{\mu}\psi_{\mu\nu}\right) = \mathfrak{I}_\mu \, ,$$

$$\frac{\partial}{\partial x_\sigma}(\varphi_{\mu\nu} + \psi_{\mu\nu}) + \frac{\partial}{\partial x_\mu}(\varphi_{\nu\sigma} + \psi_{\nu\sigma}) + \frac{\partial}{\partial x_\nu}(\varphi_{\sigma\mu} + \psi_{\sigma\mu}) = 0 \, .$$

For the energy-momentum tensor one ⟨could⟩ might try setting up the symmetric expression:

$$T_{\mu\nu} = \varepsilon\varphi_{\mu\alpha}\varphi_{\nu\alpha} + \frac{3}{4}\delta_{\mu\nu}\varphi_{\alpha\beta}^2$$

$$- \frac{1}{\mu}\psi_{\mu\alpha}\psi_{\nu\alpha} + \frac{1}{4\mu}\delta_{\mu\nu}\varphi_{\alpha\beta}^2 \qquad \left(\delta_{\mu\nu} = \left\{\begin{matrix}1\\0\end{matrix}\right\}\right)$$

$$- \frac{1}{\mu}(\varphi_{\mu\alpha}\psi_{\nu\alpha} + \varphi_{\nu\alpha}\psi_{\mu\alpha}) \, .$$

For, at rest one obtains

1) $\varepsilon\varphi_{\mu\alpha}\varphi_{\nu\alpha} + \frac{3}{4}\delta_{\mu\nu}\varphi_{\alpha\beta}^2$

$$= \begin{vmatrix} \varepsilon e_x^2 - \frac{1}{2}\varepsilon e^2 & \varepsilon e_x e_y & \varepsilon e_x e_z & 0 \\[2mm] \varepsilon e_x e_y & \varepsilon e_y - \frac{1}{2}\varepsilon e^2 & \varepsilon e_y e_z & 0 \\[2mm] \varepsilon e_x e_z & \varepsilon e_y e_z & \varepsilon e_z - \frac{1}{2}\varepsilon e^2 & 0 \\[2mm] 0 & 0 & 0 & \frac{1}{2}\varepsilon e^2 \end{vmatrix} \, ,$$

2) $-\frac{1}{\mu}\psi_{\mu\alpha}\psi_{\nu\alpha} + \frac{1}{4\mu}\delta_{\mu\nu}\psi_{\alpha\beta}^2$

$$= \begin{vmatrix} -\dfrac{1}{\mu}(b_y^2 b_z^2) + \dfrac{1}{2\mu}b^2 & \dfrac{1}{\mu}b_x b_y & \dfrac{1}{\mu}b_x b_z & 0 \\[2mm] \dfrac{1}{\mu}b_x b_y & - & - & 0 \\[2mm] - & - & - & 0 \\[2mm] 0 & 0 & 0 & \dfrac{1}{2\mu}b^2 \end{vmatrix},$$

3) $\dfrac{1}{\mu}(\varphi_{\mu\alpha}\psi_{\nu\alpha} + \varphi_{\nu\alpha}\psi_{\mu\alpha})$

$$= \begin{vmatrix} 0 & 0 & 0 & \dfrac{i}{\mu}(e_y b_z - e_z b_y) \\[2mm] 0 & 0 & 0 & \dfrac{i}{\mu}(e_z b_x - e_x b_z) \\[2mm] 0 & 0 & 0 & \dfrac{i}{\mu}(e_x b_y - e_y b_x) \\[2mm] \dfrac{i}{\mu}(e_y b_z - e_z b_y) & - & - & 0 \end{vmatrix},$$

One could be satisfied with this result; nevertheless, I would like to demonstrate that the asymmetric expression by Minkowski ⟨likewise⟩ seems to relate much more *naturally* to the field equations. For this purpose I first multiply the current I_α with $\varphi_{\mu\alpha} + \psi_{\mu\alpha}$. Then the result is

$$(\varphi_{\mu\alpha} + \psi_{\mu\alpha})I_\alpha = (\varphi_{\mu\alpha} + \psi_{\mu\alpha})\frac{\partial}{\partial x_\nu}\left(\varepsilon\varphi_{\alpha\nu} + \frac{1}{\mu}\psi_{\alpha\nu}\right)$$

$$= -\frac{\partial}{\partial x_\nu}(\varepsilon\varphi_{\mu\alpha}\varphi_{\nu\alpha}) + \varepsilon\varphi_{\nu\alpha}\frac{\partial\varphi_{\mu\alpha}}{\partial x_\nu} + \varphi_{\mu\alpha}\frac{\partial}{\partial x_\nu}\left(\frac{1}{\mu}\psi_{\alpha\nu}\right)$$

$$-\frac{\partial}{\partial x_\nu}\left(\frac{1}{\mu}\psi_{\mu\alpha}\psi_{\nu\alpha}\right) + \frac{1}{\mu}\psi_{\nu\alpha}\frac{\partial\psi_{\mu\alpha}}{\partial x_\nu} + \psi_{\mu\alpha}\frac{\partial}{\partial x_\nu}(\varepsilon\varphi_{\alpha\nu})$$

$$= -\frac{\partial}{\partial x_\nu}\left(\varepsilon\varphi_{\mu\alpha}\varphi_{\nu\alpha} - \frac{\varepsilon}{4}\delta_{\mu\nu}\varphi_{\alpha\beta}^2\right) - \frac{1}{4}\frac{\partial}{\partial x_\mu}(\varepsilon\varphi_{\alpha\nu}^2)$$

$$-\frac{\partial}{\partial x_\nu}\left(\frac{1}{\mu}\psi_{\mu\alpha}\psi_{\nu\alpha} - \frac{1}{4\mu}\delta_{\mu\nu}\psi_{\alpha\beta}^2\right) - \frac{1}{4}\frac{\partial}{\partial x_\mu}\left(\frac{1}{\mu}\psi_{\alpha\nu}^2\right)$$

$$-\frac{\partial}{\partial x_\nu}\left(\varepsilon\varphi_{\nu\alpha}\psi_{\mu\alpha}+\frac{1}{\mu}\varphi_{\mu\alpha}\psi_{\nu\alpha}\right)$$

$$+\varepsilon\varphi_{\nu\alpha}\frac{\partial}{\partial x_\nu}(\varphi_{\mu\alpha}+\psi_{\mu\alpha})+\frac{1}{\mu}\psi_{\nu\alpha}\frac{\partial}{\partial x_\nu}(\varphi_{\mu\alpha}+\psi_{\mu\alpha}).$$

If one therefore sets

$$T_{\mu\nu}=-\varepsilon\varphi_{\mu\alpha}\varphi_{\nu\alpha}+\frac{3}{4}\delta_{\mu\nu}\varphi_{\alpha\beta}^2$$

$$-\frac{1}{\mu}\psi_{\mu\alpha}\psi_{\nu\alpha}+\frac{1}{4\mu}\delta_{\mu\nu}\varphi_{\alpha\beta}^2$$

$$-\left(\varepsilon\varphi_{\nu\alpha}\psi_{\mu\alpha}+\frac{1}{\mu}\varphi_{\mu\alpha}\psi_{\nu\alpha}\right),$$

one thus obtains an elegant form for the energy-momentum equation:

$$\left(\frac{\partial T_{\mu\nu}}{\partial x_\nu}=(\varphi_{\mu\alpha}+\psi_{\mu\alpha})I_\alpha+\frac{1}{4}\frac{\partial}{\partial x_\mu}\left(\varepsilon\varphi_{\alpha\beta}^2+\frac{1}{\mu}\psi_{\alpha\beta}^2\right)\right)\qquad(!)$$

$$-\varepsilon\varphi_{\nu\alpha}\frac{\partial}{\partial x_\nu}(\varphi_{\mu\alpha}+\psi_{\mu\alpha})-\frac{1}{\mu}\psi_{\nu\alpha}\frac{\partial}{\partial x_\nu}(\varphi_{\mu\alpha}+\psi_{\mu\alpha}).$$

The third term of the expression $T_{\mu\nu}$ thereby yields the components, at rest:

$$\varepsilon\varphi_{\nu\alpha}\psi_{\mu\alpha}+\frac{1}{\mu}\varphi_{\mu\alpha}\psi_{\nu\alpha}$$

$$=\begin{vmatrix}0 & 0 & 0 & i\varepsilon(e_y b_z-e_z b_y)\\ 0 & 0 & 0 & i\varepsilon(e_z b_x-e_x b_z)\\ 0 & 0 & 0 & i\varepsilon(\text{------------})\\ \frac{i}{\mu}(e_y b_z-e_z b_y) & \frac{i}{\mu}(\text{--------}) & \frac{i}{\mu}(\text{--------}) & 0\end{vmatrix}.$$

Esteemed Professor! How do you choose?—between whether
1) $T_{\mu\nu}$ should be symmetric, or

2) $\dfrac{\partial T_{\mu\nu}}{\partial x_\nu}$ should assume a natural, mathematically (as well as probably also

physically) so elegant a form!

In great respect, yours most sincerely,

Jun Ishiwara.

Best regards to your gracious wife!

423. From Sergei F. von Oldenburg[1]

[Petrograd, 18 January 1923][2]

[Not selected for translation.]

424. To Edgar, Else, and Edgar Michel Meyer[1]

Colombo, 20 January 1923

Dear Meyers altogether!

Japan was magnificent. Intelligent, fine fellows and in addition fine painting and architecture. Then Southeast Asia,[2] now Ceylon with interesting experiences. It's onward now to Palestine, Spain, and then back home. Warm regards, yours,

A. Einstein.

425. "On the General Theory of Relativity"

[*Einstein 1923e*]

COMPLETED 22 January 1923
PRESENTED 15 February 1923
PUBLISHED 12 March 1923

IN: *Preußische Akademie der Wissenschaften* (Berlin). *Sitzungsberichte* (1923): 32–38.

[p. 32]   § 1. Generalities. Formulation of the Field Equations

The general theory of relativity's mathematical structure was originally completely based upon the metric, i.e., on the invariant[1]

$$ds^2 = g_{\mu\nu}dx_\mu dx_\nu.$$ (1)

The quantities $g_{\mu\nu}$ and their derivatives described the metric field and gravitational field. Compared to these, the components $\phi_{\mu\nu}$ of the electric field were essentially alien constructs. The wish to understand the gravitational field and the electromagnetic field as one fundamental entity dominated the endeavors of theoreticians during the last few years.[2]

A mathematical finding that we owe to Levi-Civita and Weyl rewarded these efforts:[3] The derivative of the Riemann tensor of curvature, which is fundamental in the general theory of relativity, is based most naturally on the parallel-shift law of vectors ("affine relation")

$$\delta A^{\mu} = -\Gamma^{\mu}_{\alpha\beta} A^{\alpha} dx_{\beta} \ . \tag{2}$$

This may be traced back to (1) by means of the postulate that a vector's value does not change upon parallel displacement; yet such a reduction is not logically necessary. This H. Weyl first recognized; he based on this finding a generalization of Riemannian geometry that in his opinion yields the theory of the electromagnetic field.[4] Weyl does not assign invariance to the value of a line element or of a vector, resp., but only to the *relation* between the values of two line elements or vectors sharing the same starting point. The parallel displacement (2) must be designed so as to leave this relation unchanged. The basis of this theory can be described as semimetric. In my opinion, this is not the way to arrive at a physically useful theory.[5] Even from a purely logical point of view, it should surely appear more satisfying to base the theory *solely* on (2), provided one feels inclined to drop the invariant (1) as the theory's basis.

Eddington did this and noticed that, on the contrary, a metric invariant of type (1), whose physical existence cannot be doubted, can be based on (2). Since, from (2) follows the existence of the fourth-order Riemannian tensor[6] [p. 33]

$$R^{i}_{k,lm} = -\frac{\partial \Gamma^{i}_{kl}}{\partial x_{m}} + \Gamma^{i}_{\tau l}\Gamma^{\tau}_{km} + \frac{\partial \Gamma^{i}_{km}}{\partial x_{l}} - \Gamma^{i}_{\tau m}\Gamma^{\tau}_{kl},$$

and thence, by reducing the indices i and m, follows the existence of the second-order Riemannian tensor

$$R_{kl} = -\frac{\partial \Gamma^{\alpha}_{kl}}{\partial x_{\alpha}} + \Gamma^{\alpha}_{k\beta}\Gamma^{\beta}_{l\alpha} + \frac{\partial \Gamma^{\alpha}_{k\alpha}}{\partial x_{l}} - \Gamma^{\alpha}_{kl}\Gamma^{\beta}_{\alpha\beta}, \tag{3}$$

whose fundamental importance in gravitational theory is well known.

$$R_{kl}dx_{k}dx_{l}$$

is therefore an invariant of the line element that Eddington regards as a metrical invariant.[7]

The R_{kl}'s, for any choice of $\Gamma^\alpha_{\mu\nu}$, which are subject only to the symmetric condition

$$\Gamma^\alpha_{\mu\nu} = \Gamma^\alpha_{\nu\mu} \tag{4}$$

do not form any symmetric tensor. If the tensor (R_{kl}) is broken down into a symmetric and antisymmetric one according to the equation

$$R_{kl} = g_{kl} + \phi_{kl}, \tag{5}$$

whereby

$$\phi_{kl} = \frac{1}{2}\left(\frac{\partial \Gamma^\alpha_{k\alpha}}{\partial x_l} - \frac{\partial \Gamma^\alpha_{l\alpha}}{\partial x_k}\right), \tag{6}$$

then it suggests itself to set the tensor (g_{kl}) equal to the metric tensor g_{kl} but to regard the tensor (ϕ_{kl}), which satisfies the relation

$$\frac{\partial \phi_{kl}}{\partial x_m} + \frac{\partial \phi_{lm}}{\partial x_k} + \frac{\partial \phi_{mk}}{\partial x_l} = 0, \tag{7}$$

as the tensor for the electromagnetic field.[8]

First a comment in support of the limiting symmetry condition (4). From (2) follows the displacement law for the covariant vector through the natural stipulation that the scalar product of a contravariant and a covariant vector does not change upon parallel displacement. From this follows the law[9]

$$\delta B_\mu = \Gamma^\alpha_{\mu\beta} B_\alpha dx_\beta.$$

From this follows, in the familiar manner, the tensorial character of

$$\frac{\partial B_\mu}{\partial x_\nu} - \Gamma^\alpha_{\mu\beta} B_\alpha.$$

[p. 34]   From this and from the tensorial character of $\frac{\partial B_\mu}{\partial x_\nu} - \frac{\partial B_\nu}{\partial x_\mu}$, the tensorial character of

$\Gamma^\alpha_{\mu\nu} - \Gamma^\alpha_{\nu\mu}$ can then be concluded. From this and also from the foregoing, it then follows that

$$\frac{\partial B_\mu}{\partial x_\nu} - \Gamma^\alpha_{\nu\mu} B_\alpha$$

also has a tensorial nature. Therefore, the symmetry condition (4) is necessary if the unequivocal character of the vector's covariant extension is to be preserved.

In Eddington's theory the 40 quantities $\Gamma^\alpha_{\mu\nu}$ appear as unknown functions of the x_ν's, similar to the 14 quantities $g_{\mu\nu}$ and ϕ_μ in the original theory of relativity. Now, the unsolved problem by Eddington involves finding the equations necessary for determining these quantities. Hamilton's principle offers the most comfortable method for this. Let $\mathfrak{H}$ be a scalar density dependent on the Γ's and their first derivatives, then for each continuously vanishing variation of the $\Gamma_{\mu\nu}^{\alpha}$'s at the edge of the integration region

$$\delta\left\{ \int \mathfrak{H}\, d\tau \right\} = 0 \tag{8}$$

should hold. The field equations that, because of the tensorial character of $\delta\Gamma_{\mu\nu}^{\alpha}$ are likewise tensorial in character, then read

$$0 = \mathfrak{H}_{\alpha}^{\mu\nu} = \frac{\partial \mathfrak{H}}{\partial \Gamma_{\mu\nu}^{\alpha}} - \frac{\partial}{\partial x_{\sigma}}\left(\frac{\partial \mathfrak{H}}{\partial \Gamma_{\mu\nu,\sigma}^{\alpha}}\right), \tag{9}$$

where

$$\frac{\partial \Gamma_{\mu\nu}^{\alpha}}{\partial x_{\sigma}} = \Gamma_{\mu\nu,\sigma}^{\alpha}.$$

Thereby it is assumed that $\mathfrak{H}$ is an (algebraic) function of the $R_{k,lm}^{i}$'s. Our main task involves choosing this function.

Tensor densities exist that are *rational* second-order functions of the $R_{k,lm}^{i}$'s; they can be obtained by means of the tensor density δ^{iklm}, the components of which are equal to 1 or -1, depending on whether *iklm* is an even or odd permutation of 1, 2, 3, 4. One such tensor density is, e.g.,

$$R_{k,lm}^{i} R_{i,\sigma\tau}^{k} \delta^{lm\sigma\tau}.$$

I consider it correct, however, to confine ourselves to the tensor densities formed out of the reduced tensor R_{kl}, or, resp., out of S_{kl} and ϕ_{kl},[10] because we are only inclined to ascribe physical meaning to these quantities. Then we have to permit irrational functions, as we are already accustomed to doing from the previously developed general theory of relativity (e.g., $\sqrt{-g}$). Then, too, there are various options, among which the following seems the most interesting to me:

$$\mathfrak{H} = 2\sqrt{-|R_{kl}|}, \tag{10}$$

which is analogous to the volume's tensor density and *is formed* from R_{kl} *without being broken down into the symmetric and antisymmetric parts.* This Hamiltonian proves to be usable, consequently the theory achieves in an ideal way the unifica-   [p. 35] tion of gravitation and electricity under a single concept; not only the same Γ's determine both kinds of fields, but also the Hamiltonian is a thoroughly uniform one, whereas it had hitherto been composed of logically independent summands.[11]

In the following the usefulness of the theory is made plausible.

§ 2. The New Theory's Relationship with the Earlier Results of General Relativity Theory

First a comment about equation (5). $g_{kl}dx_k dx_l$ represents the metric invariant for a "cosmic" measuring rod. If $g_{kl}dx_k dx_l$ is supposed to describe a squared unit length of human dimensions, then one has to set

$$\lambda^2 R_{kl} = g_{kl} + \phi_{kl}, \tag{5a}$$

where λ is a very large number.[12] One therefore has, according to (3),

$$\frac{1}{\lambda^2}g_{kl} = \frac{\partial \Gamma^\alpha_{kl}}{\partial x_\alpha} + \frac{1}{2}\left(\frac{\partial \Gamma^\alpha_{k\alpha}}{\partial x_l} + \frac{\partial \Gamma^\alpha_{l\alpha}}{\partial x_k}\right) + \Gamma^\alpha_{k\beta}\Gamma^\beta_{l\alpha} - \Gamma^\alpha_{kl}\Gamma^\beta_{\alpha\beta} \tag{11}$$

$$\frac{1}{\lambda^2}\phi_{kl} = \frac{1}{2}\left(\frac{\partial \Gamma^\alpha_{k\alpha}}{\partial x_l} - \frac{\partial \Gamma^\alpha_{l\alpha}}{\partial x_k}\right). \tag{12}$$

We now execute the variation indicated in (8) under the general assumption that $\mathfrak{H}$ is a function of g_{kl} and ϕ_{kl}, left undetermined for the time being. Then[13]

$$\delta\mathfrak{H} = \frac{\partial \mathfrak{H}}{\partial g_{kl}}\delta g_{kl} + \frac{\partial \mathfrak{H}}{\partial \phi_{kl}}\delta\phi_{kl} = \mathfrak{g}^{kl}\delta g_{kl} + f^{kl}\delta\phi_{kl}, \tag{13}$$

where $\mathfrak{g}^{kl}$ signifies a symmetric tensor density, and f^{kl} an antisymmetric one. Taking (11), (12), and (13) into account, (8) assumes the form

$$0 = \int d\tau \delta\Gamma^\alpha_{kl}\left\{\mathfrak{g}^{kl};_\alpha - \frac{1}{2}\delta^k_\alpha \mathfrak{g}^{l\sigma};_\sigma - \frac{1}{2}\delta^l_\alpha \mathfrak{g}^{k\sigma};_\sigma - \frac{1}{2}\delta^k_\alpha \frac{\partial f^{l\sigma}}{\partial x_\sigma} - \frac{1}{2}\delta^l_\alpha \frac{\partial f^{k\sigma}}{\partial x_\sigma}\right\}. \tag{14}$$

Since we interpret ϕ_{kl} as the electromagnetic field's covariant tensor, we should regard f^{kl} as the electromagnetic field's contravariant tensor density and

$$i^l = \frac{\partial f^{l\sigma}}{\partial x_\sigma} \tag{15}$$

as the density of the current. In (14), $\mathfrak{g}^{kl};_\alpha$ means the covariant extension of $\mathfrak{g}^{kl}$ according to the formula

$$\mathfrak{g}^{kl};_\alpha = \frac{\partial \mathfrak{g}^{kl}}{\partial x_\alpha} + \mathfrak{g}^{\sigma l}\Gamma^k_{\sigma\alpha} + \mathfrak{g}^{k\sigma}\Gamma^l_{\sigma\alpha} - \mathfrak{g}^{kl}\Gamma^\sigma_{\alpha\sigma}. \tag{16}$$

From (14) follows

$$0 = \mathfrak{g}^{kl};_\alpha - \frac{1}{2}\delta^k_\alpha \mathfrak{g}^{l\sigma};_\sigma - \frac{1}{2}\delta^l_\alpha \mathfrak{g}^{k\sigma};_\sigma - \frac{1}{2}\delta^k_\alpha i^l - \frac{1}{2}\delta^l_\alpha i^k. \tag{17}$$

By combining this equation with the ones obtained from reducing the indices α [p. 36]
and l,

$$0 = 3 \mathfrak{s}^{\; l\sigma}_{\; \; ;\sigma} + 5 i^l.$$ (18)

Finally, as a general result of our variation consideration, it follows that[14]

$$0 = \mathfrak{s}^{\; kl}_{\; \; ;\alpha} + \frac{1}{3}\delta^k_\alpha i^l + \frac{1}{3}\delta^l_\alpha i^k.$$ (19)

These are 40 equations from which the Γ values are computable. For this purpose
we introduce the tensors s_{kl} and s^{kl}, resp., which belong to the tensor density $\mathfrak{s}^{kl}$;
thereby, these tensors share the same interrelationship as the one between covariant
and contravariant fundamental tensors ($g_{\mu\nu}$ and $g^{\mu\nu}$) in general relativity theory.
Thus the following equations may hold.

$$\mathfrak{s}^{\; kl} = s^{kl}\sqrt{-|s_{ik}|}$$ (20)

$$s_{\alpha i} s^{\beta i} = \delta^\beta_\alpha.$$ (21)

Furthermore, we set

$$i^l = \sqrt{-|s_{lk}|} \qquad i^l = \sqrt{-s}\, i^l$$ (22)

$$i_l = s_{l\tau} i^\tau.$$ (23)

Then we obtain from well-known calculations in general relativity theory:

$$\Gamma^\alpha_{kl} = \frac{1}{2} s^{\alpha\beta}\left(\frac{\partial s_{k\beta}}{\partial x_l} + \frac{\partial s_{l\beta}}{\partial x_k} - \frac{\partial s_{kl}}{\partial x_\beta}\right) - \frac{1}{2} s_{kl} i^\alpha + \frac{1}{6}\delta^\alpha_k i_l + \frac{1}{6}\delta^\alpha_l i_k.$$ (24)

One must imagine these values for Γ plugged into (11) and (12). Because from
(13), $\mathfrak{s}$ and $\mathfrak{f}$ are expressible as g and ϕ through the choice of Hamiltonian, therefore
after substitution the equations (11) and (12) suffice to determine the unknown
functions. Now, in order to recognize the physical legitimacy of the Hamiltonian
choice reached in (10), let us next look at the case of a missing electromagnetic
field. According to (10) and (13), then

$$\mathfrak{s}^{\; kl} = g^{kl}\sqrt{-g}$$

$$\mathfrak{f}^{\; kl} = 0,$$

where g^{kl} and g correspond to g_{kl} in the conventional relation from general relativity theory. Equation (24) then takes on the well-known form

$$\Gamma^\alpha_{kl} = \frac{1}{2} g^{\alpha\beta}\left(\frac{\partial g_{k\beta}}{\partial x_l} + \frac{\partial g_{l\beta}}{\partial x_k} - \frac{\partial g_{kl}}{\partial x_\beta}\right),$$ (24a)

which equation, together with (11), exactly yields the vacuum equation of the gravitational field in the general theory of relativity with a vanishing electromagnetic field, taking the cosmological term into account. This is a strong argument in favor of our choice of Hamiltonian as well as of the usefulness of the theory overall.

[p. 37]   We now move on to the case where the electromagnetic field does not vanish. To start with, from (12) and (24) generally follows:

$$\frac{1}{\lambda^2}\phi_{kl} = \frac{1}{6}\left(\frac{\partial i_k}{\partial x_l} - \frac{\partial i_l}{\partial x_k}\right).$$

(25)

The insight here is that at absolutely vanishing current density no electric field is possible. However, the extraordinary smallness of $\frac{1}{\lambda^2}$ implies that finite ϕ_{kl}'s are only possible with tiny, practically vanishing covariant current densities. Hence, with the exception of singular spots, the current density practically vanishes. Consequently the equations

$$\frac{\partial f^{kl}}{\partial x_l} = 0 \ldots.$$

(26)

$$\frac{\partial \phi_{kl}}{\partial x_\sigma} + \frac{\partial \phi_{l\sigma}}{\partial x_k} + \frac{\partial \phi_{\sigma k}}{\partial x_l} = 0$$

(27)

hold there very approximately, the latter equation of which holds strictly, with (12) taken into account. The relation between the ϕ's and f's, with our Hamiltonian choice, is determined in that the quantities

$$\mathfrak{r}^{kl} = \mathfrak{g}^{kl} + f^{kl}$$

are the underdeterminants of the r_{kl}'s multiplied by the root from the negatively taken determinants r of

$$r_{kl} = g_{kl} + \phi_{kl}.$$

For, if those standardized underdeterminants are called r^{kl}, then you get

$$\delta r = r r^{kl} \delta r_{kl}$$

and consequently,

$$d\mathfrak{H} = \delta(2\sqrt{-r}) = \frac{1}{\sqrt{-r}}\delta(-r) = \sqrt{-r}\, r^{kl}\delta r_{kl} = \sqrt{-r}\, r^{kl}(\delta g_{kl} + \delta\phi_{kl}),$$

from which the postulate follows.

The approximate computation of the f^{kl}'s is thus simple in the important case that the r_{kl}'s differ only infinitesimally little from the constant values δ_{kl} ($= 1$ or $= 0$, resp.). In this case, to first approximation, it is—whereby the time coordinate is chosen in the conventional way to be imaginary—

$$f^{kl} = \phi_{kl}.$$

This result, in combination with (26) and (27), therefore shows that to first approximation Maxwell's equations for empty space hold (for weak enough fields).

Whether our theory also covers the elementary electric charge can only be decided by a rigorous calculation of the centrally symmetric static field.[15] In any event, equation (25) shows that finite values for current density i^l are only possible [p. 38] if at the same time the i_l's diminish to order of magnitude $\dfrac{1}{\lambda^2}$; thus singularity-free electrons would be conceivable. The remarkable thing is that according to this theory, positive and negative electricity clearly can differ more than merely in sign.[16]—

The above analysis demonstrates that Eddington's general idea, combined with Hamilton's principle, leads to a theory virtually free of arbitrariness that does justice to our current knowledge about gravitation and electricity and unifies the two types of field in a truly perfect way.

Haruna Maru, January 1923.[17]

426. To Nippon Puroretaria Domei[1]

[On board SS *Haruna Maru*, 22 January 1923][2]

[See documentary edition for the original English.]

427. From Chaim Weizmann

Zionist Organisation, [London] W. C. 1, 77 Ct. Russell St., 4 February 1923

Hearty welcome to the land of Israel [*Eretz Israel*][1] whose regeneration you will largely contribute [to]. Feel sure that happy impressions you will receive will link you still closer to our hopes. Your visit will be great inspiration to settlement [*Yi[s]huv*].[2] Would be glad of telegraphic communication, your plans near future. Leaving twentyfourth America

Weizmann

Translator's note: Original telegram written in English.

428. From Federigo Enriques[1]

Rome, 8 February 1923

[Not selected for translation.]

429. To Chaim Weizmann

[Haifa, 11 February 1923][1]

Dear Mr. Weizmann,

Yesterday and today I visited the technical high school and the technical college in Haifa[2] and got a very favorable impression of the educational work up to now. It would be of great benefit if the means could be found to get the teaching started at the technical college, because everything is ready and the need is great. The problems here are great but the mood is positive and the work admirable.

Cordial regards and good success, yours,

Einstein.

430. To Arthur S. Eddington

14 February 1923

[Not selected for translation.]

431. From Heinrich Lüders[1]

Berlin, 15 February 1923

[Not selected for translation.]

432. From Nicholas M. Butler[1]

[New York,] 26 February 1923

[See documentary edition for the English letter.]

433. To Jun Ishiwara

[Spain, after 26 February 1923, or Berlin, after 21 March][1]

[Not selected for translation.]

434. From Arthur Biram

Haifa, 1 March 1923

[Not selected for translation.]

435. From Gano Dunn

[New York,] Thursday, 1 March 1923

[See documentary edition for the original English.]

436. To Wilhelm Westphal

Madrid, 9 Lealtad Street, 2 March 1923

Dear Colleague,

I hurry to answer your kind letter of the 20th of Feb., which just reached me today. The invitation by the Mexican government pleases me exceedingly, particularly also the fact that my going along is not necessarily being demanded.[1] I am not a practicing astronomer so my participation on the expedition would be worthless; besides, I am now, after all those strenuous travels and lectures, yearning for a more static existence. So in your reply please thank the Mexican government cordially also in my name and mention my nonparticipation in the expedition. I would welcome it very much if Mr. Ludendorff, with his experience, led it.[2] As he would be bearing the main responsibility, the choice of the other participants should also be left mainly to him.

Regarding the tower telescope, I believe that Mr. Stumpf's abilities and energy for independent research in this difficult field do not suffice,[3] not even with Mr. Freundlich's collaboration.[4] However, I console myself with the thought that Mr. Grotrian is a capable mind with leadership qualities[5] and Mr. Stumpf would be a useful and willing coworker of Mr. Grotrian. As you do not mention Mr. Grotrian, I fear that this matter has come to nothing. Mr. Stumpf is not useful, however, as an *independent* researcher. All of you do not need to wait for me with your decisions, because, when I am finished here, I still want to be in Zurich for a few days, visiting my boys. If you wish to take my opinion into account, then this [is] can happen on the basis of the above-mentioned.

Best regards from your

A. Einstein.

437. From Mauricio David[1]

Madrid, 2 March 1923

[Not selected for translation.]

438. To the Spanish Academy of Sciences

[*Einstein 1923d*]

DATED 4 March 1923
PUBLISHED 1923

IN: *Discursos pronunciados en la sesión solemne que se dignó presidir S. M. el Rey el día 4 de marzo de 1923 celebrada para hacer entrega del diploma de académico corresponsal al profesor Albert Einstein.* Real Academia de Ciencias Exactas, Fisicas y Naturales. Madrid: Talleres Poligráficos, 1923, pp. 19–20.

YOUR MAJESTY:[1]

Respected Colleagues, please allow me to express my most profound gratitude and satisfaction at your having elected me as your correspondent of your Academy. Bonds such as those we have established today demonstrate anew that the intellectual forces that unite peoples cannot be permanently destroyed by the political storms of our times.

Your words, dear Mr. Cabrera,[2] have touched the very bottom of my heart, not because they convey for me the honor of this great recognition, but because they demonstrate how deliberately and affectionately you have studied my life's work, echoing the poet's words: "We wish to receive less praise and, instead, may we be read industriously." You also touched upon the weak point of the theory of light quanta, an arduous subject for our generation of physicists. I believe that those difficulties can only be overcome by a theory that not only fundamentally modifies the principle of energy, but perhaps also expands that of causality.[3] Just a while ago, Tetrode has pointed precisely to such possibilities.[4] Even though the principles for the solution to this basic problem have not yet become clear, nevertheless the new impetus towards the unification of all the forces of nature, born in the cradle of the theory of relativity, promises satisfactory results. The method employed in this venture is purely mathematico-speculative, characteristic of Levi-Civita, Weyl, Eddington.[5] In this way one can completely relieve the foundation of physics from the disturbing dualism summed up by those two words, *gravitation* and *electricity*.

I have found the words you have spoken very significant – a reflection of your optimistic hope in the development of science in Spain. Moments of active participation in the global progress of understanding depend upon external conditions that have now been realized in your country. I believe that a tormented and imperiled Europe can turn its eyes full of hope towards this people, which is now heading down the road to scientific work after having produced such grand things in the arts for humanity.

439. From Michael I. Pupin[1]

[New York,] 4 March 1923

[Not selected for translation.]

440. From Carl Brinkmann

Berlin-Grunewald, 9 March 1923

[Not selected for translation.]

441. To Hermann Anschütz-Kaempfe

Madrid, 10 March [1923][1]

How enthusiastic I am to see what has happened with the design![2] To a happy reunion—hopefully soon—yours,

A. Einstein.

442. From Maja Winteler-Einstein

Florence, Colonnata Sesto, 11 March 1922 [1923][1]

My dear Albert,

You're probably celebrating your birthday at home and so I don't want to miss being part of the chorus of congratulators, one weak voice . . . .

May you stay healthy and be content! I wish you whatever you could possibly wish for yourself. You've achieved everything that one dreams about in youth, and you deserve it! That must be a fine feeling.

From the newspapers I learned that now you finally got the Nobel Prize. My warmest congratulations for that, too. Now you're relieved of the endless money worries as well, because of Miza and the boys.[2]

I didn't hear much from you both during your fairy-tale voyage but, because of this, thought the more often of you. I even imagined the good idea would occur to you to drop by from Spain and see how we're doing here. After such a trip that would have been a short detour; but I can also understand that you were anxious to get home. Did everything go smoothly as well? Itini[3] sent me a newspaper clipping in which I read that you had delivered a talk in Jerusalem in Hebrew.[4] Whence did you reap that science? You're an absolute conjurer!

We're living in paradise here. Our marriage is as harmonious again as ever[5] and I would be completely content if I didn't constantly have the feeling that you were somehow peeved. You mustn't think us stupid before you've seen it with your own eyes. One can't make such things clear by letter.

We didn't pay the 1,000 francs back to you yet[6] because we thought that after having received the Nobel Prize, it's not so urgent for you now anymore and we can make payments for the house with it. But next year you'll certainly get it, that is, this fall. Paul will be sending you an accounting about the remainder.

Don't you want to come once with your boys to spend the spring or summer vacation with us? I would so much like to get to know the younger boy, you know. We have enough room and it would be such a great joy for me.

Pauli is in excellent health; he didn't get a cough all winter and his nerves are much better as well. Country living works wonders. And since my illness I haven't come down with anything either.

When you have some time, or if Elsa does, do write me at length about the voyage. I would be a very appreciative audience.

Greetings to all, an affectionate birthday kiss to you, from your

Maja.

443. From Zionistische Vereinigung für Deutschland (Betty Frankenstein[1])

Berlin, 14 March 1923

Esteemed Professor,

We received news from Munich that at meetings of the National Socialist Party it has been repeatedly alleged that you, esteemed Professor, denied your German origins during your stay in Paris[2] in order to win over sympathetic listeners in France.

We would consider it very good if you, esteemed Professor, would deny this tendentious report and would be obliged if you would send us your denial for dissemination.

In great respect,

Zionistische Vereinigung für Deutschland, Secretariat

Frank

444. From *Vossische Zeitung*

[Berlin,] 15 March 1923

[Not selected for translation.]

445. From Svante Arrhenius

Stockholm, Experimentalfältet, 17 March 1923

Esteemed Colleague,

I received your kind letter of 10 January in February and thank you very much.[1]

The Nobel Foundation deposited the money, at my suggestion, in a bank that pays 3% interest, with a fortnight's withdrawal notice. The information that the award of the prize was supposed to take place in June is not correct. The local German ambassador, Mr. Nadolny,[2] has already received the pert. documents. Now you could naturally attend the next prize conferral on Dec. 10, if you wish to experience such a ceremony.[3] I do believe, though, that you and your wife would prefer to see Sweden, for which December, as the darkest month, is not very suitable. I have a suggestion that may perhaps meet your approval. This summer a large Scandinavian exhibition will be held in Gothenburg[4] on 10 May–Sept. There will also be a (Scandinavian) scientific convention taking place between 9–14 July. It would be splendid if you then gave a talk comprehensible to a general audience. You can choose the topic—one would certainly be extremely grateful for a talk about your theory of relativity. At the exhibition you can see Sweden in concentrated form.

From there we may travel via Christiania[5] to Stockholm and perhaps to Dalekarlien[6] or farther north. We could set the itinerary in Gothenburg. If you write me about this, I shall register you (and your esteemed wife, if she so wishes) as participants at the scientific convention. Do also please inform me of the topic of your talk. I shall then have everything arranged for you in Gothenburg.

Gratefully looking forward to your kind reply at your next convenience, I remain with my best compliments to you, your wife, and Mrs. Grete Hamburger, yours very truly,

Svante Arrhenius.

446. To Albert Karr-Krüsi

[Zurich,] 20 March 1923

[Not selected for translation.]

447. To Pierre Comert

Zurich, 21 March 1923

Dear Mr. Comert,

I recently gained the conviction that the League of Nations has neither the energy nor the good intention to fulfill its great cause.[1] As a serious pacifist I therefore do not consider it right to be connected with the same in any way. I ask you please to remove my name from the list of members on the committee.[2]

In great respect,

Albert Einstein.

448. From Paul Winteler

22 March 1923

[Not selected for translation.]

449. To Svante Arrhenius

Berlin, 23 March 1923

Esteemed Colleague,

Please accept again my sincere thanks for your nice capitalistic solicitude.[1] I very much approve of your proposal of coming to Cotenburg[2] in July instead of to Stockholm in the winter and am very willing to hold the requested lecture. As far as the latter is concerned, I only regret that my new theory about the basic unity of gravitation and electromagnetism cannot be presented in popular form.[3] Perhaps the lecture could be structured into a popular part and a more scientific part. This just as a suggestion; I shall naturally guide myself entirely according to your wishes.

I shall not be taking my wife along, who has a very long trip behind her, of course, but perhaps my eldest son, who is currently studying.[4] Whatever else we should like to undertake I leave, with your kind permission, to the provident care of the gods.

With friendly greetings, I am very sincerely yours.

450. To Carl Brinkmann

Berlin, 23 March 1923

Dear Mr. Brinkmann,

When the substitute issue arose, you were the first man I thought of.[1] I was only held back at the time by the consideration that through your connections to the Foreign Office you could perhaps be perceived more as a politician than a scholar, if not, indeed, an organ of the German government. This I believed should definitely be avoided, as the other men on the committee likewise have no professional ties to their governments (as far as I know).

Now, however, the whole question is moot. My annoyance about the impotence and moral dependence of the League of Nations influenced me to write a final resignation to the League of Nations and to underscore it by publicizing it in the press.[2] This may have been undiplomatic; but my gut feeling compelled me to take this step.

With my best wishes for your activities in Riga[3] and kind regards, I am yours.

451. To Zionistische Vereinigung für Deutschland

Berlin, 23 March 1923

Dear Sir,

Just returned from my travels, I hasten to reply to your letter of the 14th of this mo.[1] and authorize you to issue a denial in the following form: Neither in Paris nor elsewhere did I deny that I was born in Ulm as the son of German parents and that I became a citizen of Switzerland through immigration. I can furthermore attest that I never tried to ingratiate myself anywhere.

Very respectfully.

452. From Michele Besso

[Bern, 23 March 1923][1]

Dear Albert,

You now fear the other half as little as you feared the first half when you took your stand against the 93.[2] Even so, it seems to me that you had a better opportunity to get to know those people you did not want to go along with then than the ones whose goodwill you now thought it necessary to dispute, in public even. The objection that the L[eague] of N[ations][3] is far too centralized in structure for it to be possible to attribute any *single* will to it, was your contribution. Yet I hear you answer: "That's not what counts; what I mean is clear anyway"; roughly thus: "The proportion of stupidities imposed on this country is too great. It simply *will* have to defend itself somehow; I don't want to contribute in the least to these restrictions that the others must try to lay in its path, as soon as they don't happen to succeed. I am throwing my little bit of weight into the balance along these lines: the victorious cause pleased the gods, but the vanquished pleased Cato [*victrix causa diis placuit sed victa Catoni*]."[4] Let there not have been any distortion of the photographic plate—where it hinges on a matter of 10^{-4} cm![5] and none of the safeguards required for verification, by means of external "experimental results," be forgotten! Maybe when you have been closer to the one again, you will dislike the others less (again). And the slow twisting approach by the Americans through all the internal difficulties[6] will make you optimistic again about this entity, to which Alb[ert] Thomas from the Labor Organization and Lord Robert Cecil belong.[7]

What will the coming months bring, I wonder?

Warm greetings, yours,

Michele.

What has been done cannot be undone! [*Facta infecta fieri nequeunt!*][8]—Just for the sake of perspective, I also add that your decision is being used as an indication of a weakening of the position of the Sw[iss] state (in favor of an unautonomous social democracy, the "Federation for the Independence of Switzerland" and similar doubtful creations).[9]

On the other hand, I cannot judge whether it is entirely insignificant for the internat. position of Jewry and whether perhaps an improvement in individual countries opposes it.

But the main thing is, of course, not the initially arising consequences but the deeper justification: and about that I expressed my doubts above. Again, yours,

M.

Whoever reads the forthcoming report—in mid-March 1923—by Gonzague de Reynold[10] about the meetings—of August 1922!—will surely understand better why it did not <give you a homey feeling> make you feel at home being among that company. Patience!, I would probably have said to you if you had complained to me about it. Now, I say: Patience! to myself—the world doesn't look very charming on either side, though

15 April 1923

Dear Albert, I don't quite know what these lines have been waiting for: They didn't turn out the way ⟨I⟩ that would have been right; but *everything* goes a bit wrong. I believe, though, that I was still secretly waiting for some report from you, perhaps still unconsciously continuing to wait for your transit through Switzerland.

Lately there was a report in the daily paper according to which, during your absence recently, an Academy colleague presented a paper of yours that was supposedly particularly important.[11]

I've heard nothing about the findings by the eclipse expedition that was meant to detect something.[12]

I'll enjoy hearing a report from you.—That is by no means to say that you have to address the subject of the first part of this message in any way. This much I do know about you, just as about myself: We just do it as best we can and know.

Yours,

Michele.

We, Mr. Chavan and I, duly received your beautiful cards from Japan. Chavan also wanted to congratulate you on the Nobel but didn't know where to write.[13] I . . . didn't know what to say. To many people, the worries about the money wouldn't be troubling—to you, it will surely have brought not merely advantages.[14] It's enough to drive one nuts ["*Embetieren*,"] how unreasonable people are. One should and *wants* to do so much, but it doesn't suffice anyway to get anywhere [*niene hi*].[15]

A few days after your trip through Bern, my mother died. Then two of my close cousins in Milan.[16] A short while ago, Vero's father-in-law (a man of my age group—simple as that!).[17] Strange how life is!

M.

By the way, I too in the meantime had been sick for two weeks; a sort of influenza with fever. Already long gone. Not the chronic laziness [*Infaulentia*].[18]

453. To [Ilse Einstein][1]

[Berlin, before 24 March 1923][2]

1) Muraour[3] should return books. Please write.

2) Tax [authorities] want itemization of capital assets as of the end of 1922.

3) Ministry alleges that I acquired Prussian citizenship upon entering employment in 1914.[4] Please ask Moszkowski[5] whether this is binding, considering that I was not treated as a Prussian during the war. Furthermore, what other kinds of consequences this would have. Whether my boys are therefore also Prussians.

4) What steps are necessary about compensation for damage to apartment (Moszk.)

454. To Heinrich Lüders

Berlin, 24 March 1923

Esteemed Secretary,

With reference to your esteemed letter of the 15th of Feb. of this yr.,[1] I take the liberty of informing you of the following. When my appointment to our Academy was being envisioned in 1913, my colleague Haber called my attention to the fact that my appointment would have the consequence that I would become a citizen of the Prussian State. Because I considered it important that absolutely nothing change with regard to my citizenship, I made my acceptance of an eventual appointment conditional upon satisfaction of this provision, which was then also granted accordingly.[2] I do not doubt that this matter can be verified in the ministerial files; furthermore, I know that this situation is known to my colleagues Haber and Nernst.[3]

In great respect,

A. Einstein.

455. To Sergei F. von Oldenburg

Berlin, 24 March 1923

Highly esteemed Colleague,

I happily and thankfully accept the election as corresponding member of your Academy.[1] I followed with admiration how, with so much love and success, scientific research was kept alive in your severely tested country.

In utmost respect,

A. Einstein.

456. From Paul Ehrenfest

[Leiden,] 27 March 1923

Dear Einstein

First of all, cord. greetings on your return back home from the exotic world.[1]— I am sorry to hear that your wife isn't well. Give her my sincere regards—a speedy recovery!

I was extremely pleased that you were still able to see my wife[2] and was touched straight to my heart that you spoke spontaneously about coming to Leyden; I hadn't dared to ask you about that.—But now, naturally, I will joyfully and resolutely jump at the opportunity: I beg you please to choose: *Around 11 April* or *around 20 May* (= Whit Sunday)—Later I am going to have to conduct such an enormous number of examinations, partly in Leyden, partly in Delft, that any contact between us would be impossible.– "Around" gives you some freedom of choice:– I entreat you: (A) To inform me as soon as possible what you choose, with exact indications of the limiting dates; (B) To then also—forgive the impudence of the request!—really keep to it.– I'd prefer most if *Ilse* replies to me!!!–

I just heard: From 13 May until circa 5 June Lorentz will be in England. From 5 May–27 May De Sitter likewise.[3] Around 11 April—therefore, e.g., from 7 April onwards you'll encounter everyone!

Cord. regards, yours,

P. Ehrenfest

457. From Richard Stern[1]

Berlin, 28 March 1923

[Not selected for translation.]

458. From Hermann Anschütz-Kaempfe

Kiel, 23 Heikendorfer Way, 31 March 1923

Dear, esteemed Professor Einstein,

I do not know whether these lines will find you in Berlin already or will be waiting for you; from your two messages from Japan and Spain[1] I gather, though, that it should be right about now.

We, i.e., my wife and I, have been here for 14 days and want to stay another 4 weeks or so; then all sorts of urgent things summon us back to Lautrach. It would be nice if you came while we are here; and there is enough interesting work to do. The new compass, in its first model, has meanwhile been thoroughly tortured the entire winter and behaved itself quite splendidly; we are just about to put the second sphere, in a slightly improved form, onto the testing stand.

In my letter to you at the University of Tokyo I reported to you about your apartment; now everything aside from a few little details is ready there and it is awaiting your arrival;[2] it even has 2 entrances, one for the gentry and one, as Kossel[3] expressed it, for deliverers and experimental physicists.

At the present time we are thinking about an axleless reversing motor; that is, a relay bobbin and armature, the rotor of which is centered by means of electro-inductive repulsion; I think that must work and the sensitivity must then become very great.

Would Berlin, resp., your friends over there, not be very cross if you wanted to leave again so soon? But it would be preferable if you came soon so that we could install you here personally. And in May/June we shall be away and in August your room in Lautrach is already reserved for you and your boys.

With cordial regards from both of us and the whole company to you and Mrs. Einstein, yours,

Anschütz-Kaempfe

INDEX

References are collected under the appropriate English heading. Certain institutions, organizations, and concepts that have no standard English translation are listed under their foreign designation (with cross-references from an English translation). For the meaning of abbreviations, see the List of Abbreviations in the documentary edition.